T5-AQQ-849

Handbook of Lipid Research   5

The Phospholipases

Handbook of Lipid Research

Editor: Donald J. Hanahan
The University of Texas Health Center at San Antonio
San Antonio, Texas

Handbook of
Lipid Research

5

The Phospholipases

Moseley Waite

Bowman Gray School of Medicine
Wake Forest University
Winston-Salem, North Carolina

Plenum Press · **New York and London**

Library of Congress Cataloging in Publication Data

Waite, Moseley.
  The phospholipases.

  (Handbook of lipid research; v. 5)
  Includes bibliographies and index.
  1. Phospholipases. [DNLM: 1. Phospholipases. QU 85 H236 1978 v. 5]
QP751.H33   vol. 5   574.19′247 s   87-14132
[QP609.P55]   [574.19′247]
ISBN 0-306-42621-8

QP
609
.P55
W34x
1987

© 1987 Plenum Press, New York
A Division of Plenum Publishing Corporation
233 Spring Street, New York, N.Y. 10013

All rights reserved

No part of this book may be reproduced, stored in a retrieval system, or transmitted
in any form or by any means, electronic, mechanical, photocopying, microfilming,
recording, or otherwise, without written permission from the Publisher

Printed in the United States of America

This book is dedicated to my two mentors

Salih J. Wakil

and

Laurens L. M. van Deenen

I owe much for the efforts they gave

And to my family

May they always find the white water to paddle

BIOM 2//7/8%

Preface

Phospholipases are a class of ubiquitous enzymes that have in common their substrate and the fact that they are all esterases. Beyond that, they are a diverse group of enzymes that fall into two broad categories, the acyl hydrolases and the phosphodiesterases. The former group is made up of the phospholipases A_1 and A_2, phospholipase B, and the lysophospholipases. On the other hand, the phosphodiesterases are the phospholipases C and D. The scheme indicates the site of attack of each type of phospholipase.

The lysophospholipases, not shown, have in some cases properties similar to phospholipase B and are known to attack the acyl ester at either position 1 or position 2 of the glycerol backbone. Furthermore, some of the phospholipases C and D do not hydrolyze phosphoglycerides but use sphingomyelinase as their substrate. These phospholipases C are also referred to as sphingomyelinases. The products of that reaction are phosphocholine plus ceramide.

The recommended nomenclature for the phospholipases follows the pattern of differentiating between the acyl hydrolases and the phosphodiesterases (*Enzyme Nomenclature,* Academic Press, New York–San Francisco–London, 1979):

Recommended name(s)	EC number
Phospholipase A_1	3.1.1.32
Phospholipase A_2	3.1.1.4
Phospholipase B (lysophospholipase)	3.1.1.5
Phospholipase C	3.1.4.3
Sphingomyelin phosphodiesterase (phospholipase C)	3.1.4.12
Phospholipase D	3.1.4.4
Lysophospholipase D (alkylglycerophosphoethanolamine phosphodiesterase)	3.1.4.39
Sphingomyelin phosphodiesterase D	3.1.4.41

In addition to the differences in substrate specificity, considerable differences exist between the molecular and catalytic properties, as would be expected. Within a given class of phospholipases, however, a remarkable degree of homology is found. This is particularly true of the phospholipases A_2 from venoms and the mammalian pancreas. Considerable information on this homology and the evolution of the enzymes is available, and it is covered in some detail in this book. A final determination of the extent of homology in a given class of enzymes will, in most cases, require much additional work.

Clearly, the functions of phospholipases are diverse. Basically, two general types of function can be considered, even though this consideration is somewhat too simplistic. First, there is the action of phospholipases in digestion, which exists at many levels. Digestion of extracellular phospholipids can serve as a mechanism by which bacteria can derive a source of phosphate, while digestion of dietary neutral lipid and phospholipid is dependent on the action of phospholipases in the intestine; this action therefore provides the essential fatty acids required for normal metabolism.

Regulatory phospholipases have low catalytic activity and are present in small quantities in cells relative to the digestive enzymes. In addition, they are under stringent regulation that limits their activity, thereby protecting the cell from membrane degradation and cytolysis. These phospholipases have come under intense investigation of late with the recognition that the formation of a number of bioactive lipids such as the eicosanoids (e.g., prostaglandins and leukotrienes) is dependent on the action of phospholipases. Likewise, the very fundamental acyl composition of the phosphoglycerides that constitute the bilayer of membranes is dependent on phospholipase action in a deacylation–reacylation cycle.

This book attempts to cover many aspects of the basic enzymology of phospholipases in addition to some functional aspects of physiologic action, especially in Chapter 11. While the book is primarily organized phylogenetically, there are some notable exceptions to this rule. This is more evident toward the later portion when mechanisms of action and function are covered. The first chapter is oriented toward the basic considerations of the nature of the substrate and methods of enzyme assay. In all cases, useful references to more detailed descriptions of the material are given so that the reader may pursue those areas of interest beyond the scope of this book.

Moseley Waite

Acknowledgments

Many people contributed to this book, people to whom the author is deeply indebted. A number of the concepts and conclusions reached here have been gleaned in some fashion from others, for there are a number of excellent scientists working in one fashion or another on phospholipases. However, in a book such as this, some inaccuracies, omissions, and overstatements exist; those shortcomings clearly are the responsibility of the author.

Specific acknowledgments for all those who responded to correspondence and provided data are indicated in the text; I am greatly appreciative of this help. I would like to give particular recognition to the following, who spent considerable time in discussion of phospholipases with me: Drs. H. van den Bosch, A. Slotboom, G. H. deHaas, G. Scherphof, A. Bangham, R. Verger, S. Gatt, P. Elsbach, R. Franson, M. Wells, W. Parce, M. Thomas, P. Cullis, K. Chepenik, B. Smith, R. Lumb, E. Dennis, R. Heinrikson, S. Jackowski, C. Rock, P. Majerus, R. Snyderman, G. Weissmann, M. Jain, and A. Abdel-Latif. Special thanks go to Dr. D. Hanahan, who conned me into writing this book but who paid the price by reading rough drafts; his comments, insights, and encouragements were invaluable.

My own work on phospholipases has benefited from the work of a number of people at the Bowman Gray School of Medicine. For their efforts in the past 19 years, a toast: P. Sisson, L. King, V. Roddick, M. Kennedy, D. Greene, R. Franson, J. Newkirk, W. Parce, R. Morton, R. El-Maghrabi, G. Beaudry, J. Belcher, F. Chilton, L. Daniel, R. Wykle, F. Cochran, H. Griffin, G. Kucera, L. Kucera, C. Krebs, J. Parker, P. Lippiello, J. Damen, C. Miller, J. O'Flaherty, M. Robinson, D. Smith, L. Swendsen, H. Mueller, M. Thomas, M. Samuel, L. DeChatelet, L. Miller, G. Charles, S. Britt, G. Cansler, D. McLaughlin, and C. Cunningham.

This book would not exist without the heroic efforts of Tommye Campbell. From hundreds of pages of handwritten manuscript and duplicate copies of figures and tables, she carved out the finished copy. Likewise, she handled permissions correspondence and kept references in order. I will not be able to show adequate appreciation other than to simply say thanks and it's done.

Much of the initial writing was done in the Department of Biochemistry at the University of British Columbia; Drs. Jean and Dennis Vance were wonderful hosts in a magnificent environment.

M.W.

Contents

Chapter 4
Other Microorganisms

Chapter 5
Plant Phospholipases

Chapter 6
Cellular Phospholipases A$_1$ and Lysophospholipases of Mammals

Chapter 10

Mechanism of Phospholipase A_2 Action

Chapter 11

Function of Phospholipases –

Chapter 1

Assay of Phospholipases

The purpose of this chapter is to describe briefly some of the most commonly used assays for the different classes of phospholipases. Also, some less frequently used assays are described that potentially could be of use under certain conditions. In the following chapters, mention is made of the specifics of the assay used in a particular study and reference to Chapter 1 can be useful as a general reference. Some chapters focus on substrate–enzyme interaction and therefore give better insight into the molecular mechanism of the assays used; this is particularly true of Chapter 10.

1.1. General Considerations and Choice of Assay

It is of interest to note changes in the analytical methods used over a period of some 40 years and to imagine what the state of the field would be without the vast improvements in assays that have occurred. In her classical early studies on the toxin of *Clostridium welchii*, Macfarlane (Macfarlane and Knight, 1941) wrote the following: "A typical hydrolysis was carried out as follows: 830 mg lecithin, emulsified in H_2O, 1.5 ml 0.1 M $CaCl_2$ and 5 mg dry *Cl. welchii* toxin (phospholipase C) were mixed in a total volume of 15 ml adjusted to pH 7.4 with 0.1 N NaOH and placed in a thermostat at 37°; a further 2 mg toxin were added after 5 hr and the mixture was titrated back to pH 7.4 with 0.1 N NaOH at intervals until no further liberation of acid took place, by which time a layer of fat had risen to the surface." This was then assayed by chemical hydrolysis of the product phosphocholine followed by choline and phosphorus determinations. Furthermore, the phosphocholine was crystallized and elemental analysis performed. The neutral lipid (diacylglycerol) was assayed for its saponification value and the freed fatty acids separated into saturated and unsaturated fatty acids based on the solubility of the Pb^{2+} salts. While this rigorous approach was essential at that time for the characterization of the products of phospholipase action, it would hardly lend itself to the types of study ongoing today, such as enzyme purification, or to the action of phospholipases within membranes.

The assay of phospholipases is somewhat complex owing to the water-

1

insoluble nature of the substrate employed. While it has been shown that phospholipase A_2 can hydrolyze phosphatidylcholine in the monomeric form, optimal activity is reached only above the critical micellar concentration (CMC) (Roholt and Schlamowitz, 1961; Pieterson *et al.,* 1974). Because phospholipases preferentially hydrolyze substrate above their CMC, it is essential in the study of phospholipases to know the physical properties of the substrate. To simplify the understanding of the substrate, most investigators have employed model systems as substrates such as micelles or liposomes. Figure 1-1 illustrates three phases of lipid aggregates to be discussed as model substrates (Cullis and Hope, 1985). Even though these usually are not the physiological substrates, the properties of model membranes are simple relative to the complexities of the natural substrates, that is, biological membranes. As our understanding of phospholipase–model substrate interaction increases, undoubtedly more emphasis will be placed on the molecular mechanisms of phospholipase–membrane interaction. Indeed, some very elegant studies involving the activity of phospholipase–red cell membrane interaction have been carried out that have led to a better understanding of the molecular packing and architecture of phospholipids in this membrane.

The method of assay has depended on the type of phospholipase and its catalytic turnover rate. Many of the extracellular phospholipases (venom, bacterial, pancreatic) have specific activities higher than 1000 μmoles/min/mg.

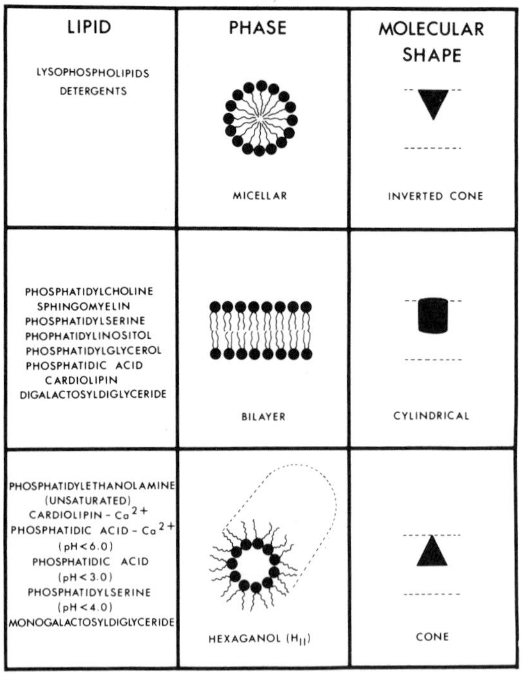

LIPID	PHASE	MOLECULAR SHAPE
LYSOPHOSPHOLIPIDS DETERGENTS	MICELLAR	INVERTED CONE
PHOSPHATIDYLCHOLINE SPHINGOMYELIN PHOSPHATIDYLSERINE PHOPHATIDYLINOSITOL PHOSPHATIDYLGLYCEROL PHOSPHATIDIC ACID CARDIOLIPIN DIGALACTOSYLDIGLYCERIDE	BILAYER	CYLINDRICAL
PHOSPHATIDYLETHANOLAMINE (UNSATURATED) CARDIOLIPIN - Ca^{2+} PHOSPHATIDIC ACID - Ca^{2+} (pH < 6.0) PHOSPHATIDIC ACID (pH < 3.0) PHOSPHATIDYLSERINE (pH < 4.0) MONOGALACTOSYLDIGLYCERIDE	HEXAGANOL (H_{II})	CONE

Figure 1-1. The polymorphic phases of phospholipids are cartooned in this figure. A major factor dictating the molecular shape and therefore the phase is the relative cross-sectional area of the polar head group and the hydrocarbon acyl chains. (From Cullis and Hope, 1985.)

It is quite convenient to assay these enzymes using a titration procedure measuring the proton liberated during hydrolysis. On the other hand, workers studying the less active cellular phospholipase often employ radio tracer techniques that are extremely sensitive. These techniques generally suffer from being much more laborious, however. Recently, Aarsman *et al.* (1976) have developed a sensitive colorimetric assay for phospholipases A using model compounds with thiol acyl esters that has some attractive features including sensitivity and simplicity. The product of the reaction contains a free sulfhydryl that can be coupled with 5,5′-dithiobis-2-nitrobenzoate (Elman's reagent). It is not certain at this point that these model compounds are suitable substrates for all phospholipases A_2, and the kinetic values obtained with these are usually different from those obtained with the natural oxy acyl ester substrates. Some less commonly used or limited types of assay that have been reported will not be covered even though for the particular problem under investigation they were quite well suited. These include charring of product on thin-layer chromatography plates or the interesting new variation of this using flame ionization detection of phospholipids and hydrolysis products separated on coated quartz rods (Ackman, 1981). Bioluminescent (Ulitzur and Heller, 1981), fluorimetric (Gatt *et al.*, 1981; Hendrickson and Rauk, 1981), and hemolytic (Vogel *et al.*, 1981) assays have been described that could become quite valuable in this area of research once their generality has been established. Likewise, gas chromatography or high-pressure liquid chromatography of fatty acids and glyceride products will not be covered. The procedures are described elsewhere in *Methods in Enzymology* (Patten *et al.*, 1981; Porter and Weenen, 1981) or in Kates' detailed chapter entitled "Techniques in Lipidology" (Kates, 1972).

1.1.1. Physical Form of Substrate

A major factor regulating phospholipase action is the initial interaction of the enzyme with bulk-phase substrate prior to the binding of the substrate molecule into the active site of the enzyme. For this reason, knowledge of the aggregated state of the substrate is essential. The following sections are brief descriptions of various forms of substrate used for phospholipase assays, lipid preparations, and methods of assay. For more details of phospholipase–substrate interaction, the reader should refer to Chapters 9 and 10 and to Verger (1980), Verger and deHaas (1976), Dennis (1983), or Verheij *et al.* (1981*b*).

1.1.2. Liposomes

Undoubtedly, liposomes are the most commonly used substrates for measurement of phospholipase activity. While liposomes vary in complexity and size, they share a common property described by Bangham and coworkers over 20 years ago; namely, they are osmotically active bimolecular leaflets (Bangham *et al.*, 1965). In this form, the phospholipids are organized in a manner closely resembling biological membranes. There are two basic types of liposome, ones with a single bimolecular leaflet and those with multilamellar leaflets. The size of the single-shell vesicles can range from approximately 25 nm (Huang and Lee, 1973) to 1.2 μm or greater (Kim and Martin, 1981); multilamellar vesicles can be even larger. Most methods of liposome preparation yield a range of vesicle sizes and, in some cases, a mixture of single and multilamellar vesicles. The characteristics of the liposome are dependent on the phospholipid composition, the solute composition, and, to a great extent, on the method of vesicle preparation. Certain pure phospholipids will not form single-shell vesicles; for example, phosphatidylethanolamine will not form vesicles unless it is mixed with nearly equal amounts of another phospholipid, usually phosphatidylcholine. Indeed, recent evidence obtained by nuclear magnetic resonance measurements indicates that phosphatidylethanolamine does not form the usual bilayer structure but rather forms a structure similar to hexagonal type II (Fig. 1-1; Cullis and deKruijff, 1978).

An extremely important question arises in the preparation of liposomes concerning the effective amount and concentration of phospholipid available to the enzyme at the lipid–water interface. This question exists under any conditions in which the substrate is not present in true solution. For example, phospholipid can be added to an aqueous solution such that the preparation has 1 μmole of phospholipid per milliliter of solution. While this concentration would be considered 1 mM, the effective concentration of the substrate on the surface of the liposome essentially would be pure phospholipid plus water of hydration. This would be on the order of 7.5–11.0 M assuming the approximate dimensions of 2.0 nm by 0.6 nm^2 for the phospholipid molecule in the bilayer (Cullis and Hope, 1985). The actual area is dependent on the acyl and polar head groups as well as the ionic composition of the medium. This consideration is significant since many, if not all, phospholipases bind to the liposome and effectively are active on substrate at this very high concentration. For this reason, classical Michaelis–Menten kinetics should not be employed as is done with soluble substrates. While there are many examples of K_m and V_{max} values being given for phospholipases, these serve only to demonstrate a relative affinity constant and maximal velocity under the conditions employed.

Figure 1-2 is a representation of the type of enzyme–liposome interaction that is thought to occur with many phospholipases, in particular, the A$_2$ type. This cartoon, drawn to the approximate scale of limited size single-lamellar vesicle and the porcine pancreatic phospholipase A$_2$ (2.2 × 3.0 × 4.2 nm) (Dijkstra *et al.*, 1978), has four rate factors that need consideration in addition to the catalytic rate of the phospholipase: (1) the rate at which the enzyme attaches to and leaves the liposome (on–off rate); (2) the rate at which the products of the reaction, monoacylphospholipid and fatty acid, diffuse from the enzyme in the outer monolayer; (3) the rate at which the product monoacylphospholipid and/or fatty acid "flip-flops" into the inner monolayer, possibly exchanging with a substrate diacylphospholipid; and (4) the rate at which the substrate diacylphospholipid diffuses to contact the phospholipase. Another possible factor that could influence the rate of hydrolysis is the exchange of phospholipid between particles, although this has been shown

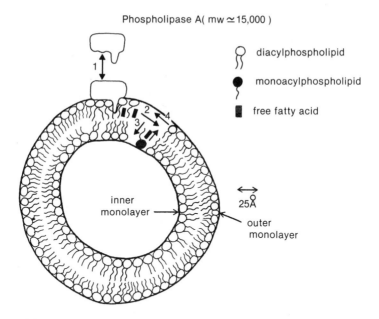

Figure 1-2. This cartoon shows the reversible binding of a phospholipase to a bilayer vesicle (step 1). Following hydrolysis, products (step 2) and substrate (step 4) diffuse away from the enzyme and bilayer "flip-flop" of lipid occurs (step 3).

to be extremely slow for micelles (Kornberg and McConnell, 1971). Two of the basic characteristics of the liposome that can influence the on–off rate of an enzyme are the electrokinetic surface properties (Bangham and Dawson, 1962) and the molecular packing of phospholipid molecules within the liposome (Hauser *et al.*, 1981). The motions of substrate, product, and enzyme in the liposome are interdependent events and combined, they could influence the overall observed rate of catalysis. This model does not take into account the activity of the enzyme of monomeric (soluble) molecules or the diffusion of product into the aqueous phase. Although these events can occur, probably they are negligible in magnitude. Indeed, Jain *et al.* (1980) and Hunt and Jones (1984) demonstrated that under conditions of limited hydrolysis, the bilayer structure is maintained. When the extent of hydrolysis exceeds one-half (outer monolayer) then the "flip-flop" of product and substrate can become a major factor in the limitation of activity. This becomes even a greater consideration when multilamellar vesicles are used as substrates. Potentially, the size of the vesicle becomes a consideration when one realizes that the total surface area ranges from nearly 2000 nm^2 to 4.5 μm^2 (diameter of liposome 25 nm to 1.2 μm). This 2200-fold difference in surface area could be of considerable importance in contrasting the relative contribution of the enzyme on–off rate (1) with diffusion rates (2–4). Also, the difference in surface area is related to the radius of curvature and hence the packing characteristics of the phospholipid molecules. A theoretical consideration of diffusion-controlled reactions in planar structures (two dimensions) has been developed by

Hardt (1979). Likewise, the process of association–dissociation of a solute (e.g., phospholipase) combined with diffusion in a plane has been described (Schranner and Richter, 1978).

The action of the phospholipases D would in general be considered similar to the phospholipases A since the product, phosphatidic acid, maintains the bilayer structure. On the other hand, phospholipase C produces diacylglycerol that does not maintain bilayer structures but forms droplets in the bilayer (Verkleij *et al.*, 1973).

The motion of the lipid in the plane of the monolayer increases dramatically above the phase transition of the lipid. It is of considerable importance in understanding phospholipase–liposome interaction that the pancreatic phospholipase which penetrates and degrades liposomes poorly either above or below the phase transition catalyzes extensive hydrolysis at the phase transition (op den Kamp *et al.*, 1974; Wilschut *et al.*, 1978). These workers concluded that the enzyme is better able to penetrate the liposome when two phases are present and a physical derangement has occurred between the two phases.

1.1.3. Emulsions

Often detergents are used to disperse phospholipid substrates for the assay of phospholipases. The use of these detergents in general has been empirical in nature since there is no single rule governing the substrate preference of all phospholipases. As a result, only a few workers have characterized the physical structures of the emulsions used for substrates. While it is adequate for projects such as the purification of phospholipases, only limited knowledge of the mechanism of action of the phospholipases is obtained with uncharacterized detergent emulsions. Some notable exceptions to this generality will be given in the section on the snake venom and pancreatic phospholipases (Chapter 10) where detailed studies of emulsified substrates have been carried out.

Two general points should be considered in using detergents with phospholipids: (1) the charge characteristics of the detergent and (2) the influence the detergent will have on the aggregated state of the phospholipid (i.e., micelle or vesicle). The charge characteristic is important since many if not all phospholipases are markedly influenced by the surface (Zeta) potential of the emulsion. By using detergents of varying charge and determining the Zeta potential of the emulsion or liposome using microelectrophoresis (Bangham and Dawson, 1959), one can approach the problem of phospholipase–substrate charge interaction in a more quantitative fashion. Neutral detergents are employed more widely than those with charge since the problem changes if substrate charge and ion binding are avoided.

The structure produced by the addition of detergent to phospholipids will be influenced both by the structure and the quantity of the detergent used. Many detergents will form micelles that are, in the simplest form, solid spheres or ellipsoids as opposed to the liposome that has an aqueous center surrounded by the bimolecular leaflet(s) [for a complete discussion of comparison of the structure of micelles and liposomes, see Volume 4 of this series (Small, 1986) or *The Hydrophobic Effect* (Tanford, 1980)]. With sufficient detergent, the bilayer structure is lost and the phospholipid is dispersed in a micellar emulsion. Because of the amphipathic nature of the phospholipids, however, the polar head group will be at the aqueous interface and thus exposed to the phospholipase under investigation. Deems *et al.* (1975) have termed mixtures of phospholipid and detergent (Triton X-100 in this case) *mixed micelles*. The concentration of the substrate phospholipid in the surface is dictated by the ratio of detergent to phospholipid. As a consequence, phospholipase activity will be decreased as the ratio of detergent to phospholipase is increased owing to surface dilution of the substrate.

In general, phospholipids in emulsions would not be considered models for a natural substrate since most phospholipases act on the bilayer of biological membranes. However, a notable exception is the digestive pancreatic phospholipase A_2 which has as its natural substrate bile salt–phospholipid emulsions in the intestine (Johnston, 1977).

1.1.4. Monomolecular Films

One of the important parameters of phospholipase action is the influence of surface pressure or molecular packing on enzymatic activity. The most popular and useful approach to this study is the use of monomolecular films of phospholipid. A monomolecular film (monolayer) of phospholipid is formed at an air–water interface when a solution of phospholipid in a volatile organic solution is introduced to the aqueous solution (Fig. 1-3). As depicted on the left, the phospholipase A cannot bind or penetrate the monolayer at high pressure. However, when the pressure is lowered (right), the enzyme can bind to the film and a molecule(s) of substrate can enter the active site (Verger *et al.*, 1973). In theory, but not always in fact, the phospholipases C and D would be less dependent on penetration since the phosphate ester attached by these enzymes extend into the aqueous phase.

Generally, assays on monolayers are carried out in a Teflon trough in which the surface area can be varied with precision (Fig. 1-4). As depicted, the area is regulated by a movable barrier that can compress the film and increase the surface pressure. The surface pressure is monitored by a wire or plate suspended in the film that is coupled to a microbalance. The microbalance in turn can regulate the surface area by activating the barrier drive

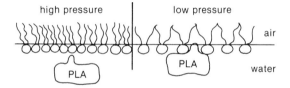

Figure 1-3. The packing of phospholipid at the air–water interface at different surface pressures is cartooned in this figure. At low but not high pressure, the phospholipase can bind or penetrate and initiate catalysis.

the organization of a monomolecular film with phospholipase (PL) penetration

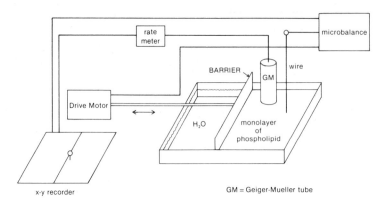

Figure 1-4. A scheme of an automated monolayer system. As described in the text, the barrier is driven by a motor that is controlled by a microbalance to maintain constant pressure. The Geiger–Mueller rate meter separately records the change in radioactivity as a function of time. See Verger and deHaas (1973) for the comparison of a *zero-order* and *first-order* trough.

to maintain a constant pressure and therefore a constant number of molecules per area—a constant substrate concentration. When radiolabeled phospholipid films are used, the surface radioactivity can be monitored by a Geiger–Mueller tube.

The enzyme, introduced to the subphase, produces a radiolabeled product that diffuses into the subphase thereby reducing the surface radioactivity which is monitored by the Geiger–Mueller tube. It is essential in this technique to have a substrate that produces water-soluble products upon hydrolysis. Variations on this theme have been described by various workers, each having a feature or features that adapt this apparatus to the particular requirements of the study under investigation (Verger, 1980).

Related to the monolayer system is a two-phase system often used with the phospholipases of snake venom (Hanahan, 1952; Marinetti, 1964). In this case, an ether–water interface replaces the previously described air–water interface. The phospholipase present in the aqueous subphase is thought to act on lipid present at the interface.

1.1.5. Natural Cellular Membranes

The use of natural membranes as substrates for phospholipases has the obvious advantage of being the source of phospholipids that this class of enzyme encounters *in situ*. Weighed against this advantage are the problems of heterogeneity of the phospholipids present, possible competing enzymatic activities such as reacylation systems, and the lack of the possibility to regulate the physical properties of the lipids described earlier. If radio tracer techniques are to be used for the assay, the labeling procedure is quite straightforward. For example, *E. coli* have been grown on radiolabeled fatty acids and used as substrate for phospholipases purified from leukocytes (Elsbach *et al.*, 1979) and platelets (Jesse and Franson, 1979). Likewise, membrane-associated phospholipases can be investigated in cell suspensions or in culture by prelabeling cells with fatty acids, $[^{32}PO]_4^{-3}$, glycerol, or the base moiety prior to initiation of the lipolytic activity. The phospholipids in organelles from tissues of animals can be labeled by injection of animals with the appropriate phospholipid precursor prior to the isolation of the organelle. However, it is not essential to radiolabel membranes for such studies; it is possible to quantitate chemically the hydrolysis products (fatty acid, monoacylphospholipid, choline, diacylglycerol, etc.). These quantitative procedures are much more laborious and are not employed generally. Such analytical procedures are important, however, in establishing the quantitation of radio tracer techniques.

1.1.6. Lipoproteins

The use of lipoproteins as substrates for phospholipases is rather unusual and would not be employed except in specific situations. Two enzymes that degrade neutral glycerides in lipoproteins, the lipoprotein and hepatic lipases, have been shown to degrade the phospholipids of various lipoproteins under appropriate conditions (Scow and Egelrud, 1976; Waite and Sisson, 1976a; Stocks and Dalton, 1980). These cell surface enzymes might play a significant role in the degradation of phospholipids from the surface of lipoproteins as they are removed from circulation. Also, venom phospholipases have been used to modify lipoproteins and to probe their structure (Reman *et al.*, 1978). The general classification of the lipoproteins is given in Table 1-1.

1.2. Methods of Assay

1.2.1. Titration

The use of titration is applicable to all types of phospholipase since a proton is liberated by the action of each. There are two major limitations that must be recognized though; it is rather insensitive and it is limited in the range of pH that can be used. A minimum enzyme activity that can be detected is about 100 nmoles of $[H^+]$ liberated per minute. This clearly eliminates the use of titration for the assay of intracellular phospholipases. The lowest pH that can be used for assaying is about 6.0. Below that pH the pK_a of the fatty acid is approached and the proportion of the undissociated fatty acid becomes too high. Practically, this is not a major problem, however, since most phospholipases with sufficient activity for titration have optimal activity above pH 6.0. Prior to using titration as an assay for phospholipase activity, it is essential to establish the type of activity (A_1, A_2, etc.) and the total extent of molecular degradation. For example, an enzyme preparation may contain more than one phospholipase that results in multiple sites of attack on the substrate.

The basic procedure involves an apparatus that automatically titrates an alkaline solution and maintains a constant pH. In a sense, the titrator serves as the buffer; therefore, no buffer is included in the reaction mixture. Perhaps

Table 1-1. Composition of Serum Lipoproteins

Lipoprotein	Size (nm)	Density	Phospholipid	Free	Esterified	Triglyceride	Protein	Major apoproteins
				Cholesterol				
Chylomicra	100–1000	<0.95	7	2	5	85	2	A-I,B,C-I,C-II,C-III,E
Very-low-density lipoprotein (VLDL)	30–75	0.950–1.006	18	8	15	50	9	A-I,B,C-I,C-II,C-III,E
Low-density lipoprotein (LDL)	20–25	1.019–1.063	20	9	36	10	25	B
High-density lipoprotein (HDL)	10–15	1.063–1.210	28	5	22	5	40	A-I,C-I,C-II,C-III,D,E

Percent composition (approximate) — Lipid

one of the most straightforward titration assays to be employed was described by deHaas *et al.* (1968). They homogenized an egg yolk in water, added deoxycholate and $CaCl_2$, and measured the uptake of alkali as the pH was maintained at 8.0. The reaction rate was adjusted such that the reaction was linear for 3–4 min which shows the value of this procedure when kinetics are under consideration. It was interesting that whole egg yolk was a better substrate than purified egg lecithin primarily due to the acidic phospholipids present in the egg-yolk emulsion.

1.2.2. Colorimetric and Fluorimetric Assays

Naturally occurring phospholipids do not contain chromophores or fluorophores and any assay of this nature must either couple a product with another agent for detection or the substrate must be synthesized with a component that undergoes a spectral change upon hydrolysis. Until recently, these types of assay were relatively insensitive and often lacked specificity or were limited by interference with other compounds [see van den Bosch and Aarsman (1979) for review]. An early colorimetric assay for the release of fatty acid was based on a modification of the procedure originally described by Duncombe (1963). The basic principle relies on the interaction of Cu^{2+} with released fatty acid, extraction of the complex into $CHCl_3$, and quantitation of the Cu^{2+} by complexing it with diphenylcarbazide in the presence of diphenylcarbazone (Itaya, 1977). This procedure was claimed to measure as little as 10 nmoles of a series of long-chain fatty acids although van den Bosch and Aarsman (1979) could not obtain reproducible results with less than 40 nmoles of product. An interesting variation of this approach was developed which employed radiolabeled $^{60}Co^{2+}$ or $^{63}Ni^{2+}$ to complex the released fatty acid (Ho and Meng, 1969; Ho, 1970). The use of $^{63}Ni^{2+}$, a β emitter, was preferred to $^{60}Co^{2+}$, a γ emitter, for reasons of safety. This procedure was shown to be able to detect as little as 1 nmole of fatty acid. A major problem with these approaches is the interference by phospholipid that interacts with the metal and extracts into the organic phase. This can be circumvented by the removal of an extracted phospholipid with silicic acid (dePierre, 1977).

A more sensitive colorimetric assay was developed by Aarsman *et al.* (1976) who synthesized a thiol-containing glycol-containing analog of lecithin, thioglycolecithin. The thiol group released by hydrolysis was reacted with Elman's reagent to give a continuous recording of as little as 1 nmole product formed per minute. A major advantage of this analog is its property of forming rather clear and stable emulsions, owing to its having a single fatty acyl chain. Dithioacylphospholipids are now used by a number of groups [see Farooqui *et al.* (1984)]. The use of water-insoluble substrates requires an optically clear substrate dispersion. Volwerk *et al.* (1979) synthesized phosphatidylcholines that contained short-chain fatty acyl groups esterified to the thioglycerol phosphocholine. Indeed, these proved to be better substrates for the phospholipase A_2 from pancreas in that the enzyme had a K_m 5–10 times lower and V_{max} 2–5 times higher than dipalmitoyllecithin with oxyesters (Aarsman and van den Bosch, 1979). Another approach was to use mixed micelles of long acyl chain phospholipids and Triton X-100 (Hendrickson and Dennis, 1984).

Recently, fluorescently labeled phosphatidylcholines were synthesized (Sunamoto *et al.*, 1980; Hendrickson and Rauk, 1981) and used as substrates for phospholipase A_2 from snake venom and porcine pancreas above its CMC (7 μM). 1,2,-Bis[10-(4-pyrano)butyryl]phosphatidylcholine had an excimer emission at 480 nm when excited at 332 nm while the products monopyranobutyryl glycerophosphocholine and free pyranobutyrate (Fig. 1-5) had monomer emission maxima at 382 and 400 nm. This difference in emission properties of the products and substrate provided a very convenient assay that could measure the production of the activity of 10 nmoles of substrate hydrolyzed per minute. This is 10–100 times more sensitive than the titration assay normally used for this enzyme. In competition studies, it was shown that the pyrene-containing analog was as good a substrate as didecanoyl glycerophosphocholine. While not as sensitive as the assay employing the thiol derivatives (Aarsman and van den Bosch, 1979), it has the advantage of not requiring a thiol reagent in the assay mixture that could interfere with thiol-containing enzymes. Another fluorescent probe that can be used is parinaric acid that can be incorporated into phospholipids (Wolf *et al.*, 1981), although its lack of stability has limited its use. The tight binding of the parinaric acid

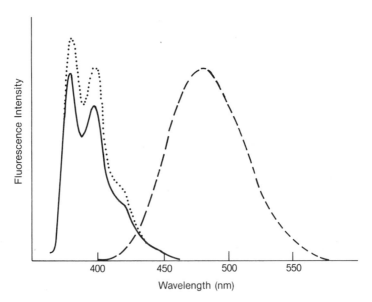

Figure 1-5. Uncorrected emission spectra of 22.3 μM di[10-(4-pyrano)butyryl]-phosphatidylcholine, ---; 0.9 μM mono[10-(4-pyrano)butyryl]phosphatidylcholine, ···; 0.9 μM 4-(1-pyrano)butanoic acid, —. Excitation at 332 nm; excitation and emission slit widths, 0.5 mm; fluorescence intensities (arbitrary units) for mono[10-(4-pyrano)butyryl]phosphatidylcholine and 4-(1-pyrano)butanoic acid are normalized; di[10-(4-pyrano)butyryl]phosphatidylcholine is at a higher gain. (From Hendrickson and Rauk, 1981.)

by albumin upon release gives a marked change in the fluorescent depolarization that detects activity in the micromolar range.

Trinitrophenylaminolauric acid was incorporated into a wide range of lipids for use as a (phospho)lipase substrate (Gatt *et al.*, 1981). This procedure, however, required the extraction of the product.

1.2.3. Assay Employing Radiolabeled Substrates

The use of radiolabeled substrates probably is the most popular approach at present to the assay of phospholipases that have a low turnover number. There are two main advantages to these substrates: (1) The investigator can quite readily establish the type of phospholipase under investigation and (2) they afford the greatest sensitivity. The main disadvantages are the expense of purchasing or preparing substrates plus the extraction and often the chromatography of products.

There are two general approaches to the preparation of radiolabeled phospholipids: (1) chemical synthesis and (2) biosynthetic labeling. While there are numerous procedures for the chemical synthesis of phospholipids, one must take into consideration that very small quantities are involved if the specific radioactivity of the product is to be kept high. [See Rosenthal (1975) for a good review of phospholipid synthesis.] For this reason, chemical syntheses must be limited to one or two steps. A straightforward procedure is to acylate glycero-3-phosphocholine with radiolabeled acyl chloride (Baer and Buchnea, 1959). It is possible to replace the fatty acid at either positions 1 or 2 of the glycerol by cleaving the phosphatidylcholine with phospholipases A_1, A_2, or a lipase with phospholipase A_1 activity followed by reacylation with the desired acyl chain. In the deacylation–reacylation steps, care has to be taken to minimize migration of the acyl chain from one position to the other (Pluckthun and Dennis, 1982*a*). Other phospholipids can be prepared from this radiolabeled phosphatidylcholine using the base exchange properties of phospholipase D from plants (Comfurius and Zwaal, 1977; Eibl and Kovatchev, 1981). In this and other preparations, the specific radioactivity is most conveniently based on the phosphorus content of the lipid.

Another convenient chemical synthesis of phosphatidylcholine involves demethylation of the choline followed by methylation with $[^{14}C]CH_3I$ as described by Stoffel (1975). Treatment of phosphatidylcholine with sodium benzene thiolate or 1,4-diazobicyclo-(2,2,2)-octane yields dimethyl phosphatidylethanolamine. This product is converted in good yield to the radiolabeled phosphatidylcholine in an overnight reaction. In both procedures, the products are purified by chromatographic procedures prior to use.

Other phospholipids such as phosphatidylinositol, cardiolipin, and the less frequently used bacterial and plant lipids generally are prepared biosynthetically. Potentially, however, phospholipase D could be used to synthesize cardiolipin from phosphatidylglycerol (Stanacev *et al.*, 1973).

The approaches to biosynthetic preparation of substrates range from presenting an intact organism or group of cells with a radiolabeled precursor to incubation of the immediate precursors for the product desired with an appropriate enzyme source. Sometimes the approach to be used is almost arbitrary and is heavily influenced by the systems handled most commonly by the investigator. A few frequently used approaches are described below.

1.2.4. Preparation of Fatty Acyl-Labeled Phospholipids

This approach is most easily conducted by using the capacity of liver preparations to form acyl CoAs as well as to acylate monoacyl lipids with the acyl CoA (Waite and van Deenen, 1967). The lysophospholipid with the fatty acyl group in position 1 of the glycerol (1-acyl lysophospholipid) is easily formed by the action of a phospholipase A_2 on the diacylphospholipid. It is important to select a phospholipase A_2 with the correct substrate specificity for the phospholipid to be synthesized (to be discussed later with the individual enzymes). As mentioned earlier, a phospholipase A_1 or a lipase can be used to prepare the lysophospholipid with the fatty acyl group in position 2 of the glycerol (2-acyl lysophospholipid). However, an easier approach is to use a plasmalogen that contains a vinyl ethyl in position 1. Brief exposure of the plasmalogen to acid causes degradation of the vinyl ether bond yielding the lysophospholipid and an aldehyde. Caution must be exercised, however, as rapid migration of the acyl chain from position 2 to position 1 can occur, especially under acidic conditions. The lysophospholipid is then incubated with the radiolabeled fatty acid, ATP, CoA, $MgCl_2$, and fresh microsomes. The product, once purified by some chromatographic procedure, should be degraded with a phospholipase A_2 to determine the positional specificity of the radiolabeled acyl chains (Waite and van Deenen, 1967). Generally, an unsaturated fatty acid when incubated with a 1-acyl lysophospholipid will incorporate almost exclusively (better than 95%) in position 2 of the glycerol. However, a saturated fatty acid does not incorporate into position 1 of the glycerol with the same degree of specificity, probably because of competition between synthesis *de novo* and reacylation (Lands and Crawford, 1976). A more systematic approach to the preparation of phosphatidylcholine labeled in position 2 of the glycerol has been described (Conner *et al.*, 1981). They were able to obtain a 90% yield of phosphatidylcholine labeled with both saturated and unsaturated acids with a high degree of positional specificity. Although relatively small amounts of phosphatidylcholine were synthesized (0.5 µmole), presumably the preparation could be scaled up proportionately. These procedures can be used for a number of phospholipids, including phosphatidylinositol, and provide an excellent way for distinguishing between phospholipases A_1, A_2, and lysophospholipase.

Uniformly labeled phosphatidylcholine has been used to assay a mixture of phospholipases A_2, C, and D by measuring the production of free fatty acid (A_2), lysophosphatidylcholine (A_2), diacylglycerol (C), phosphatidic acid (D), and water-soluble products (C and D) (Grossman *et al.*, 1974). While such an approach can be quite valuable, the picture would be rather difficult to interpret if phospholipase A_1 and/or lysophospholipase is present.

The products are easily extracted for analysis by the procedure of Bligh and Dyer (1959) or some variation thereof. The products and substrates are then separated by some thin-layer chromatographic, paper chromatographic, or high-pressure liquid chromatographic system. However, for quick routine screening, simply looking at the release of fatty acids one can preferentially extract the fatty acid using the modification of the Dole procedure described by van den Bosch *et al.* (1974). They quite correctly pointed out that this procedure also extracts some phospholipid into the heptane layer. Shakir (1981) and our own laboratory group have employed a system using isopropyl ether to extract fatty acid from an acidified reaction mixture. Both their procedure and our procedure require the use of silicic acid to remove the extracted phospholipid.

The amount of lysophospholipase activity is estimated by the difference between the total amount of lysophospholipid and free fatty acid produced while the relative activities of phospholipases A_1 and A_2 are best estimated by the amounts of 1-acyl and 2-acyl lysophospholipid recovered. This is a minimal estimate of phospholipase activity since a portion of the product lysophospholipid is further degraded. If no lysophospholipase is active, then comparison of either the lysophospholipids or fatty acids should give the same determination of phospholipases A_1 and A_2.

It should be recognized that the biosynthetically prepared substrates are mixtures of a class of phospholipids that vary in their acyl composition (molecular species). For that reason, problems could arise to complicate the analysis and give skewed results. For example, the acyl composition can influence phospholipase activity and therefore phospholipids with certain acyl groups are preferentially degraded.

Substrates that are useful for the phospholipases C and D contain the radiolabel in the polar head group, or in the case of phospholipase C, the phosphate. In this way, the investigator can conveniently measure release of radiolabel into the polar phase of an organic extraction of the reaction mixture. For example, Low (1981) described the preparation of inositol-labeled phosphatidylinositol using rat liver microsomes that was convenient for the assay of phospholipase C. Other phospholipids can be prepared in a similar fashion using the appropriate precursor. *E. coli* sonicates, when fractionated to remove the phosphatidylserine decarboxylase, can be used to produce serine-labeled phosphatidylserine (Kanfer and Kennedy, 1964; Newkirk and Waite, 1973).

Another biosynthetic approach used to radiolabel the polar portion of the molecule (phosphate, choline, ethanolamine) is to inject the labeled precursor into rats; these are efficiently incorporated into liver membranes within 90 min (Bjørnstad, 1965; Waite *et al.*, 1969). A good source for phospholipids containing [^{32}P] is yeast that is particularly rich in phosphatidylinositol (Irvine *et al.*, 1978). *E. coli* can yield relatively high levels of [^{32}P] cardiolipin when grown under the appropriate conditions; cardiolipin increases when the cells enter the stationary phase [for review, see Raetz (1978)]. Also, Wurster *et al.* (1971) demonstrated that the cardiolipin content of *E. coli* increases appreciably when levorphanol is present in the culture medium.

The procedures cited here are satisfactory for obtaining lipid-labeled membranes as well as for the isolated radiolabeled phospholipids. The choice of precursor is rather important, however, and its metabolic fate must be known. For example, when radiolabeled ethanolamine is injected into rats, the radiolabeled phosphatidylethanolamine produced will be converted to phosphatidylcholine to an appreciable extent (Bjørnstad, 1965). However, if all phospholipids in a membrane are to be labeled, precursors such as glycerol (if glycerol kinase is present in the system), glucose, fatty acids, or phosphate are all good precursors. The labeling of membrane is most helpful when a membrane-associated phospholipase is under investigation.

The variety of procedures described in this chapter allows the investigator a number of options to choose from in his or her studies of phospholipases. Many examples of the particular use of a technique will be given in the following chapters. The examples given are varied and chosen according to the nature of the enzyme under investigation. Also, a better understanding of the importance of the physical state of the phospholipid should become more apparent through a number of detailed studies presented.

Chapter 2

Bacterial Acyl Hydrolases (Phospholipases A, B, and Lysophospholipases)

Chapter 1 described how phospholipases are assayed. In this chapter, the description of individual phospholipases begins with those found in the simplest organisms, the bacterial phospholipases. In this and subsequent chapters, emphasis is placed on the isolation and structures of these enzymes and how they interact with substrate and metal ion cofactors. When known, the function(s) of the phospholipase is covered.

2.1. General Considerations

Etiologically, it is of interest that few phospholipases A from prokaryotics have been described, relative to the number of phospholipases C and D. A broad distinction can be made here between the acyl hydrolases and the phosphodiesterases; in general, the former are cell-associated while the latter are soluble, often secreted from the bacterium. This chapter covers acyl hydrolase as a unit. Often phospholipases A have a broad specificity and show a lack of absolute positional specificity. Here a few of the better characterized prokaryotic phospholipases A are described, including enzymes that have high levels of lysophospholipase activity.

2.1.1. Escherichia coli

One of the earliest observations of the deacylation of phospholipids by *E. coli* was actually made in a study on *E. coli* sonicate acylation of 1-acyl lysophosphatidylethanolamine (Proulx and van Deenen, 1966). Subsequently, it was found that different phospholipases A could be differentiated somewhat on the basis of their pH optimum, ion requirement, and inhibition by detergents (Proulx and Fung, 1969). A predominant phospholipase A_1 activity was detected at acid and alkaline pH values, whereas phospholipase A_2 activity was only in the alkaline pH range. The phospholipase A_1 activity at pH 5.0 was inhibited by sodium dodecyl sulfate (SDS) and could be similar to or the same as the detergent-sensitive phospholipase A_1 studied later by Doi *et al.* (1972). This report was followed by the active pursuit of acyl hydrolases in *E. coli*. It was correctly perceived that such a system should give a clearer picture of the factors that regulate membrane structure and function, especially since the system is easily manipulated by genetic means (Scandella and Kornberg, 1971). This interest in membrane structure and function led Scandella and Kornberg to the classic study of the phospholipases A_1 of *E. coli* outer membrane. Table 2-1 lists some of the strains of *E. coli* that have been demonstrated to have acyl hydrolases.

2.1.1.1. E. coli Detergent-resistant (DR) Phospholipase A_1

Key to the success in purifying this phospholipase A_1 some 5000-fold was the observation that the enzyme was stable to treatment with SDS and organic solvents (Scandella and Kornberg, 1971). In all probability, the enzyme they purified was similar to or the same as that described by Okuyama and Nojima (1969) and Fung and Proulx (1969) who used detergents in their assay systems. Later, this purified enzyme was termed the detergent-resistant (DR) phospholipase A_1. As will be described later, the enzyme, although preferential for the ester at position 1 of the glycerol, is not absolutely specific for the site of attack. The procedure of Scandella and Kornberg (1971) used a number of steps designed to select for proteins with a high affinity for lipid. The final step, preparative electrophoresis, required prior extraction of the phospholipid with butanol. As a consequence, the enzyme tended to aggregate and conventional purification techniques such as column chromatography could no longer be used. It was quite stable, however, at neutral pH, even in the presence of SDS. Because of these properties of the enzyme, it was possible to examine thoroughly the properties of this enzyme. Indeed, it was feasible to assay the preparations after SDS–PAGE and to determine molecular weight to be 29,000 daltons.

The DR phospholipase A_1 was further investigated by Nishijima *et al.* (1977) who reported the molecular weight to be 21,000 daltons for the native enzyme and 28,000 daltons for the enzyme inactivated by heating at 98°C in SDS, mercaptoethanol, and urea. This apparent molecular weight change was attributed to an unfolding phenomenon. The question of the correct molecular weight was pinpointed from the nucleotide sequence of the encoding gene to be from 30,809 daltons (Homma *et al.*, 1984a) to 29,946 daltons (deGeus *et al.*, 1984). Nishijima *et al.* (1977) also found the pI of the enzyme to be near 5.0.

Detergents such as Triton X-100 and SDS had some interesting effects on hydrolysis that related to the hydrophobicity of the detergent (Scandella and Kornberg, 1971). There was an inverse relation between the chain length of the alkyl sulfate detergent and the concentration of the detergent required for inhibition. For example, 700 μM octyl sulfate was required for 50% inhibition, whereas only 40 μM tetradecyl sulfate caused a comparable effect.

Table 2-1. Phospholipases A of E. coli[a]

Strain	Positional specificity[b]	Cell localization	Metal ion required	pH optimum	Effect of detergent	Reference
015	A_1	N.D.	Ca^{2+} (Mg^{2+})	5.0	Sensitive	Proulx and Fung (1969)
015	A_1	N.D.	Ca^{2+} (Mg^{2+})	8.3	Resistant	Proulx and Fung (1969)
015	A_2	N.D.	Ca^{2+}	N.D.	N.D.	Proulx and Fung (1969)
015 and 0118	A_1	N.D.	N.D.	N.D.	Resistant	Fung and Proulx (1969)
K19	A_1 (?)	—	—	—	—	Audet et al. (1974)
Bfad	A_1	Extracellular	Ca^{2+}	N.D.	Resistant	Patriarca et al. (1972)
W	N.D.	Spheroplasts	N.D.	6.0	Resistant	Patriarca et al. (1972)
W	N.D.	Spheroplasts	N.D.	9.0	Resistant	Scandella and Kornberg (1971)
B	A_1	Outer membrane	Ca^{2+}	8.4	Resistant	Doi et al. (1972)
K12	A_1	Outer membrane	Ca^{2+}	7.0	Resistant	Doi and Nojima (1976)
K12	A_1	Cytosol	Ca^{2+}	7.0	Sensitive	Tamori et al. (1979)
K12	Lyso PL	Membranes and cytosol	N.D.	9.7	Sensitive	Doi et al. (1972)

[a] Abbreviations: N.D., not determined; PL, phospholipid.
[b] From Scandella and Kornberg (1971).

On the other hand, Triton X-100 at 0.5 mg/ml increased the K_m for phosphatidylglycerol 40-fold from 0.34 to 15 μM. At this concentration of Triton, the substrate would be in a mixed micelle rather than a bilayer liposome (Dennis, 1983). Scandella and Kornberg (1971) recognized the difficulty in interpreting such data, pointing out that the detergents could have an effect on the enzyme directly, an effect on the substrate, or likely, a combination of the two. It was reported that Triton X-100 did not alter the pH optimum, pH 8.4, nor the requirement for Ca^{2+}. Subsequent work by Tamori *et al.* (1979) demonstrated that detergents as well as phospholipids above their CMC actually protected the enzyme against inactivation. Curiously, in the absence of detergent, the enzyme was irreversibly inactivated at 37°C but was stable between about 45 and 70°C. Detergent and phospholipid protection was pronounced at 37°C. Presumably, the phospholipase had a hydrophobic region exposed at 37°C that was unstable in the absence of lipid. It might be expected that at elevated temperatures a conformational change occurred that buried this active hydrophobic region, leading to a more stable structure.

The purified phospholipase A_1 hydrolyzed all the major phospholipids found in *E. coli*, that is, phosphatidylglycerol, phosphatidylethanolamine, lysophosphatidylethanolamine, and cardiolipin (both acyl groups in position 1 of the glycerols). Furthermore, the fatty acid composition had only a slight effect on the rate of hydrolysis. The enzyme, however, was shown to lack activity on *sn*-1 phospholipid and on trioleoylglycerol (Scandella and Kornberg, 1971).

The question of positional specificity of the enzyme is interesting and somewhat unsettled. Scandella and Kornberg (1971) did not go into the details of their results but indicated that the phospholipase A was at least 90% specific for the acyl ester in position 1 of the glycerol in chemically synthesized phosphatidylcholine. This question was reopened by Nishijima *et al.* (1977) who studied the catabolism of biosynthesized phosphatidylethanolamine and phosphatidylcholine. Using an enzyme preparation quite similar to that described by Scandella and Kornberg (1971), they concluded that the enzyme had A_1, A_2, as well as lysophospholipase activity directed toward both positions. Owing to the complexity of the system, they made no attempt to quantitate the relative activities. However, based on the low amount of [^{14}C]lysophosphatidylethanolamine produced from 1-[^{14}C]palmitoyl phosphatidylethanolamine, it would appear that the A_2 activity is considerably lower than the A_1 activity. Neither group investigated the complications caused by acyl migration. Studies by Abe *et al.* (1974) in which DR phospholipases A_1 and A_2 were simultaneously transduced by phage PI suggested that the hydrolysis of the two acyl esters is catalyzed by a single protein and the enzyme could more accurately be defined as a phospholipase B.

A relation was shown between phospholipase activity and T_4 phase infection, measured by the release of [^{32}P] lysophosphatidylethanolamine from [^{32}P]$_i$ prelabeled cells (Scandella and Kornberg, 1971). This activity, localized in the outer membrane fraction, was not specifically activated by the phage, however. Heat and other treatments of the cell likewise induced lysophosphatidylethanolamine formation. This deacylation process was found to be in competition with the reacylation process since an ATP-generating system reduced the amount of lysophosphatidylethanolamine recovered.

Scandella and Kornberg (1971) made some rather interesting calculations in an attempt to visualize the phospholipase's activity *in situ*. They calculated that the turnover number for the enzyme *in vitro* was five times lower than that of the activity *in situ*. While the reason for this higher estimated activity of the enzyme in the membrane is not obvious, it appears evident that diffusion of phospholipid within the membrane is not rate limiting. It remains a problem, however, to define the regulation of this enzyme within the cell and indeed to determine how much of the turnover of cellular phospholipid is the result of this particular phospholipase. Some subsequent work with mutants lacking phospholipases A sheds some light on this problem.

2.1.1.2. E. coli Detergent-Sensitive (DS) Phospholipase

The second major phospholipase of *E. coli* was described shortly after the purification of the DR phospholipase A. Doi *et al.* (1972) demonstrated that another phospholipase was present in cell sonicates that was inactivated by detergents and organic solvents and preferentially hydrolyzes phosphatidylglycerol. This phospholipase is termed detergent sensitive (DS). A number of *E. coli* strains were found to be deficient in the DR phospholipase, thus minimizing competitive hydrolytic reactions in the study of DS enzyme (Doi *et al.*, 1972). In addition to its substrate specificity, it differed from the DR enzyme in its subcellular localization; nearly 80% was localized in the 100,500-*g* supernatant fraction. The positional specificity of the DS enzyme has not been established. Table 2-2 summarizes the differences between the two phospholipases A.

The distribution of phospholipase found confirms the reports of Scandella and Kornberg (1971) and Bell *et al.* (1971). The latter group localized a number of lipid-metabolizing enzymes and found that the synthetic enzymes

Table 2-2. Comparison of the Properties of Detergent-Resistant and Detergent-Sensitive Phospholipase A[a]

	Detergent-resistant[b] phospholipase A	Detergent-sensitive[b] phospholipase A
Detergents	Stable or activated	Inactivated
Organic solvents	Activated	Inactivated
Preincubation at 60°C	Stable	Inactivated
Requirement for Ca²⁺	+	±
Substrate specificity	Phosphatidylethanolamine, phosphatidylglycerol	Phosphatidylglycerol
Localization	105,000-*g* Precipitate	105,000-*g* Supernatant

[a] From Doi *et al.* (1972).
[b] The detergent-resistant activity was measured in the presence of 0.5% Triton X-100 or 50% methanol; the detergent-sensitive activity was assayed without detergent or organic solvent.

are localized primarily in the cytoplasmic membrane. The differences in the distribution of the catabolic and anabolic systems might represent a regulatory system. If such is the case, one would expect that there would be a net flow of lipid from the cytosolic membrane (the site of synthesis) to the outer membrane (the site of catabolism). The role of the cytosolic enzyme would be an additional factor to consider in such a model. It is possible that the phospholipid in the cytosolic membrane could be catabolized to some extent by the DS phospholipase localized in the cytosolic fraction. Indeed, this enzyme might partition into the cytosolic membrane; roughly 20% of its activity was recovered in the 105,000-g pellet (Doi *et al.*, 1972).

·The recent success in cloning the gene responsible for the synthesis of the DR phospholipase A has given considerable new insight into the structure as well as its mode of synthesis and function within the *E. coli* (Homma *et al.*, 1984*a*, 1984*b*; deGeus *et al.*, 1983, 1984). Earlier work (Abe *et al.*, 1974; Doi and Nojima, 1976) identified the gene region lacking in mutants of *E. coli* K-12 responsible for the synthesis of the DR phospholipase A and termed this region pldA (*phospholipid degradation*). Some questions concerning the details of gene region remain, however. Plasmids containing the pldA gene were introduced into *E. coli* and stains were obtained that produced the DR phospholipase A in quantities 40–60 times that of the wild type. A proenzyme was produced with a normal leader sequence of 20 amino acids that was proteolytically processed and inserted into the outer membrane. The overproduction of the DR phospholipase did not cause any detectable phenotypic changes in the cells, including the phospholipid composition of the membrane. These observations, coupled with the fact that mutants defective in the DR phospholipase have a normal phenotype, leave open the question concerning its function.

The hydrophobic profile of the enzyme was calculated from the amino acid sequence deduced by Homma *et al.* (1984*a*) (Fig. 2-1). As can be seen, there is no region of extensive hydrophobicity, which suggested to Homma *et al.* (1984*a*) that the binding of the DR phospholipase to the outer membrane of the *E. coli* could not be explained by simple hydrophobic bonding. The amino acid composition of this enzyme was quite different from that of the known venom and pancreatic phospholipases A_2 (see Chapter 9) (Table 2-3). The *E. coli* enzyme has no cysteine residues, whereas most pancreatic and venom phospholipases A_2 have seven disulfide bonds. However, there was some suggestion that, like the venom and pancreatic phospholipases A_2, the *E. coli* enzyme has a histidine in the active site. Furthermore, recent structural analysis of the enzymes by Maraganore *et al.* (1986*b*) suggests that the DR phospholipase has a Ca^{2+}-*binding loop* analogous to venom phospholipases (see Fig. 10-5).

The sequence deduced by Homma *et al.* (1984*a*) was identical to that reported by deGeus *et al.* (1984) for amino acids in the 14–269 region of the mature enzyme. Differences noted in the amino terminal, however, led to a different molecular weight; deGeus *et al.* (1984) calculated molecular weight = 29,946 daltons from the DNA sequence which is similar to that found by SDS–PAGE of the purified phospholipase A, while Homma *et al.*

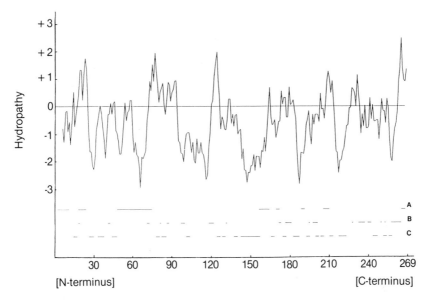

Figure 2-1. Hydropathy profile and expected secondary structure of detergent-resistant phospholipase A. Hydropathy was calculated according to Kyte and Doolittle (1982) at a span setting of 7. The positions above the midpoint line represent the hydrophobic region. The average of 19 amino acid residues from the most hydrophobic region, from 71 to 90, was calculated to be 1.01. This is lower than that of the membrane-spanning segment, 1.6, suggesting that this region does not traverse the lipid bilayer. The unexpected secondary structure was predicted by the method of Nagano (1977). The lines A, B, and C represent the α-helix, β-turns, and β-structure regions, respectively. (From Homma *et al.*, 1984*a*.)

(1984*a*, 1984*b*) calculated a molecular weight = 30,809 daltons. The difference in molecular weight resulted from a difference in nine amino acids.

2.1.1.3. *E. coli Lysophospholipases*

E. coli also contains two lysophospholipases, one cytosolic and one that is membrane associated (Albright *et al.*, 1973; Doi and Nojima, 1975). The cytosolic lysophospholipase has been purified and shown to be distinct from the phospholipases based on its properties as well as its presence in phospholipase-deficient mutants (Doi and Nojima, 1975). Since it preferentially hydrolyzes 1-acyl lysophospholipids, it was termed lysophospholipase L_1. This enzyme, purified some 1500-fold from the cytosolic fraction, had monoacylglycerol-hydrolyzing activity equal to its lysophospholipase activity. It did not hydrolyze di- or triacylglycerols, however, and is therefore distinct from the lipase found in *E. coli* that hydrolyzed monoacylglycerol (Nantel *et al.*, 1978). Even though this lysophospholipase was fully active on monomeric lipid molecules, it did not hydrolyze *p*-nitrophenylacetate. While it was claimed to be active on 2-acyl lysophospholipid, caution must be taken in interpretation of such experiments because of the migration (Pluckthun and Dennis, 1982*a*). The molecular weight of this enzyme (39,500 daltons) again clearly differentiates

Table 2-3. Amino Acid Compositions of E. coli Detergent-Resistant Phospholipase A and Bovine Pancreatic Phospholipase $A_2{}^a$

	Detergent-resistant phospholipase A	Bovine pancreatic phospholipase A_2
Lys	10 (3.7 mole %)	11 (8.9 mole %)
His	5 (1.9)	2 (1.6)
Arg	12 (4.5)	2 (1.6)
Asp	16 (5.9)	9 (7.3)
Asn	20 (7.4)	16 (13.0)
Glu	15 (5.6)	5 (4.1)
Gln	11 (4.1)	3 (2.4)
Thr	17 (6.3)	4 (3.3)
Ser	19 (7.1)	10 (8.1)
Pro	11 (4.1)	5 (4.1)
Gly	22 (8.2)	6 (4.9)
Ala	16 (5.9)	6 (4.9)
Cys/2	0 (0.0)	14 (11.4)
Val	16 (5.9)	4 (3.3)
Met	5 (1.9)	1 (0.8)
Ile	9 (3.3)	5 (4.1)
Leu	25 (9.3)	8 (6.5)
Tyr	22 (8.2)	7 (5.7)
Phe	9 (3.3)	4 (3.3)
Trp	9 (3.3)	1 (0.8)
Amino acid residues	269	123

[a] From Homma *et al.* (1984*a*). The amino acid composition of detergent-resistant phospholipase A was deduced from the nucleotide sequence of its structural gene, pldA, and that of bovine pancreatic phospholipase A_2 was calculated from the results of Puijk *et al.* (1977).

it from the membranous phospholipase. Presumably, this enzyme could act in concert with the cytosolic DS phospholipase to bring about the complete degradation of membrane phospholipid.

The lysophospholipase that preferentially hydrolyzes at position 2 of the glycerol moiety, termed L_2, has only recently been studied in detail. It is different from the soluble lysophospholipase L_1 and is associated with the inner membrane. Karasawa *et al.* (1985) purified it approximately 700-fold. Since it was solubilized by KCl extraction, it probably is a peripheral membrane protein. Nucleotide sequence analysis of genes, which encodes for the lysophospholipase L_2, the pldB gene, showed that the deduced amino acid sequence agreed with that obtained for the 15 residues in the NH_2 terminus and that molecular weight of the enzyme is 38,500 daltons (Kobayashi *et al.*, 1985*a*). Furthermore, there was no homology between this amino acid sequence and that of other *E. coli* acyl hydrolases. The genes coding for DR phospholipase A_2 and lysophospholipase L_2 were cloned together to study the gene organization (Kobayashi *et al.*, 1985*b*). By constructing an appropriate plasmid, pK01, expression of the genes in the maxicell system showed that the two genes were separated by a 3-kilobase region.

The lysophospholipase activities of *E. coli* K-12 were also investigated by the isolation of two mutants, one that overproduced lysophospholipases and one deficient in lysophospholipase activity (Kobayashi *et al.,* 1984). Comparison of different mutant colonies showed a variable ratio of lysophospholipase $L_1 : L_2$ activity, an observation that further supports the contention that the two activities are catalyzed by distinct proteins. There was, however, a constant ratio of lysophospholipase L_2 activity and an activity that transacylates the acyl group from 2-acyl lysophosphatidylethanolamine to the glycerol moiety of phosphatidylglycerol. From these data, it was suggested that the transacylase and lysophospholipase L_2 activities could be catalyzed by a single enzyme. The plasmid responsible for lysophospholipase L_2 activity, pK01, was identified and used to transform an *E. coli* mutant deficient in lysophospholipase L_2 activity (Homma *et al.,* 1983). This resulted in a transformed cell that then overproduced lysophospholipase L_2. This hybrid plasmid also contained the pldA gene, leading to the conclusion that the pldB gene, coding for lysophospholipase L_2, is between the pldA and metE genes.

While the functions of these phospholipases A_1 are not fully understood, in all probability this class of enzyme should play a significant function in bacterial membrane metabolism especially under conditions subject to stress. On the other hand, no absolute requirement for a phospholipase has been established since mutants of *E. coli* deficient in each of the two major phospholipases A as well as the double mutant lacking both DS and DR phospholipases A_1 have been isolated and studied (Doi and Nojima, 1976). These workers found that double mutants totally lacked activity on exogenous diacyl substrates. However, all mutants had normal lipid compositions and growth characteristics even with temperature changes that were shown to stimulate phospholipid turnover. When autolysis of membranous phospholipid was investigated, lipolysis was noted with the wild type but less with the mutants, in particular the double mutant lacking both phospholipases. Audet *et al.* (1974) likewise found that a strain of *E. coli*, Bfad, that lost membrane lipid and had a more fragile membrane when perturbed, released considerably more phospholipase A into the medium. Further evidence that the outer membrane phospholipase is involved in membrane phospholipid degradation was provided by Vos *et al.* (1978) when they found that wild type *E. coli* S-15 degraded membrane lipid extensively upon perturbation of the cells, compared with phospholipase-deficient mutants. Therefore, it would appear that the phospholipases A of *E. coli* membranes can degrade cellular phospholipid upon cellular damage, thereby providing potential repair mechanisms via the deacylation–reacylation pathway for the outer membranes that lack the capacity to synthesize phospholipid *de novo* (Vos *et al.,* 1978). A strain of *E. coli*, envC, known to have defective membranes, was shown to have an active phospholipase A capable of degrading exogenous phosphatidylethanolamine (Michel and Starka, 1979). These results demonstrated that an unregulated phospholipase in the bacterial membrane is deleterious. Taken together, these various reports indicate that phospholipases A are normally regulated by factors that are yet to be described. These studies therefore showed that while the mutants lacked phospholipases, the mutant cells had normal membrane turnover, leaving open the questions as to how many other phospholipases might be present and what the functions are of the two phospholipases A, DS and DR.

2.1.2. Bacillus megaterium

Kornberg's group became interested in the phospholipase activity of *B. megaterium* when it was found that during sporulation there was a decrease in the phospholipid content of the cell as well as marked changes in the membranous structures contained within the cell (Kornberg *et al.*, 1968; Bertsch *et al.*, 1969). These changes plus the cellular lysis that occurs during the liberation of the spore suggested important roles for phospholipases in reproduction of the bacteria (Raybin *et al.*, 1972). Phospholipase A_1 activity was found to be maximal at the time of maximal sporangial production, just prior to the release of free spores. [The review by Kornberg *et al.* (1968) provides a background on the biochemistry of sporulation.] There was a decline in the measurable phospholipase as the spores became resistant to sonication—the result of becoming cryptic. The enzyme was released from resistant spores upon their semination, yielding a preparation with activity five times higher than the sporangia. Isolation of the enzyme from spores therefore gave distinct advantages in the purification of the enzyme. It was of interest but perhaps disappointing to the authors that spore semination was not dependent on the phospholipase A_1, although the activity of other phospholipases was not examined.

The 170-fold purification of the phospholipase A_1 was facilitated by its water-soluble character. The purified enzyme was found to have a molecular weight of 26,000 or 29,000 daltons, depending on the method of analysis. Evidence was obtained by sedimentation analysis of a crude preparation of the enzyme that the enzyme probably associates with other cellular components *in situ*. Experiments employing chemically synthesized phosphatidylcholine showed its action to be oriented almost, if not totally, exclusively toward the acyl ester at position 1 of the glycerol. The activity, which was optimal at pH 6.5, was not affected by EDTA, Ca^{2+}, or Mg^{2+}. Detergents, however, had a marked influence on the activity of the purified phospholipase A_1. When phosphatidylglycerol was used as substrate, the enzyme had an absolute requirement either for a neutral (Triton X-100) or anionic (taurocholate) detergent. Other acidic phospholipids such as cardiolipin, phosphatidate, or 1-acyl lysophosphatidylglycerol were poor substrates under any conditions employed. On the other hand, neutral or slightly acidic phospholipids, phosphatidylcholine, phosphatidylethanolamine, and glucosaminyl phosphatidylglycerol were hydrolyzed rapidly in the presence of taurocholate but not Triton X-100. As emphasized by the authors, it would be of considerable interest to separate the charge effects of the detergent from the true specificity of the enzyme for recognizing the structure of the polar head group. The finding that the crude sporangial preparation was less dependent on detergent led to the suggestion that the cell might well contain a natural, detergentlike activator.

The purified phospholipase was also capable of degrading the phosphatidylglycerol and phosphatidylethanolamine in the plasma membranes of *E. coli,* albeit at a rate 1% of that obtained with pure phosphatidylglycerol. While this slow rate of hydrolysis was attributed to a charge phenomenon, other possibilities do exist such as molecular packing of the lipid molecules within the membrane. Such a suggestion might be reasonable, based on the stimulation of hydrolysis by Triton X-100 when pure phosphatidylglycerol was the substrate. Based on these characteristics, this phospholipase A_1 is clearly distinct from that purified from *E. coli* (Scandella and Kornberg, 1971).

The studies by these authors, plus newer knowledge of the mechanism of phospholipase action, should open new interest in bacterial systems. Such studies offer a number of advantages to the student of phospholipases including the relative ease of purifying and characterizing the phospholipase, the possibility of genetic manipulation of the system, and the synchronized biological activity of the cells.

2.1.3. Mycobacterium phlei

As with many of the phospholipases that have been purified and characterized, the initial interest was stimulated by studies that demonstrated a difference in the rate of turnover of phospholipids of *M. phlei* (Akamatsu *et al.,* 1967). The group of Nojima (Ono and Nojima, 1969) found that the cardiolipin deacylating activity in disrupted cells was localized in the membrane fraction primarily. For reasons that are not yet understood, heating of the membrane fraction at 60°C for 15 min led to a three-fold activation of the deacylation. The kinetics of deacylation were studied using the membrane preparation in an attempt to demonstrate the sequence of deacylation steps involved. While they did demonstrate the accumulation of some intermediates, these were not characterized as to the number of acyl residues remaining or their position of attachment to the glycerol moieties. Comparison of the data in this paper with a later paper characterizing a 500-fold purified phospholipase A_1 (Nishijima *et al.,* 1974) suggests that more than one phospholipase was present in the membrane fraction. This was indicated by the finding that the hydrolysis of cardiolipin by the membranes had two distinct pH optima (4.0 and 5.5), while the purified phospholipase A had optimal activity of 4.5. Other explanations such as pH specific inhibitors present in the membrane could account for the dual optima in the membranes, however.

Since the phospholipase A_1 was tightly bound to the cellular membrane, extraction with detergent (Triton X-100) was essential. Extraction of the enzyme with organic solvent to remove lipid rendered the enzyme unmanageable unless SDS or a mixture of urea and BRIJ 58 was used for specified steps in purification. Four forms of the enzyme were obtained that differed in their molecular weights (45,000 and 27,000 daltons) and in their capacity to bind to DEAE columns. Only one form was studied in detail so the relation of these four forms is yet to be determined, as is the question of positional and substrate specificities. While the enzyme preparation was specific for acyl esters at position 1 of the glycerol in phosphatidylcholine and phosphatidylethanolamine, the authors reported that both 1-acyl and 2-acyl esters of lysophosphatidylethanolamine could be hydrolyzed. They recognized the possibility of other explanations, including the problems one might expect with acyl migration, and concluded that the enzyme is a "positional nonspecific lysophospholipase with phospholipase A_1 activity," a phospholipase B. The understanding of the mechanism of action of this and the similar *E. coli* K-12 enzyme will be fascinating to unravel.

The specificity for the polar head group was likewise intriguing. At acid pH values (3.0–5.0), the acidic phospholipids such as cardiolipin and phosphatidylglycerol were hydrolyzed preferentially, whereas at alkaline pH (8.0–10.0), the ethanolamine-containing phospholipids were preferentially attacked. This very complex behavior was interpreted as a requirement by the enzyme for *nonionic substrates,* but this remains to be proved. Indeed, this system would be ideal for study of the effects of substrate surface charge on catalysis. The effect of metal ions on activity (slight stimulation by Ca^{2+}, inhibition by Fe^{3+}) likewise could be in part a surface-charge phenomenon.

The presentation of the substrate was in a detergent complex. All ionic detergents tested were inhibitory, again presumably the result of a shift in the charge characteristics of the substrate complex. Triton X-100 at low concentrations (0.015–0.05%, roughly equivalent to a Triton : substrate ratio of 20 : 60) stimulated activity on phosphatidylethanolamine and lysophosphatidylethanolamine causing changes in the calculated K_m and V_{max} values. These effects, while not in the same detergent : substrate range used by Dennis, probably is a reflection of his surface dilution phenomenon (Dennis, 1973*a*), namely, dispersion and surface dilution effects.

2.1.4. Acyl Hydrolases of Genus Vibrio

Both phospholipase and lysophospholipases have been identified in *Vibrio.* Testa *et al.* (1984) found both phospholipase A_2 and lysophospholipase in the culture fluids of *V. vulnificus.* The two could be differentiated by their heat stability and isoelectric point. While not yet purified and characterized, the acyl hydrolases are distinct from the major cytolysin produced by the bacterium. Some evidence was found to suggest that one or both of the enzymes might have cytolytic activity. A lysophospholipase was purified from Triton X-100-extracted *V. parahaemolyticus* that appears to be distinct from the extracellular *V. vulnificus* (Misaki and Matsumoto, 1978), based primarily on differences in their isoelectric points.

Chapter 3

Bacterial Phosphodiesterases (Phospholipases C and D)

The bacterial phosphodiesterases differ from the acyl hydrolase covered in Chapter 2 in that many enzymes of this class are secreted from the cell in a soluble form, a characteristic that has facilitated their purification and characterization. In general, these enzymes attack cellular membranes and either directly or indirectly are involved in cell killing and tissue damage in host organism. Because of their involvement in pathogenesis, they were often identified initially as a toxin before their catalytic function was defined. Indeed, much of the initial interest in bacterial phospholipases was generated by investigators seeking to minimize the effects of bacterial infections.

3.1. General Considerations

Historically, it is of interest to note that as early as 1901 (Rnata and Caneva, 1901) it was recognized that *Vibrio eltor* was capable of degrading lecithin. Although a few bacterial phospholipases C and D had been studied in some detail, it was not until the thorough screening of gram-negative bacteria (Esselman and Liu, 1961; Kurioka and Liu, 1967) that it was appreciated that extracellular phospholipases C and D were rather common. Esselman and Liu studied some 41 species and found *lecithinase* activity in three of the five families examined. It was well established by this time that the lecithinase referred to in that study was indeed phospholipase C. Pseudomonadaceae, Spirillaceae, and Enterobacteriaceae were positive, whereas Brucellaceae and Neisseiaceae were inactive when assayed by the egg yolk procedure, an assay that measured an opaque zone of diglyceride formation in the region of the growing colony. The cleavage of the phosphodiester was demonstrated by the appearance of acid-soluble phosphorus. Table 3-1 summarizes some characteristics of bacterial phospholipases C.

Table 3-1. Some Purified Bacterial Phospholipases C

Source	Substrate(s)[a]	pH optimum	Isoelectric point	Molecular weight	Metal requirements	Zn^{2+} present
Ps. aureofaciens[b]	PE,PC,(LPE,LPC)	7.5–9.0	6.3–6.5	35,000	Ca^{2+},Mg^{2+},Mn^{2+}	Yes
Ps. fluorescens[c]	PE,PC,(LPE,PC)	7.5	—	—	Ca^{2+}	—
Ps. schuylkilliensis[d]	PC,PE	—	6.4	23,000	Ca^{2+},Mg^{2+},Zn^{2+}	—
Ps. chlororaphis[e]	—	—	6.3	54,000	?	—
Ps. aeruginosa[f]	LPC,Sph,PC	7.0–8.0	5.5	78,000	—	—
C. perfringens[g]	PC,Sph,PS,(PE,LPC)	7.0–8.0	—	30,000–41,000	$Ca^{2+}>Mg^{2+}$	Probable
C. novyi[h]	PC,Sph,PE	7.0	7.0	30,000	Ca^{2+},Mg^{2+}	Probable
C. novyi[i]	PI	7.0(?)	5.6,6.0	30,000	None	Probably not
B. cereus[j]	PC,PE,PS	6.6–8.0	6.3,6.9,7.9	23,000	Ca^{2+}	Yes
B. cereus[k]	PI,LPI	7.2–7.5	5.4	29,000	None	Probably not
B. cereus[l]	Sph	6.0–7.0	8.1	23,300	$Mg^{2+}>Ca^{2+}$	Probably not
B. thuringiensis[k]	PI,LPI	7.5	5.4	23,000	None	Probably not
Staphylococcus aureus[m]	PI	7.0–7.4	—	20,000–33,000	None	Probably not
Staphylococcus aureus[n]	Sph,LPC	7.0	9.4	26,000–38,000	Mg^{2+}	—
Streptomyces hachijoensis[o]	Sph,PC,PI,LPC,PE,CL	8.0–9.0	5.6,6.0	18,000	Mg^{2+}	—

[a] Abbreviations: CL, cardiolipin; LPC, lysophosphatidylcholine; LPE, lysophosphatidylethanolamine; LPI, lysophosphatidylinositol; PC, phosphatidylcholine; PE, phosphatidylethanolamine; PI, phosphatidylinositol; PS, phosphatidylserine; Sph, sphingomyelin.
[b] Sonoki and Ikezawa (1975, 1976).
[c] Doi and Nojima (1971).
[d] Arai et al. (1974).
[e] Lysenko (1973).
[f] Berka and Vasil (1982).
[g] Takahashi et al. (1982) and Dawson (1973).
[h] Taguchi and Ikezawa (1975).
[i] Taguchi and Ikezawa (1978).
[j] Björklid and Little (1980), Little (1981a), and Zwaal et al. 1971.
[k] Ikezawa et al. (1983).
[l] Tomita et al. (1982) and Ikezawa et al. (1978).
[m] Low and Finean (1977) and Low (1981).
[n] Mollby (1978).
[o] Okawa and Yamaguchi (1975a).

3.1.1. Genus Clostridium

The toxins of the clostridium genus have been studied with considerable interest because of their role in gangrene (Macfarlane and Knight, 1941). The indication that the α toxin produced by *C. perfringens (welchii)* had phospholipase C activity prompted efforts to purify and establish precisely the mechanism of action of this enzyme on membranes. Similar activity has been found in other species including *C. novyi* Type A, *C. haemolyticum, C. bifermentans,* and *C. sordellii* (Mollby, 1978). Since this family of bacteria produces an array of toxic proteins, it is necessary in studying the toxicity of phospholipases C to obtain pure preparations of the enzymes. The major source for this undertaking has been *C. perfringens.* A number of the approaches employed and their degrees of success are covered by Avigad (1976) and Mollby (1978).

Generally, purification involves DEAE ion-exchange chromatography, gel filtration, precipitation by $(NH_4)_2SO_4$, and isoelectric focusing. The use of Zn^{2+} and glycerol solutions to stabilize the enzyme facilitated enzyme purification. Perhaps the most novel and effective procedure has been described by Takahashi *et al.* (1981) who succeeded in purifying the enzyme approximately 2500-fold over the crude culture filtrate. The unique feature of their scheme was affinity chromatography on a column of Agarose-linked egg yolk lipoprotein. In this single step, they achieved an 80- to 90-fold purification with a yield of 65%. In all steps, the phospholipase C activity, the biologic activity (lethal toxicity and hemolytic activity), and the capacity to bind antitoxin copurified; this provided a strong argument for the action of a single protein. The purified enzyme was heterogeneous on isoelectric focusing with pI values of 5.2, 5.3, 5.5, and 5.6 (designated α_0, α_1, α_2, and α_3) (Takahashi *et al.,* 1974). The biologic and enzymatic activities of the four fractions were in the same ratios, indicating that the molecular differences in the protein did not affect its functions. Bird *et al.* (1974) previously found four components of phospholipase C activity upon electrofocusing in a polyacrylamide gel, although their values were somewhat lower (pI = 4.75, 5.15, 5.35, and 5.55). The molecular basis for this microheterogeneity is not yet understood.

The molecular weight of this phospholipase C was somewhat difficult to establish with certainty. Several workers using SDS polyacrylamide electrophoresis reported molecular weights ranging from 43,000 to around 100,000 daltons (Mollby, 1978). The most recent value, 43,000 daltons, is probably the most accurate since it was established using enzyme preparations purified to homogeneity and determined by immunodiffusion, ultracentrifugation, and electrophoresis (Takahashi *et al.,* 1974; Yamakawa and Ohsaka, 1977). A more recent purification scheme yielded a preparation of the phospholipase C free of any detectable contaminations and had the same molecular weight, 43,000 daltons, as determined by several criteria (Krug and Kent, 1984). Some of the higher molecular weights reported could have resulted from enzyme polymerization (Ikezawa *et al.,* 1976). Yamakawa and Ohsaka (1977) concluded that the apparent molecular weight of 31,000 daltons found by them and others (Mollby and Wadstrom, 1973) was artifactually low because of enzyme–dextran interaction.

It is generally accepted that the phospholipase C from *Cl. perfringens* has a broad range of substrate specificity (Stahl, 1973). Several workers reported that phosphatidylcholine is the preferred substrate but that sphingomyelin, phosphatidylserine, phosphatidylethanolamine, lysophosphatidylethanolamine, lysophosphatidylcholine, ceramides, and ceramide phosphonates were also degraded. However, these studies were not done with purified preparation of phospholipase C and therefore must be viewed with caution. The work of Krug and Kent (1984) showed that a highly purified preparation preferentially hydrolyzed phosphatidylcholine; sphingomyelin was degraded at 35% of the rate of phosphatidylcholine and lysophosphatidylcholine was not significantly attacked. Surface charge of the substrate emulsion also affected the apparent substrate specificity as shown by the pioneering work of Bangham and Dawson (1962). They found that the addition of stearylamine or high concentrations of Ca^{2+} greatly enhanced the rate of lipolysis. (This effect, along with the influence of molecular packing on phospholipase activity, is covered in greater detail in Chapter 10.)

The first suggestion that this phospholipase C, like the enzyme from *Bacillus cereus* (to be covered later), required both Zn^{2+} and Ca^{2+} for activity came from Ispolatovskaya (1971). The requirement for Zn^{2+}, difficult to show directly, was implicated in an elegant study by Sato and Murata (1973) who demonstrated that the organism, when grown in Zn^{2+}-free medium, produces an inactive proenzyme that could be activated by the addition of Zn^{2+}. Furthermore, the enzymatic activity, when inhibited by *o*-phenanthroline, could be reactivated by Zn^{2+}, Co^{2+}, or Mn^{2+} but not Ca^{2+}. These studies supported the earlier postulate that the enzyme is a metalloenzyme (Ispolatovskaya, 1971). Although Ca^{2+} cannot replace Zn^{2+}, it is required for activity. Currently, it is thought that Ca^{2+} plays two roles: (1) It functions catalytically at low concentrations (Dawson *et al.*, 1976), and (2) it was shown to produce a positive surface change that enhances activity (Bangham and Dawson, 1962).

Taguchi and Ikezawa (1975, 1977, 1978) found many similarities between the phospholipase C of *C. novyi* Type A, termed γ toxin, and the enzyme purified from *C. perfringens*. The *C. novyi* enzyme is a Zn^{2+} metalloprotein, requires Ca^{2+} or Mg^{2+} for full activity (depending on the substrate), and has a rather broad substrate specificity. Also, the molecular weight is similar to that from *C. perfringens*, 30,000 daltons, as determined by gel filtration. Two significant differences are to be noted, however: (1) The isoelectric point of the enzyme from *C. novyi* Type A is 7.1 (Taguchi and Ikezawa, 1977), and (2) a separate phospholipase C active on phosphatidylinositol was separated from the phospholipase C active on the other phospholipids (Taguchi and Ikezawa, 1978). The phosphatidylinositol-specific enzyme was found to have pI values of 5.6 and 6.0 which allowed easy separation from the other phospholipase C using isoelectric focusing. The products of hydrolysis catalyzed by the phosphatidylinositol phospholipase C was a mixture of inositol-1-phosphate and inositol-1,2-cyclic phosphate, similar to the enzyme from *B. cereus*. Since this enzyme was not inhibited by *o*-phenanthroline, it did not appear to be a metalloenzyme. Likewise, the catalytic activity was not dependent on Ca^{2+} or Mg^{2+}.

3.1.2. Bacillus cereus

The possibility that bacterial toxins could be assigned an enzymatic function proved to be a powerful driving force in the field of bacterial toxicology during the 1950s and 1960s. Despite the fact that many of the phospholipases purified were not toxins, it was recognized in the 1960s that these enzymes were valuable tools in probing biologic and model membranes. Chu (1949) expanded the investigations on bacterial phospholipases C by employing the approach of Macfarlane and Knight (1941) on *B. cereus* and *B. mycoides*. It was recognized that the culture filtrate of these organisms contained a Ca^{2+}-dependent phospholipase C that had a broad substrate specificity. Since it was found that many of the properties of the lipolytic activity were similar to those of hemolytic activity and lethal activity in mice, phospholipase C was implicated as a toxoid (this was later shown not to be the case). In the course of the study, it was noted that the enzyme was resistant to heat denaturation and could be stabilized by glycerol. Both these observations were employed later in the purification of the phosphatidylcholine-active phospholipase C (Zwaal *et al.*, 1971; Little *et al.*, 1975; Myrnes and Little, 1980). It was also noted that normal rabbit serum inhibited the enzymatic activity on emulsions of phosphatidylcholine but had no effect on the activity when the phosphatidylcholine was mixed with lipovitellenin, a protein from egg yolk. This difference led to the conclusion that "its failure to suppress the hydrolysis of lecithin bound in egg-yolk lipoproteins is another illustration of the complexity of the biological system," which is still a vexing problem in the study of phospholipases.

Two significant questions arose from the finding that phospholipase C was released from the cell into the medium: (1) Is there a cellular phospholipase C? and (2) Can the released phospholipase C degrade the lipids in the bacterial membrane? The first problem was approached by Kushner (1962) who found that roughly one-half to two-thirds of the enzymatic activity in the whole culture mixture was recovered in medium. When the isolated cells were assayed, however, there was very little activity detected unless the extracellular fluids were added. This led to the proposal by Kushner (1962) that medium contained a factor essential for the activity of the cell-associated enzyme. A number of interesting characteristics on the production of phospholipase(s) C by various strains of *B. cereus* were presented in this paper even though it was not recognized at that time that more than one phospholipase C was present in the system.

3.1.2.1. Separation of Substrate-Specific Phospholipases C

The recognition that more than one phospholipase was present in culture filtrates of *B. cereus* was a major step in our understanding of extracellular phospholipases C (Slein and Logan, 1963). Using DEAE–cellulose chromatography or paper electrophoresis, Slein and Logan separated a phospholipase C that copurified with the phosphatasemia factor from a phospholipase C that was inhibitory to phosphatasemia factor activity. The phosphatasemia factor releases alkaline phosphatase from bone slices. These authors used commercial preparations of phosphatidylcholine as substrate that fortunately were contaminated with other phospholipids. Had pure phosphatidylcholine been used, the activity associated with the phosphatasemia factor might have been missed. These authors recognized the probability that impurities in the substrate preparation influenced their results. Even though it was not recognized that the individual phospholipases C were active on different substrates in the crude mixture, these results prompted a further study which led to the recognition that three phospholipases C were present in the culture filtrate: one most active on phosphatidylethanolamine and phosphatidylcholine, one active on sphingomyelin (sphingomyelinase), and one active on phosphatidylinositol (Slein and Logan, 1965). In this paper, they were able to demonstrate that the phosphatasemia factor had all the properties of the phospholipase C active on phosphatidylinositol, although there was some indication that synergism occurred between the phosphatidylinositol-degrading enzyme and the phospholipase C that was active on phosphatidylcholine and phosphatidylethanolamine. Likewise, the phosphatidylinositol-specific phospholipase C purified from *B. thuringiensis* had phosphatase-releasing activity and recently was shown to release acetylcholine esterase from erythrocytes (Ikezawa *et al.*, 1983).

The phospholipases C from *B. cereus* do not have lethality or hemolytic activities (Ottolenghi *et al.*, 1961; Molnar, 1962). The same conclusion was reached by Johnson and Bonventre (1967) who differentially inactivated the phospholipases and the toxoid and hemolysin activities in culture filtrates. This paper, as well as a later review by the same authors (Bonventre and Johnson, 1971), describes the various studies done during this period to dissect the differences between the phospholipases and the toxins produced by *B. cereus*. Houtsmuller and van Deenen (1963) concluded in a series of studies that the phospholipase C from *B. cereus* could degrade a wide range of phospholipids in their natural state and in a purified form. The list of phospholipids degraded included the anionic phosphatidylglycerol and the cationic phosphatidylglycerol ornithine; only phosphatidic acid and cardiolipin were not degraded. This suggested that the phospholipase(s) C from *B. cereus* was not dramatically influenced by the surface charge of the phospholipids, unlike the similar enzyme from *C. perfringens* (Bangham and Dawson, 1962). It would appear therefore that the cellular phospholipase C might be responsible for the degradation of membrane phospholipids and this raised the interesting possibility that the process could be regulated by the extracellular phospholipase C activator described by Kushner (1962).

The recognition that more than one phospholipase C existed in culture filtrates of *B. cereus* led Zwaal *et al.* (1971) to purify the enzyme active on phosphatidylcholine. Their goal was to obtain a preparation of the enzyme sufficiently pure to use a probe of membrane structure. Their method included glycerol in the purification that stabilized the enzyme. They were able to demonstrate that about 70% phospholipids in red cell ghosts were degraded despite the fact that the hydrolysis of the phospholipid caused no hemolytic activity. In a similar study, Mavis *et al.* (1972) found that nearly all the phospholipids in or derived from the membranes of *E. coli* were degraded by the phospholipase C from *B. cereus*. These lipids were mainly phosphatidylethanolamine, phosphatidylglycerol, and cardiolipin. Cardiolipin, however, was degraded only some 20–40%, depending on the physical state of the lipids. It must be concluded that the physical state of the substrate regulates activity on the highly acidic phospholipids since it has recently been stated that crystallized phospholipase C from *B. cereus* does not hydrolyze phosphatidylglycerol or cardiolipin (Little, 1981*a*).

Since the initial publication of Zwaal *et al.* (1971), the phosphatidylcholine-active phospholipase C has been exhaustively studied by a group of Norwegian investigators. Otnaess *et al.* (1974) screened a series of mutants of *B. cereus* and found one (AT10987 AB-1) that produced five times more enzyme than the parental line. Based on the observation that it was cross-reactive with antibody raised against the parental line, they concluded that the phospholipase C from the mutant was similar or identical to that from the wild type. This group devised a rapid and efficient purification scheme that employed affinity chromatography on egg-yolk phosphatidylcholine bound to Sepharose as the primary procedure. This procedure yielded an enzyme preparation purified some 1000-fold with recoveries exceeding 70% (Little *et al.*, 1975). More recently, Myrnes and Little (1980) introduced a novel purification scheme that took advantage of the enzyme's heat stability and property of renaturing with Zn^{2+} in 4 M guanidinium chloride. In this procedure, they started with the culture foam since the enzyme accumulates in the surface foam, perhaps because the hydrophobic amino-terminal region of the molecule causes the enzyme to behave as a surfactant (Otnaess *et al.*, 1977).

3.1.2.2. *Phosphatidylcholine-Hydrolyzing Phospholipase C: Structure and Role of* Zn^{2+}

The availability of large quantities of highly purified phospholipase C allowed detailed studies on the nature of the enzyme and role that the bound Zn^{2+} plays in its structural stability and catalytic function. The isoelectric point of the enzyme was dependent on the Zn^{2+} content of the preparation, as might be expected. The pI ranged from 6.3 for the apoenzyme to 7.9 for the enzyme saturated with 8–9 moles Zn^{2+} per mole of enzyme. The holoenzyme with 2 moles Zn^{2+} per mole of enzyme had a pI of 6.9 (Bjørklid and Little, 1980). Using a series of amino acid blocking agents, Little's group found that two lysyl, one histidyl, one arginyl, and at least one carboxyl group were required for full activity (Aurebekk and Little, 1977*a*, 1977*b*; Little, 1977*a*; Little and Aurebekk, 1977). One of the two active lysyl groups is located at position 6 of the hydrophobic N-terminal region (Myrnes and Little, 1981). In general, it appeared that blocking the lysyl, histidyl, and carboxyl (but not the arginyl) groups did not affect the ability of the enzyme to bind Sepharose-immobilized substrate. Also, when the activities of the native and blocked enzymes were compared using soluble as well as micellar substrates, it was found that the K_m but not the $V_{\max}$ was altered. A notable exception to this was the finding that blocking the histidyl group altered the maximal activity rather than the K_m (Little, 1977a). An additional four histidyl residues became exposed when the two molecules of Zn^{2+} were removed. This led to the postulate that these four histidyl residues are involved in coordinating the two required molecules of Zn^{2+} in the enzyme (Little, 1977a). Since it was not shown that binding of the substrate or a substrate analog prevented in-activation by the blocking agents, it is not yet certain if the reactive amino acid residues are in the active site.

Zinc plays a crucial role in stabilizing the structure of this enzyme and has some regulatory role in its substrate specificity. Not only did these studies point out the crucial function of Zn^{2+} in the structure but they also provided the rationale for the novel purification scheme currently employed, as pre-viously mentioned (Myrnes and Little, 1980). Based on heat stability studies and using urea as a chaotropic agent, Little concluded that four possible states of the enzyme could be recognized. These proceeded from the native state to a hyperfolded state, then into a reversibly denatured state, and finally into a state that is irreversibly denatured (Little, 1978). The final state was not totally opened since it still contained 2 moles of Zn^{2+}. Removal of the Zn^{2+} also led to a final irreversible denaturation in the absence but not in the presence of 8-M urea. This suggested that while Zn^{2+} stabilized the structure, there was a core stabilized by urea that supercedes the effect of Zn^{2+}. There were indications that the Zn^{2+}-free enzyme was similar to the form of the enzyme obtained when the enzyme with Zn^{2+} was heated. Guanidinium chlo-ride, on the other hand, released the Zn^{2+} from the enzyme irreversibly and exposed all histidine residues (Little and Johansen, 1979). This denaturation could be reversed, however, if Zn^{2+} were added back to the heat-coagulated enzyme. Likewise, treatment with guanidinium chloride and Zn^{2+} restored activity. The stabilization by Zn^{2+} was not totally specific since substitution of Zn^{2+} with other metal ions such as Co^{2+}, Ni^{2+}, and Mn^{2+} did protect some-what if present during guanidinium-chloride-induced unfolding (Little, 1981b). Although details of these analyses were not presented (Little, 1981b), x-ray diffraction studies indicated that the two metal ions are about 0.5 nm apart. No sulfhydryl bridges are present in the enzyme which simplified the de-naturation–renaturation and metal substitution studies.

The metal content of the enzyme apparently affected the active site of the enzyme as well as influenced the stability of the enzyme. The native enzyme that contained two Zn^{2+} ions did not hydrolyze sphingomyelin. However, when Co^{2+} replaced the Zn^{2+}, sphingomyelin was degraded (Otnaess, 1980). [A more recent study indicates that the highly purified phospholipase C that contains Zn^{2+} could degrade sphingomyelin although the rate was no more than 1% of that obtained with phosphatidylcholine (Hetland and Prydz, 1982).] While it has been recognized for some time that the state of the lipid can regulate substrate specificity, the more recent finding that the enzyme with Zn^{2+} and Co^{2+} had greater activity on both erythrocyte ghosts and myelin than either the enzyme with Zn^{2+} or Co^{2+} singly forces the conclusion that very subtle changes in the enzyme–substrate fit can regulate this apparent specificity of hydrolysis (Little *et al.,* 1982). This is not unexpected, however, when one considers the similarity in the structures of phosphatidylcholine and sphingomyelin. The most obvious differences between the two substrates adjacent to the site of hydrolysis are the two carbonyls of the esters at positions 1 and 2 of the glycerol in phosphatidylcholine compared with the hydroxyl at position 3 and the amide linkage at position 2 in sphingomyelin. Apparently, the bulk around base group N is not crucial since phosphatidylethanolamine, which lacks the methyl groups on the N, is a good substrate. It is not yet known if the enhanced activity on sphingomyelin of the enzyme with bound Zn^{2+} and Co^{2+} is the result of direct interaction of the substrate with metal ions or the result of a conformational change in the enzyme structure. It has been suggested that inhibition of activity on soluble dehexanoyl phosphatidylcholine by cations was the result of cation interaction with the Zn^{2+} (Aakre and Little, 1982). If so, the enhanced activity of the enzyme with Zn^{2+} and Co^{2+} on sphingomyelin could be the result of metal–substrate interaction.

High concentrations of large anions such as HCO_3^-, Br^-, Cl^-, NO_3^-, CNO^-, and I^- inhibited the native phospholipase C action on pure phosphatidylcholine or erythrocyte ghosts (Aakre and Little, 1982). This inhibition was related to binding of the anion at a site 4.8 nm away from the nearest Zn^{2+} ($^{113}Cd^{2+}$) binding site (Aalmo *et al.,* 1984). The distance suggested to the authors that inhibition was not due to a metal halide formation but resulted from a specific interaction of the anion with an arginine residue. Arginine was shown to be required for activity in studies using 2,3-butanedione and that I^- could block this irreversible inactivation. Figure 3-1 is a map proposed to relate the two metal ion binding sites with the anion local.

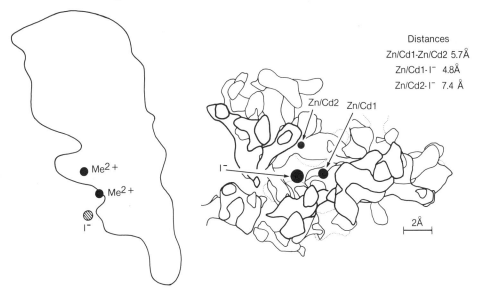

Figure 3-1. *Left:* A molecular outline of phospholipase C. A section through the two metal ions (denoted as Me^{2+}) is shown. The high occupancy I^- is indicated but is actually on a plane above that shown. *Right:* Electron density profiles through 10 sections of the electron density map at 2.8-Å resolution showing the cleft containing the two metal sites and the high occupancy I^-. The sections are at intervals of 1 Å. Interionic distances are shown. (From Aalmo *et al.,* 1984.)

A clearer picture of substrate binding to this phospholipase C was obtained by El-Sayed *et al.* (1985). These workers used a series of short acyl chain phosphatidylcholines and found that monomeric substrates must have acyl chains at least six carbons long to be bound by the enzyme. This is of interest since it suggests some hydrophobic interaction between the substrate and enzyme and that the catalytic site is buried somewhat in the protein. Evidence was also provided that the carbonyl group and the adjacent components of the substrate are also involved in binding. For example, if the acyl ester bonds were replaced by ether bonds, little hydrolysis occurred.

3.1.2.3. *Phosphatidylinositol-hydrolyzing Phospholipase C*

The phosphatidylinositol-specific phospholipase C has been highly purified from culture broth of *B. cereus* and positively identified as the phosphatasemia factor (Ikezawa *et al.*, 1976; Ohyabu *et al.*, 1978). The purification of this phospholipase C was not as successful as that used for the phosphatidylcholine-hydrolyzing phospholipase C in that the recovery of activity and the specific activity was roughly one-third that of the phosphatidylcholine-active enzyme (24 vs. 81% yield and 440 vs. 1400 units/mg protein). Very little has been established about the molecular structure of this enzyme although it differs from the phosphatidylcholine-specific enzyme in at least two important ways: (1) It does not appear to require metal ions, and (2) it can use the hydroxyl at position 2 of the inositol in place of water as the phosphate acceptor (product myoinositol 1,2-cyclic phosphate). One of the interesting aspects of this enzyme is its capacity to release alkaline phosphatase from the cell surfaces. *In vivo*, it appears that the tissues that release the phosphatase are primarily the liver, erythrocyte, and kidney, based on inhibition studies that differentiate between the organ-specific phosphatases. The release of the phosphatase specifically requires the phosphatidylinositol hydrolyzing enzyme since the addition of the phosphatidylcholine hydrolyzing phospholipase C or the sphingomyelinase did not enhance phosphatase release from kidney even though those enzymes can attack the phospholipids in a variety of membranes. It therefore appears that the binding of the phosphatase requires phosphatidylinositol (Ohyabu *et al.*, 1978).

Sundler *et al.* (1978) carried out more detailed studies on this enzyme in order to define its interaction with micelles. They used phosphatidylinositol diluted with phosphatidylcholine and Triton X-100 as substrate to determine the specificity of binding and the effect of surface dilution or *two-dimensional concentration* on the enzyme activity. A ratio of Triton X-100:phosphatidylinositol of 2.5:4 gave optimal activity, some 8- to 10-fold higher than that obtained with pure phosphatidylinositol sonicated in buffer. Higher ratios of Triton X-100:phosphatidylinositol resulted in lowered activity, an observation interpreted by Roberts *et al.* (1978) to be the result of the dilution of phosphatidylinositol in the surface of mixed micelles. However, simple surface dilution was not the sole cause of inhibition by Triton X-100, based on the finding that dilution of phosphatidylinositol with phosphatidylcholine in mixed micelles did not result in reduced activity. These results left open the question of the mode of inhibition by Triton X-100. It was proposed also that this enzyme differed from the pancreatic phospholipase A_2 in that the interaction of the phospholipase C occurred through its active site rather than the "interphase recognition site" that is proposed to be separate from the catalytic site (Verger and deHaas, 1976; see Chapters 9 and 10).

3.1.2.4. Sphingomyelin-hydrolyzing Phospholipase C (Sphingomyelinase)

The sphingomyelinase from *B. cereus* has also been purified to homogeneity and studied with emphasis on the metal and detergent requirements (Ikezawa *et al.*, 1978; Tomita *et al.*, 1982). Interestingly, the specific activity of this enzyme was similar to that of the phosphatidylinositol-specific phospholipase C, about 300–400 units/mg protein. The sphingomyelinase has a requirement for and was purified with bound Mg^{2+}. Under the appropriate conditions, Ca^{2+} and Zn^{2+} stimulated the apoenzyme somewhat. Understanding the nature of the metal ion stimulation was complex since the effect of metal ions on hydrolysis was dependent on whether or not the substrate was soluble or aggregated and on the presence of detergents. Tomita *et al.* (1982) showed convincingly that Ca^{2+} was a competitive inhibitor of Mg^{2+} activation. However, before these results can be appreciated fully, more detailed information on the physical state of the substrate–detergent–metal ion mixtures will be required. It is interesting to note that this enzyme, like other bacterial phospholipases, can degrade lysophosphatidylcholine. Mg^{2+} stimulation of lysophosphatidylcholine hydrolysis was not great and Ca^{2+} had little effect, when compared with the activity on sphingomyelin. Molecular weight determinations of the enzyme varied considerably and were dependent on the method of analysis. The value obtained by SDS–PAGE was 41,000 daltons, whereas gel filtration and sedimentation equilibrium analysis gave a value of about 24,000 daltons. Since glycoproteins are known to have artifactually high molecular weights on SDS–PAGE, it is possible that this enzyme is a glycoprotein, unlike most other bacterial phospholipases C.

3.1.3. Staphylococcus aureus

As with many of the early studies of bacterial phospholipases, the initial interest in *S. aureus* centered on the relation between the nature of the extracellular toxin and the pathogenic characteristics of the organism. It was recognized in the early work on *S. aureus* toxins that more than one toxin was produced by the bacteria and that the relative amounts of the toxins depended on the strain of the organism assayed (Glenny and Stevens, 1935). One of the toxins, termed β, was shown to have phospholipase C (sphingomyelinase) activity (Magnusson *et al.*, 1962; Doery *et al.*, 1963). Two phospholipases C were isolated from the culture fluids of the αβ hemolytic variant strain of *S. aureus:* One was active on sphingomyelin and lysophosphatidylcholine and the other was active on phosphatidylinositol and lysophosphatidylinositol (Doery *et al.*, 1965). These two enzymes were separated by curtain electrophoresis and could be distinguished immunologically and by their pH optima. The phosphatidylinositol-specific enzyme was produced by strains of *S. aureus* that produced only the α toxin as well as the αβ strain that produced both toxins. On the other hand, the sphingomyelin-specific phospholipase C (sphingo-myelinase) was not produced by the α strain. These aspects of the phospho-lipases of *S. aureus* have been reviewed by Wiseman (1975).

3.1.3.1. Phosphatidylinositol-hydrolyzing Phospholipase C

Low and Finean (1977, 1978) purified the phosphatidylinositol-specific phospholipase C from the medium of *S. aureus* cultures with the goal of determining its role in hemolysis and its effect on membrane-bound enzymes. The enzyme, purified some 6000-fold, hydrolyzed about 5 μmoles of phos-phatidylinositol/min/mg, considerably lower than the sphingomyelinase of *S. aureus* (Zwaal *et al.*, 1975; Malmqvist and Mollby, 1981). The phosphatidyl-inositol-specific enzyme had a molecular weight between 20,000 and 33,000 daltons depending on the method of analysis and could, like other phospha-tidylinositol-specific phospholipases C, release alkaline phosphatase and 5′-nucleotidase from mammalian membranes.

3.1.3.2. Sphingomyelin-hydrolyzing Phospholipase C (Sphingomyelinase)

Following the demonstration that purified β toxin was lethal when in-jected into rabbits (Gow and Robinson, 1969) and that the purified toxin had sphingomyelinase activity (Maheswaran and Lindorfer, 1967; Wiseman and Caird, 1967; Wadstrom and Mollby, 1971), considerable interest centered on the properties of the sphingomyelinase and the extent to which phospholipid degradation was related to death. While some of the basic properties of the enzyme have been characterized, little is known about its lethal activity. Ap-parently, two sphingomyelinases are produced by *S. aureus*, a minor form with a pI of 3 and the major cationic form with a pI of 9.4 (Wiseman and Caird, 1967). As is the case with most sphingomyelinases, Mg^{2+} is a required cofactor and can stabilize purified preparations of the enzyme (Mollby, 1978). The anionic species awaits further purification and characterization. Since during the purification of the cationic sphingomyelinase it was noted that the apparent recovery of activity was greater than 100%, it was concluded that an inhibitor is present in the culture medium that was removed during the purification. There is a great inconsistency concerning the specific activity and the molecular weight (26,000–38,000 daltons) of the enzyme in the literature. These differences, according to Mollby (1978), can be ascribed to the differ-ences in the strains of *S. aureus* studied as well as in the assay methods. Recently, Malmqvist and coworkers have attempted to standardize the assay procedure by hydrophobically linking sphingomyelin in a monolayer coating to beads of octyl-Sepharose (Malmqvist *et al.*, 1978; Malmqvist, 1981; Malm-qvist and Mollby, 1981). Using the substrate-coated beads, Malmqvist and Mollby (1981) demonstrated that Ca^{2+} and Zn^{2+} were inhibitors of the Mg^{2+}-stimulated sphingomyelinase. The activity obtained using this assay system was about 30-fold higher than previously reported by Wadstrom and Mollby

(1971). It is not certain if these differences are the result of assay systems used or a reflection of the purity of the enzyme preparation.

Wadstrom and Mollby (1971) demonstrated in studies on hemolytic capacity of the sphingomyelinase that their preparation of enzyme had high hot–cold lytic activity against sheep and bovine erythrocytes but low activity against human erythrocytes that have 18-fold less sphingomyelin. These results were consistent with sphingomyelinase being related to hemolytic activity. In 1971, however, Zwaal *et al.* (1971) isolated the sphingomyelin-specific phospholipase C free of hemolytic activity. The preparation of Zwaal *et al.* (1971) was stabilized by 50% glycerol and had a specific activity about 25-fold higher than that used by Malmqvist and Mollby (1981) which allowed them to use less protein in their studies and reduced any contamination that might facilitate hemolysis. The preparation retained its capacity to degrade the sphingomyelin in human erythrocytes under nonlytic conditions. These results apparently separated the hydrolytic activity of the sphingomyelinase from the hemolytic activity reported by several workers. A similar problem with studies on the lethality of the sphingomyelinase relates to the purity of the preparation used. Mollby (1978) pointed out that even minor contamination by the other membrane-altering toxins could lead to falsely ascribing lethal action to the sphingomyelin-specific phospholipase C. As will be discussed in Chapter 11, the sphingomyelinase from *S. aureus* is active at high surface pressures which, in contrast to a number of other phospholipases, allows it to act on sphingomyelin in natural membranes (Demel *et al.* 1975). The capacity to act at high surface pressures would favor a more pathologic function for this enzyme, relative to other phospholipases that function only at low pressures.

3.1.4. Acinetobacter

The hemolysin activity of several strains of *Acinetobacter calcoacetius* has been identified and related to phospholipase C activity (Lehmann, 1973*a*). Lehman found that the crude preparation from some strains demonstrated the hot–cold hemolytic activity, while others directly hemolyzed red cells. Since the preparations from the two types of strains were distinct immunologically, it would appear that two hemolytic activities were the result of distinct proteins (Lehmann, 1973*a*). Some differences in the hemolytic and phospholipase C activities were noted (Lehmann, 1973*b*), which await explanation. It is of interest that the preparation had leukolytic activity indistinguishable from hemolytic activity (Lehmann and Whg, 1972).

While detailed enzymology was not carried out, it is interesting to note that the apparent substrate specificity was pH dependent; at pH 8.3, phosphatidylethanolamine, phosphatidylcholine, phosphatidylserine, and sphingomyelin were all degraded, whereas at pH 6.3, only the two choline-containing phospholipids were attacked (Lehmann, 1972). While two enzymes with different pH optima in the preparation might be responsible, it is possible that the change in substrate specificity is the result of differences in the charge characteristics of the substrate at the two pH values. The enzyme(s), like many phospholipases C, required Mg^{2+} that is tightly bound to the protein (Lehmann, 1973*b*).

3.1.5. *Streptomyces hachijoensis*

Okawa and Yamaguchi reported the presence of both phospholipases C (Okawa and Yamaguchi, 1975*a*) and D (Okawa and Yamaguchi, 1975*b*) in the culture broth of *S. hachijoensis.* The phospholipase C activities were subfractionated and found to differ in their isoelectric points (5.4 and 6.0) even though they were identical immunologically. Although the substrate specificity was not studied in detail, it appeared to be broadly similar to that reported for the phospholipase C from *C. perfringens.* This enzyme, as well as the phospholipase D from this organism, was optimally active at 50°C. These enzymes should therefore be valuable in studying those phospholipids whose phase transitions are above physiologic temperature. A number of differences were shown to exist between the phospholipases C and D from this organism, demonstrating that they are distinct proteins (Table 3-2).

Since the phospholipase C preparations were not homogeneous, the final characterization of their activities is not certain.

3.1.6. *Genus Pseudomonas*

The most highly purified phospholipase C isolated from this genus was obtained by Arai *et al.* (1974) from *Ps. schuylkilliensis.* Although not thoroughly studied, this enzyme purified some 20,000-fold, hydrolyzed phosphatidyl-choline and phosphatidylethanolamine equally well. Its molecular weight, 23,000 daltons, was considerably smaller than that obtained from *Ps. aureofaciens,* 35,000 daltons (Sonoki and Ikezawa, 1975), *Ps. chlororaphis,* 53,000 daltons (Doi and Nojima, 1971), and *Ps. aeruginosa,* 78,000 daltons (Berka and Vasil, 1982). The highly purified phospholipase C preparation obtained from *Ps. aeruginosa* differed in substrate specificity from the other preparations of phospholipase C produced by these bacteria; lysophosphatidylcholine was hydrolyzed most rapidly and the enzyme only attacked choline-containing phospholipid (Berka and Vasil, 1982). Even though the molecular weights reported for the *Pseudomonas* phospholipases C were quite different, their isoelectric points ranged between 5.5 and 6.5.

The phospholipase C from *Ps. aureofaciens* (Sonoki and Ikezawa, 1975)

Table 3-2. Some Properties of the Phospholipases C and D from S. hachijoensis

	Phospholipase C (I)	Phospholipase D
pH optimum	8.0	7.5
Isoelectric point	6.0	8.6
Metal ion requirement	Mg^{2+}	Ca^{2+}, Mn^{2+}, Co^{2+}
Molecular weight	18,000	16,000
Activity on		
Phosphatidylethanolamine	No	Yes
Phosphatidylserine	No	Yes
Phosphatidylcholine	Yes	Yes

and *Ps. fluorescens* (Doi and Nojima, 1971) was purified using the procedures of ammonium sulfate fractionation, gel filtration, and ion-exchange Sephadex chromatography. The degrees of purity were 3600- and 2500-fold over the culture filtrates, respectively. The enzymes from these two species most effectively hydrolyzed phosphatidylethanolamine, unlike other bacterial phospholipases. Acidic phospholipids were poorly hydrolyzed, if at all. Apparently, the phospholipase C from *Ps. fluorescens* preferred the phosphatidylethanolamine that contained the more highly saturated fatty acids in that *E. coli* lipid was more rapidly hydrolyzed than substrate obtained from rat liver. Although more highly purified, the enzyme from *Ps. aureofaciens* had the lower specific activity (17 vs. 36 units/mg protein). Interestingly, Sonoki and Ikezawa (1976) studied the enzymatic activity in an ether–ethanol solution that contained only sufficient H_2O to introduce the enzyme, 5% of the total volume. Although the absolute rates of hydrolysis were not compared using the two methods of assay, the substrate specificity was changed; in the presence of organic solvent, phosphatidylcholine was the preferred substrate. In all cases, the activity required metal ions; Ca^{2+} was used in most of the studies. Although the conclusion was not drawn, it would appear that both were Zn^{2+}-containing enzymes, like the phosphatidylcholine-specific phospholipase C from *B. cereus* (see Section 3.1.2.2). Detergents were found to have some influence on activity. No real conclusion was drawn with the enzyme from either source concerning surface-charge requirements, although it would appear that the enzyme from *Ps. aureofaciens* could tolerate higher concentrations of acidic detergents than that obtained from *Ps. fluorescens*. With both, the appearance of the enzyme was maximal in midlog growth and only 3–10% of the total activity was cell associated.

More recently, Berka and Vasil (1982) purified the phospholipase C from *Ps. aeruginosa*. It was found that the enzyme could be eluted from DEAE–Sephacryl by tetradecyltrimethylammonium bromide or lysophosphatidylcholine but not by phosphocholine or choline. It was concluded that the enzyme bound to the column by both hydrophobic and specific ammonium bindings sites. The enzyme, purified 1800-fold, had no cysteine residues but was found to be rich in glycine, serine, threonine, aspartate, glutamate, and aromatic amino acids.

Particular attention has been paid to the function and molecular biology of the phospholipase C from *Ps. aeruginosa*. It is one of two hemolytic toxins that are related to the virulence of the organism (Berka *et al.*, 1981; Coleman *et al.*, 1983). The synthesis of the phospholipase C is repressed by high P_i concentrations. A series of elegant studies have appeared which implicate this enzyme in a P_i scavenging system required by the organism in the initiation of infection (Coleman *et al.*, 1983; Lory and Tai, 1983; Vasil *et al.*, 1984). It is postulated that the phospholipase C releases phosphocholine from host tissues that is further degraded to release the P_i for transport into the bacterium. Cloning experiments demonstrated that the DNA fragment containing the phospholipase C (the 6.1 BAM HI fragment) also coded for at least three (out of five) P_i repressible proteins. Also, a transfected strain of *Ps. aeruginosa* that overproduced phospholipase C also transported P_i at four times the rate of the strain with the vector alone (Vasil *et al.*, 1985). When the phospholipase-C-containing DNA fragment was inserted into *E. coli,* the cells were capable of synthesizing the *Ps. aeruginosa* enzyme (Coleman *et al.,* 1983; Lory and Tai, 1983). As with many cloned secretory proteins, *E. coli* was unable to process the foreign proteins and, as a consequence, the phospholipase C remained cell associated. Its distribution within the cell is not settled yet since one group (Coleman *et al.,* 1983) found the enzyme in the cytosol whereas another group (Lory and Tai, 1983) recovered the enzyme in the outer membrane. Also, when expressed in *E. coli,* the phospholipase appears to be constitutive rather than inducible since the P_i content in the medium did not influence phospholipase C synthesis.

3.2. Bacterial Phospholipases D

Classically, phospholipases D have been associated with the plant kingdom. Some notable exceptions to this rule are now recognized and it is quite likely that the phospholipases D will have a wide phylogenetic distribution when studied further. Table 3-3 lists some bacterial phospholipases D along with certain of their properties. As with other phospholipases, the bacterial phospholipases have diverse specificities, functions, and, undoubtedly, mechanisms of action. The phospholipase D from several bacterial sources will be described here—*Corynebacterium, streptomyces, Haemophilus parainfluenza,* and *Vibrio damsela.*

3.2.1. Genus Corynebacterium

A nonhemolytic strain of *Corynebacterium* was first suggested to produce a phospholipase D (Fossum and Hoyem, 1963). They demonstrated that free

Table 3-3. Some Purified Bacterial Phospholipases D

Species	Substrates	pH optimum	Isoelectric point	Molecular weight	Metal requirements
C. pyogenes[a]	"Lecithinase"	—	—	—	—
C. ovis[b,c,d]	Sphingomyelin	7.9	9.8	31,000	Mg^{2+}
S. chromofuscus[e]	Phosphatidylcholine and others	—	5.1	50,000	Ca^{2+}, Mg^{2+} (Zn^{2+} ?)
S. hachijoensis[e]	Phosphatidylcholine and others	—	8.6	16,000	Ca^{2+}, Mg^{2+} (Zn^{2+} ?)
H. parainfluenzae[f]	Cardiolipin	—	—	—	Mg^{2+}
E. coli[g,h] (*S. typhimurium*) (*P. vulgaris*) (*Ps. aeruginosa*)	Cardiolipin	7.0	N.D.	N.D.	Mg^{2+} (ATP)

[a] Fossum and Hoyem (1963).
[b] Linder and Bernheimer (1978).
[c] Bernheimer *et al.* (1980).
[d] Soucek and Souckova (1974).
[e] Imamura and Horiuti (1979).
[f] Astrachan (1973).
[g] Cole *et al.* (1974).
[h] Cole and Proulx (1975).

choline was liberated from egg-yolk lipid emulsions by three different non-hemolytic strains while the closely related *C. pyogenes* appeared to lack what they termed lecithinase activity. Even though neither the identification of the substrate nor the mechanism of choline release was determined, this was an important observation that spurred more detailed work a decade later.

While the phospholipase D released by *C. ovis* is lethal, it is not hemolytic to sheep erythrocytes rich in its substrate sphingomyelin. The crude preparation of enzyme was specific for sphingomyelin, the products being ceramide phosphate and free choline (Soucek *et al.*, 1971). Soucek and Souckova (1974), however, demonstrated that the enzyme preparation could degrade membrane phospholipid in erythrocytes from sensitive species of animals. The enzyme was purified to near homogeneity by Linder and Bernheimer (1978) who determined its molecular weight, 31,000 daltons, and pI, 9.8. The purified preparation, like crude preparations, protected erythrocytes from the hemolytic action of staphylococcal sphingomyelinase (phospholipase C) and hemolysis by helianthus toxin, a protein apparently lacking catalytic activity. In a series of model studies, they found that both the phospholipase D and the helianthus toxin bound to liposomes that contained sphingomyelin. Furthermore, pretreatment of the sphingomyelin-containing liposomes with the phospholipase D prevented rupture of the liposomes caused by the action of helianthus toxin. Two possibilities were given as possible mechanisms of phospholipase D cytoprotection against toxin hemolysis: (1) The phospholipase D directly blocks the binding of the toxin, or (2) the toxin is not capable of binding to the ceramide phosphate liposomes produced by the phospholipase D. The authors concluded that the latter explanation is the more likely.

Since it was known that two nonhemolytic species of *Corynebacterium* potentiate hemolysis when grown in proximity, Bernheimer *et al.* (1980) detailed the nature of this synergism in a series of clear experiments. Based on known observations, they postulated that a factor was produced by *C. equi* that acts upon the phospholipase D (from *C. ovis*) modified erythrocytes causing them to lyse. *C. equi* is a species of *Corynebacterium* that does not produce phospholipase D. In the course of their studies, they demonstrated that the factor produced by *C. equi* is a phospholipase C with broad substrate specificity, including ceramide phosphate. The hemolytic activity exhibited the hot–cold effect characteristic of phospholipase C action on membranes. The sequence of events in hemolysis appears to be, therefore, appear to be as follows:

$$\text{Sphingomyelin} \xrightarrow[\text{C. ovis}]{\text{PLD}} \text{choline} + \text{ceramide phosphate}$$

$$\text{Phosphate} + \text{ceramide} \xleftarrow[\text{C. equi}]{} \text{PLC}$$

(labile membrane)

In the simplest case, it predicted that if this model were correct the phospholipase C from *C. equi* should be hemolytic by itself. They found, however, that phospholipase C action on pure sphingomyelin required both Mg^{2+} and a detergent (0.1% Triton X-100). The detergent was not required for the action of the phospholipase D or the action of the phospholipase C on membranous ceramide phosphate. These results then explain the synergism between the two species of *Corynebacterium* and that the molecular organization of the choline-containing lipids regulates the phospholipase C from *C. equi*.

3.2.2. Genus Streptomyces

A phospholipase D was isolated from *Streptomyces chromofuscus* that is distinct from the one obtained from the *S. hachijoensis* (Imamura and Horiuti, 1979). Its molecular weight was roughly three times larger (50,000 vs. 16,000 daltons) and its isoelectric point was considerably lower (5.1 vs. 8.6) than that of the enzyme from *S. hachijoensis*. One of the most interesting aspects of this report was the experimental approach that allows the separation of the function of the lipophilic binding site from the catalytic site. These workers took advantage of the binding of the hydrophobic binding site to palmitoylated gauze to achieve a 200- to 300-fold purification with nearly quantitative recovery. The enzyme was eluted from the gauze with 0.2% Triton X-100. Following the complete purification from culture filtrates, the enzyme was 1000-fold purified with a recovery of 46% of the original activity. Sixty mg of pure protein could be obtained from 20 liters of culture fluid. This is indeed an attractive procedure that perhaps will have wide application for extracellular phospholipases.

The purified phospholipases D from both species had a broad specificity and phosphatidylcholine was attacked most readily. Metal ions had a rather complex effect and inhibition by EDTA was achieved only at high concentrations (20 mM or greater). This raises the possibility that metal ions are tightly bound to the enzyme, perhaps as Zn^{2+} is bound to the phospholipase C of *B. cereus* (Little, 1981*b*). In the absence of Triton X-100, both Ca^{2+} and Mg^{2+} stimulated activity about 2.5-fold. In the presence of Triton X-100, however, only Ca^{2+} stimulated and other metal ions were inhibitory. Both deoxycholate and Triton X-100 stimulated the activity of the enzyme in the absence of added Ca^{2+}, but in the presence of 10 mM Ca^{2+} detergents caused inhibition at low concentrations (0.8 mM Triton X-100) that was overcome by higher concentrations of detergent (5–6 mM; 2 mM substrate, phosphatidylcholine). While it was concluded that mixed micelles were the preferred substrate, the Ca^{2+} effect is still unclear.

In the course of their studies, it was found that the enzyme was protected from inactivation by absorption to the palmitoylated gauze and that the enzyme was partially active on added substrates while bound to the gauze (Imamura and Horiuti, 1979). The latter observation suggested that the hydrophobic binding site and the catalytic sites were distinct although the low level of activity (10% of control) clouds the conclusion somewhat. A similar study with surface-immobilized phospholipase A_2 from *Naja naja naja* venom indicated that protein–protein interaction was required for optimal activity (Chapters 9 and 10) (Lombardo and Dennis, 1985b). Phospholipid, fatty acids, and detergents displaced the enzyme from the palmitoylated gauze which indicates the nonspecific nature of the hydrophobic binding site. In a related experiment, these investigators found that most of the compounds that affected binding also interacted with the enzyme to alter its heat stability.

3.2.3. Haemophilus parainfluenzae

Detailed studies have been carried out on the positional specificity of the phospholipase D partially purified from the membranes of *H. parainfluenzae* (Ono and White, 1970a, 1970b; Astrachan, 1973). Ono and White postulated that this enzyme, specific for cardiolipin and requiring Mg^{2+}, played a key role in the synthesis of other phospholipids from cardiolipin. One product of hydrolysis, phosphatidylglycerol, could be recycled to cardiolipin while the phosphatidic acid could be converted to phosphatidylethanolamine and phosphatidylglycerol. The specificity of the hydrolysis catalyzed by the phospholipase D was determined by Astrachan (1973). The cardiolipin was first degraded by the phospholipase D of *H. parainfluenzae*. Astrachan then treated the phosphatidylglycerol isolated from the phospholipase D reaction with the phospholipase C from *B. cereus*. The glycerol phosphate derived from this reaction was not a substrate for the L-glycero-3-phosphate dehydrogenase, demonstrating that the glycerol of the phosphatidylglycerol was linked via position 1. The phospholipase D is therefore specific for position 3' of the center glycerol.

3.2.4. Other Gram-negative Bacteria

Like the previously described phospholipase D of *H. parainfluenzae*, a number of gram-negative bacteria were shown by the group of Proulx (Cole *et al.*, 1974; Cole and Proulx, 1975) to contain a phospholipase D specific for cardiolipin. The group studied included *Escherichia coli, Salmonella typhimurium, Proteus vulgaris*, and *Pseudomonas aeruginosa*. Other systems known to contain cardiolipin including yeast and mitochondria did not have a comparable phospholipase D. They found that this phosphodiesterase was maximally synthesized in the late log phase of growth. ATP stimulated hydrolytic activity, especially in the stationary phase when cellular energy levels are low (Cole and Proulx, 1977). The ATP effect is not yet clear; on one hand, it is not specific for ATP and can be replaced by a number of triphosphonucleotides and by ADP, while on the other hand, pretreatment of the enzyme with ATP and subsequent removal of the excess nucleotide by enzyme isolation left the enzyme in an activated form. The latter observation led to the suggestion that a group transfer occurred from ATP. This mechanism will be of further interest since ADP can substitute for ATP. Although the complexity of multiple activation states precluded purification of the enzyme, key characteristics were established such as the pH optimum (pH 7.0), metal ion requirement (5–6 mM Mg^{2+}), and nucleotide stimulation (0.1–0.2 mM ATP).

Based on these and other studies, Audet *et al.* (1975) proposed a polyglycerophosphatidate cycle:

(1)   2 Phosphatidylglycerol $\xrightarrow[\text{cardiolipin synthase}]{Mg^{2+}}$ cardiolipin + glycerol

(2) Cardiolipin $\xrightarrow[\text{PLD}]{}$ phosphatidate + phosphatidylglycerol

Sum phosphatidylglycerol → phosphatidic acid + glycerol

Such a cycle is compatible with metabolic patterns established in *E. coli*. Two key points should be made about this cycle, however. First, even though the phospholipase D can reverse reaction (1) using glycerol as a phosphatidyl acceptor, this enzyme is distinct from the cardiolipin synthase. Second, this appears to be the means by which phosphatidylglycerol is catabolized to phosphatidate since the phospholipase D is incapable of acting on phosphatidylglycerol directly.

3.2.5. Vibrio damsela

A potent extracellular hemolysin has been purified to homogeneity and shown to have phospholipase D activity (Kothary and Kreger, 1985; Daniel *et al.*, 1986*a*). This phospholipase D was capable of degrading the lipids of erythrocytes from mice, rats, damselfish, and rabbits but not those from 10 other species. Since this phospholipase D has a broad substrate specificity, it is not readily obvious why certain species of erythrocytes escape lysis while others are destroyed. Equally puzzling is the observation that Chinese hamster ovary cells were lysed while the enzyme had limited capacity to degrade the phospholipids of the membrane of the Madin–Darby canine kidney (MDCK) cells. The failure to attack the MDCK cells was not due to the ability of the enzyme to hydrolyze the phospholipid mixture, since extracted phospholipids in liposomes were rapidly degraded.

Rather large quantities of the enzyme could be isolated from the growth medium. For example, 30 mg of pure enzyme were obtained from 1100 ml of medium in a 50% yield. This enzyme had a maximal specific enzyme activity of 2200 units/mg when assayed on phosphatidylcholine liposomes. This corresponded to 2000 hemolytic units/mg protein, one of the highest values reported. The enzyme has a molecular weight of 69,000 daltons with a pI of 5.6. Amino acid compositional analysis shows the enzyme to be rich in aspartate, serine, glutamate, and lysine. Since large quantities of this enzyme with a broad specificity can be obtained easily, it could be an important phospholipase for the study of structure–function relation and as a membrane probe.

3.3. Summary

In this chapter and Chapter 2, the phospholipases of bacteria have been covered. Their character and functions are diverse. Also, many have been found to be useful tools in studies in elucidating both chemical and membrane structures. References to a number of these enzymes will be given in the following chapters, especially Chapters 10 and 11.

Chapter 4

Other Microorganisms

Relatively few phospholipases have been studied in microorganisms other than prokaryotic bacteria. As detailed in Chapters 2 and 3, a major interest in bacterial phospholipases has emphasized their relation to toxins. Since many toxins are bacterial in origin, the interest in toxins has not sparked similar studies on phospholipases in higher microorganisms. In this chapter, emphasis is placed on the mechanism of action of the rare phospholipase B, one of the few known examples of a single enzyme that removes both acyl groups.

4.1. General Comments

Most of the phospholipases described thus far in eukaryotic microorganisms are cellular and in most cases are acyl hydrolases. Exceptions to the generality of the cellular distribution of phospholipases are known. Barrett-Bee *et al.* (1985), for example, found a correlation between the acyl hydrolase activity in the culture fluid of a group of yeasts and their pathogenicity. A stain of pathogenic *Candida albicans* with marked pathogenicity in mice had nearly 10 times the phospholipase A activity of nonpathogenic *Saccharomyces cerevisiae* and *Candida parapsilosis*.

On the other hand, the study of eukaryotic microorganisms holds great promise in elucidating the role that phospholipases play in cellular function. It is predicted that the studies reported here will lead to further studies on the cellular function of these enzymes. The unusual presence of highly active phospholipases B makes their function even more intriguing.

4.2. Penicillium notatum

Perhaps one of the most exhaustively studied acyl hydrolases in microorganisms is the phospholipase B in *P. notatum*, first recognized by Fairbairn in 1948. Early studies showed that the enzyme preparation produced only free fatty acids and a water-soluble phosphate compound (glycerophosphocholine) (Beare and Kates, 1967). This system is still under active investigation, primarily by Saito and coworkers. Perhaps one reason for the continued interest is the early recognition that a single protein might catalyze the hydrolysis of both ester bonds, thereby having both phospholipase activity and lysophospholipase activity. This concept faced the continued criticism that the dual activities could be the result of multiple enzymes present in the preparations used, namely, the combined activities of a phospholipase A and a lysophospholipase. Despite this possibility, the group of Saito undertook the task of characterizing the individual activities catalyzed by the enzymes that constitute phospholipase B activity, that is, phospholipases A_1 and A_2 and lysophospholipase. This work stands as the true first example of a phospholipase B.

Crucial to this work was the purification of the enzyme, first achieved in 1973 (Kawasaki and Saito, 1973). While the procedures employed were rather straightforward, instability of the enzyme during the later stages of purification was a problem. For reasons that remain unexplained, these workers found that the presence of EDTA helped to maintain enzyme activity. The final preparation, purified some 2300-fold over the original extract, had an apparent molecular weight of 113,000 daltons, as determined by gel filtration, an isoelectric point of 4.0, optimal activity on phosphatidylcholine and lysophosphatidylcholine at pH 4.0, and did not require metal ions. It was recognized that the molecular weight estimation could be erroneous owing to the high carbohydrate content (Kawasaki and Saito, 1973). Indeed, later studies using SDS–PAGE indicated that the molecular weight was 90,000 daltons (Okumura *et al.*, 1980). Two later modifications of the purification increased the yield roughly five- to seven-fold (8% vs. 40–60%) and increased the specific activity nearly two-fold up to 5200 units/mg protein (Imamura and Horiuti, 1980; Okumura *et al.*, 1980). The latter group applied affinity chromatography with phosphatidylserine–AH Sepharose while the first group employed hydrophobic chromatography on palmitoyl cellulose. The studies with affinity chromatography gave some interesting insights into the lipid binding vis-à-vis the catalytic activity of the enzyme. Based on the lack of binding to other phospholipid affinity columns such as phosphatidylethanolamine–CH Sepharose and glycerophosphate–CH Sepharose, these workers concluded that both the base moiety and acyl groups might be essential in the binding of the enzyme to the phospholipid. The observation that the enzyme did not deacylate any of the phospholipid Sepharose gels suggested that the lipid binding domain was distinct from the catalytic site. A similar two-site binding mechanism has been proposed for snake venom and pancreatic phospholipases (Verger *et al.*, 1973; Pluckthun and Dennis, 1982*b*).

A series of detailed studies were carried out to understand the nature of the reactions catalyzed by the purified phospholipase B. Initially, it was determined that the maximal activity was 100 times higher on pure lysophosphatidylcholine than on egg-yolk phosphatidylcholine (Kawasaki and Saito, 1973). The rate of hydrolysis of either lysophosphatidylcholine or phosphatidylcholine was inversely proportional to the acyl chain length when the acyl group was varied from decanoyl to octadecanoyl chains (Saito and Kates, 1974). Also, unsaturated lipids were deacylated more rapidly than the corresponding saturated lipid. Since it was found that the effect of the chain length on the rate of the hydrolysis was minimized by the presence of 10% ether in the reaction mixture, it would appear that physical organization of the lipid molecules markedly influences enzymatic activity (Table 4-1).

The enzyme also showed specificity for the polar head group. With the exception of cardiolipin, the highly acidic phospholipids were degraded more rapidly than the neutral phospholipids phosphatidylcholine and phosphatidylethanolamine (at pH 4.0, the pH of the assay, phosphatidylethanolamine is near its isoelectric point, as determined by microelectrophoresis). While the reactions were run in 10% ether, Saito and Kates (1974) reported that the clarity of the substrate dispersion followed the order of hydrolysis; thus, dispersion of the substrates was recognized as a problem.

A hypothesis for the mechanism of action of this enzyme was proposed in this work that served as the experimental model for future studies on the phospholipase B. Saito and Kates (1974) proposed that the enzyme has two binding sites, Site I for phosphatidylcholine and Site II for lysophosphatidylcholine. The following sequence of steps was visualized:

1. Binding of phosphatidylcholine to Site I.
2. Transfer of the acyl group at position 2 to the catalytic site (hydrolysis) and concomitant binding of lysophosphatidylcholine to Site II. (Preliminary evidence, later verified, indicates that the acyl chain at position 2 is removed prior to that at position 1.)
3. Transfer of the acyl group originating from position 2 to H_2O.
4. Transfer of the acyl group from position 1 to the catalytic site (II).
5. Transfer of the acyl group originating from position 1 to H_2O.

The hydrolysis of lysophosphatidylcholine would involve steps 3–5 only.

Probing the active site(s) of the enzyme with chemical modifiers failed to distinguish between the two proposed sites. *N*-bromosuccinimide and 2-hydroxy-5-nitrobenzyl bromide, thought to react with tryptophan residues and phenylglyoxal, reactive with arginine, inhibited the phospholipase B and lysophospholipase activity in a comparable fashion (Sugatani *et al.*, 1980).

Table 4-1. *Effect of Substrate and Solvent on Phospholipase B Hydrolysis*[a]

| | Substrate[b] | | | |
| | Phosphatidylcholine | | Lysophosphatidylcholine | |
Chain length	+ Ether	− Ether	+ Ether	− Ether
C10:0	44	55.0	6620	5550
C14:0	25	15.0	6100	4810
C18:0	11	5.3	3510	2300
C18:1	14	6.3	4140	5030

[a] From Saito and Kates (1974). The hydrolysis of phosphatidylcholine and lysophosphatidylcholine by the phospholipase B from *P. notatum* was run in the presence (+) or absence (−) of diethylether.

[b] The values given are nmoles hydrolyzed per 5 min per μg protein (Saito and Kates, 1974).

Until more is known about the mechanism of action of the enzyme, these studies can only raise the possibility that there is a commonality in a portion of the two proposed active sites. These latter authors postulated that a covalent bond could exist between the acyl groups and the enzyme. This would suggest the possibility that the enzyme could catalyze a transacylation reaction, such as found with the hepatic lipase [monoacylglycerol acyl transferase (Waite and Sisson, 1973a)] and the lysophospholipase from the cytosol of rat liver (van den Bosch *et al.*, 1965; Erbland and Marinetti, 1965a). However, from the studies reported thus far, the question of covalent acyl binding remains open.

The finding that Triton X-100 alters the activity of the enzyme on lysophosphatidylcholine relative to phosphatidylcholine was useful in verifying that the enzyme first removes the acyl group from position 2, followed by the hydrolysis at position 1 (Kawasaki *et al.*, 1975). The addition of Triton X-100 allowed the use of greater effective concentrations of phosphatidylcholine even though the apparent affinity for the substrate was decreased (K_m increased from 1.5 to 25 mM in the presence of Triton X-100). Using optimal phosphatidylcholine and Triton X-100 concentrations, the apparent V_{max} was 180 units/mg protein compared with 220 units/mg protein using lysophosphatidylcholine under the same conditions. The activity on lysophosphatidylcholine was inhibited by Triton X-100 so when comparison was made of the hydrolysis of phosphatidylcholine with Triton X-100 to that of lysophosphatidylcholine in the absence of Triton X-100, a ratio of 1:16 was obtained. By using Triton with 1-[^{14}C]stearoyl-2-[^{14}C]oleoyl glycerophosphorylcholine as substrate, sufficient lysophosphatidylcholine accumulated to show a 10-fold enhancement of [^{14}C]stearoyl over [^{14}C]oleoyl lysophosphatidylcholine, demonstrating a preference for position 2 of diacylphosphatidylcholine.

Thus, the following scheme was proposed:

$$\text{Phosphatidylcholine} \xrightarrow{\text{"}PLA\text{"}_2} \text{1-acyl glycerophosphorylcholine} + \text{fatty acid}$$

Fatty acid + glycerophosphorylcholine $\xleftarrow{\text{"}lysophospholipase\text{"}}$

The sum of the "phospholipase A_2" and "lysophospholipase" activity would therefore be the overall phospholipase B activity. Further studies with other substrates supported this scheme. For example, 1-acyl glycerophosphorylcholine was hydrolyzed at least 15 times faster than 2-acyl glycerophosphorylcholine (Sugatani *et al.*, 1978). Also, phospholipase A_2 activity was demonstrated using 1-*O*-alkyl- or 1-*O*-alk-1′-enyl-2-acyl glycerophosphorylcholine even though the diacyl compound was degraded somewhat more rapidly. Of the various neutral glycerides studied, only monoacylglycerol was hydrolyzed, although at low rate. They suggested that the difference in the hydrolytic rates of the two-positional isomers of lysophosphatidylcholine "would be due to different binding sites on the enzyme." Most likely, 1-acyl glycerophosphorylcholine is binding at Site I, the site with lower affinity for the diacylphospholipid.

The various activities of the enzyme were further dissected using 2,3-

dipalmitoyl-*sn*-glycero-1-phosphorylcholine that was degraded only at position 3, hence, phospholipase A_1 activity (Sugatani *et al.,* 1980). The substrates that were used to study the various activities of the phospholipase B are:

Activity	Substrate
Phospholipase B	1,2-Diacyl-*sn*-glycero-3-phosphorylcholine
Phospholipase A_1	2,3-Diacyl-*sn*-glycero-1-phosphorylcholine
Phospholipase A_2	1-Alkyl (or 1-alk-1′-enyl)-2-acyl glycerophosphorylcholine
Lysophospholipase L_1*	1-Acyl-glycerophosphorylcholine
Lysophospholipase L_2*	2-Acyl-glycerophosphorylcholine

*Designates position of attack, similar to the phospholipases.

The two model compounds used for the phospholipase A_1 and A_2 activities were degraded at roughly the same rate, approximately at one-half the rate of the corresponding diacylphosphatidylcholine. Since the rate of removal of the acyl chain in Site I is about equal for positions 1 and 2, it would appear that the lower rate of degradation of the two model compounds would be at the level of substrate–enzyme interaction (Michaelis-complex formation) rather than at the level of catalysis or the liberation of product.

During the course of purification using the phosphatidylserine affinity column, Okumura *et al.* (1980) found that modification of the enzyme can occur as the result of proteolysis by an endogenous protease. The cleavage(s) was thought to occur at a few sites only and may not result in the loss of small peptide fragments. Unless the enzyme preparation was electrophoresed with β-mercaptoethanol, no difference was detected in the weight of the enzyme (90,000 daltons). The proteolytically cleaved preparation had, in addition to the parent protein, peptides of 68,000, 38,000, and 33,000 daltons and proteolysis could be prevented if phenylmethylsulfonyl fluoride was added during cell disruption. The activity of the modified enzyme on lysophosphatidylcholine was not altered, whereas the attack on phosphatidylcholine was reduced some four-fold in the presence of Triton X-100 and 10-fold in the absence of the detergent. The evidence thus far supports their conclusion that both phospholipase A_1 and A_2 activity remains intact but that the ability of the enzyme to interact with substrate at the lipid–water interface is impaired. Since micellar substrate is more easily penetrated than bilayer vesicles, proteolytic modification has little effect on the hydrolysis of lysophosphatidylcholine or of dioctanoyl glycerophosphorylcholine which are known to be cylindrical micelles. These workers propose that the endogenous proteolytic cleavage might serve as a regulatory process in the cell although direct evidence for this is lacking.

A role for the carbohydrates of the enzyme has likewise been proposed to have an influence on the action of the enzyme. Okumura *et al.* (1981) postulated that proteolytic modification of the enzyme exposed more hydrophilic carbohydrate, thereby hindering the enzyme's interaction at the lipid–water interface. This postulate was based on the finding that about 55%

of the carbohydrate can be removed from proteolytically modified enzyme while only 30% is removed from the native enzyme by endoglycosidase H treatment. The number of sugars removed from the protease-cleaved enzyme was roughly 10 and reduced the molecular weight from 90,000 daltons to about 78,000 daltons. Although the site(s) of removal was not determined, it is known that all the protease-modified peptides contained carbohydrate (Okumura *et al.*, 1980). The role of carbohydrate in regulating the interaction with the lipid–water interface was supported by the finding that hydrolysis of phosphatidylcholine, but not lysophosphatidylcholine, was stimulated by removal of carbohydrate, and this was most pronounced with the proteolytically cleaved enzyme.

Clearly, this system has a number of exciting possibilities for further study, especially in the realm of enzyme–substrate interaction. Based on what has been established, it would be predicted that the details of the mechanism of action of this enzyme should be forthcoming in the near future, that is, near future recognized in the timetable of lipid enzymology; it took nearly three decades of work on this enzyme to bring it to its current status.

4.3. Saccharomyces cerevisiae

One might believe that an organism as useful to the researches, breads, and beverages of humans as *S. cerevisiae* would be better understood with regard to its lipid metabolism in general and its phospholipases in particular. Some of the earliest reports on the presence of phospholipases in *S. cerevisiae* were in the early 1960s. While there is only a scanty idea of the various lipolytic activities present in *S. cerevisiae*, some progress has been made in the purification of one enzyme, a deacylase with apparent phospholipase B activity.

4.3.1. Acyl Hydrolases (Phospholipase B)

In 1963, two groups reported that extracts of *S. cerevisiae* could degrade phosphatidylcholine. One group claimed that the extracts contained a phospholipase C (Harrison and Trevelyan, 1963), while a second group (Kokke *et al.*, 1963) provided evidence that a phospholipase A coupled with a lysophospholipase or a phospholipase B led to complete deacylation of phosphatidylcholine. Subsequent work by van den Bosch *et al.* (1967) verified the deacylation pathway by using [^{32}P]phosphatidylcholine and [^{32}P]lysophosphatidylcholine. Both substrates gave rise only to glycero-[^{32}P]phosphorylcholine.

Some important work has been done to characterize a phospholipase from the plasma membrane with B-type activity (Ichimasa *et al.*, 1984; Witt *et al.*, 1984a). Two procedures were given for the purification from membrane vesicles; one is a two-step procedure that yields 15-fold purified preparation with 70% recovery when assayed using dipalmitoyl phosphatidylcholine as substrate (Witt *et al.*, 1982). Key to these procedures was extraction of the vesicles with the zwitterionic detergent SB 12 or cholate. Some properties of the enzyme preparation have been characterized; it has a molecular weight of 145,000 daltons, a broad substrate specificity, is strongly stimulated by anionic detergents, is inhibited by a number of cations, and has optimal activity at pH 3.0–4.5. The latter two properties are similar to those of the lysosomal phospholipases from rat liver, although the yeast enzyme was localized in the plasma membrane. Since the ratio of lysophospholipase to phospholipase activity decreased during purification, a designation as to the type of phospholipase(s) under investigation cannot yet be made. Studies on the level of this enzyme in glucose deprivation should be of interest since the level of glucose is known to regulate phospholipid catabolism (Angus and Lester, 1975).

A subsequent study reported the separation of two phospholipase B activities, B_1 and B_2. B_1 had an apparent molecular weight of 220,000 daltons, whereas B_2 was 145,000 daltons. Removal of the carbohydrate on the enzymes yielded proteins of 52,000 and 67,000 daltons, respectively (Witt *et al.*, 1984*b*). The two were devoid of activity when the carbohydrate was removed and it was proposed that the carbohydrates of the two were different. Furthermore, the protein structure was different, as shown by the lack of cross-reactivity of antibodies to the two enzymes. It will be of interest to see what relation, if any, these enzymes have to the phospholipase B of *P. notatum*.

A phospholipase B similar to the plasma membrane enzyme was found to be secreted for *S. cerevisiae* (Witt *et al.*, 1984*b*) during log-phase anaerobic growth. The only significant difference between this enzyme and the membrane enzymes found was the carbohydrate content. These interesting enzymes warrant detailed study and comparison in mechanism of action and function during cell growth.

4.3.2. Phosphodiesterases (Phospholipases C and D)

Angus and Lester (1972) provided additional insight into phospholipid catabolism in *S. cerevisiae*, in particular the degradation of phosphatidylinositol. The conclusion that deacylation was the pathway of phosphatidylinositol degradation was based on the observation that yeast grown on $[^{32}P]_i$ and $[^3H]$inositol liberated glycerophosphorylinositol with a constant isotopic ratio. The possibility was raised that a phospholipase C was active and that the phosphorylinositol could be reincorporated directly into one of the other inositol-containing lipids. Such a possibility would be of interest to pursue, especially in light of the earlier report of a phospholipase C in this organism (Harrison and Trevelyan, 1963).

The catabolism of cellular phospholipid took on additional significance in the subsequent work of Angus and Lester (1975) when they reported that the accumulation of glycerophosphorylinositol, as well as other glycerophosphoryl bases, was inhibited when the energy source, glucose, was removed. This activity was thought to reside in the plasma membrane, at least in part, since similar results were obtained when exogenous lipid was the precursor of glycerophosphorylinositol. An interesting effect was noted in the absence of glucose: Free inositol accumulated rather than glycerophosphorylinositol.

While no direct evidence was provided, the authors suggest that the activity responsible for the release of inositol was a phospholipase D. The addition of glucose abruptly reversed these activities; within 10–20 min the accumulation of glycerophosphorylinositol was resumed and no further increase in free inositol was found. The effect of glucose deprivation could be blocked after an hour, indicating that protein synthesis could be a factor involved, in addition to the direct effect of glucose on the system. As with a number of other systems that completely deacylate phospholipids, only small amounts of lysophosphatidylinositol accumulated as was predicted by their finding that lysophospholipase activity was greater than that of the phospholipase A.

A role for a phospholipase D in yeast was also described by Grossman *et al.* (1973), who investigated the disappearance of mitochondria during glucose repression. The process of mitochondrial decay, measured by the loss of succinate dehydrogenase activity, was inhibited by cycloheximide but not by chloramphenicol. These findings suggest that a factor(s) is synthesized extramitochondrially that is responsible for this decay. They demonstrated that a phospholipase D activity could be induced some 3 hr following the addition of glucose, similar to the kinetics of decrease in succinic dehydrogenase activity. Comparable to the effects on succinic dehydrogenase, cycloheximide, but not chloramphenicol, inhibited the induction of phospholipase D activity. While it appears that the enzyme was synthesized extramitochondrially, it was associated with the mitochondrial fraction. It required, however, a heat-stable factor from the cytosol for full activity. The lipid product of hydrolysis, phosphatidic acid, was rapidly degraded although the mechanism was not established. The findings of these two groups are of considerable significance and provide a good system to study the function of phospholipases in cells as agents responsible for the removal of effete organelles.

4.4. Amoebic Phospholipases

4.4.1. Naegleria fowleri

Recently, interest in this water-borne amoeba centers on its association with a fatal meningoencephalitis. Pathogenic strains of organism produced a factor(s) that degraded sphingomyelin (Chang, 1979) and phosphatidylcholine (Cursons *et al.*, 1978). Hysmith and Franson (1982*a*, 1982*b*) undertook a more detailed investigation of the nature of the lipolytic activities released from pathogenic strains of the amoebas. While to date the enzymes described have not been completely purified, their work does demonstrate that there appear to be several rather active lipolytic enzymes released from a pathogenic, but not nonpathogenic, strain. These enzymes include, at a minimum, three phospholipases A, a lysophospholipase, and sphingomyelinase that varied in their time of release from the cells, pH optimum, and influence of Ca^{2+}. Similar enzymatic activities were recovered in cell extracts; however, subcellular localization was not determined. Interestingly, the lipolytic activities in the cell resemble those reported in heart, liver, and other tissues.

Media from cultures grown 72 hr were capable of degrading the major phospholipids in human myelin. This led to the speculation that the lipolytic enzymes secreted by the virulent *N. fowleri* could be involved in amoebic meningoencephalitis.

4.4.2. Acanthamoeba

There is a possibility that extracellular phospholipases might play a role in meningoencephalitis caused by *A. culbertsoni*. Lal and Garg (1979) found that this organism decreased the phospholipid content in brains of mice infected with the organism. It had been reported earlier that *A. castellanii* had activated phospholipase(s) in all cell fractions (Victoria and Korn, 1975a). Although the mitochondria had the highest specific activity of phospholipases A_1 and A_2, some activity was associated with the plasma membrane. Furthermore, the plasma membrane and the cytosol contained an active lysophospholipase, relative to the other cell fractions. With the exception of these two lysophospholipases, little further characterization was done (Victoria and Korn, 1975b). The soluble phospholipase appeared to be most active on monomeric rather than micellar lysophosphatidylcholine, based on the observation that activity increased by increasing substrate concentrations only up to the CMC, 39 μM. Above 39 μM, no further increase in activity was noted. Different and more complex kinetics were found for the plasma membrane-associated lysophospholipase, undoubtedly the result of complications introduced by the membrane lipid.

While work on the phospholipases of amoeba is still in its infancy, these organisms, similar to yeast, offer many advantages in studying the role that phospholipases play in cellular function. In addition, the finding that phospholipases might in themselves be toxins adds additional significance to the study of amoebic phospholipases. Since little is known about the pathogenic mechanisms of these organisms, such efforts certainly seem warranted.

4.5. Tetrahymena pyriformis

Considerable information exists on the lipid metabolism of *Tetrahymena*, especially with respect to membrane changes that occur upon temperature shifts. The work of G. A. Thompson, Jr., and his colleagues has implicated a major role for phospholipases in the rapid shift of the acyl composition of *Tetrahymena* membranes (Dickens and Thompson, 1982; Ramesha and Thompson, 1982; Ramesha *et al.*, 1982). This work led Nojima and coworkers to study systematically the phospholipases of this organism (Arai *et al.*, 1985). They found at least three distinct phospholipases that differed in their pH optima, Ca^{2+} dependency, and substrate and positional specificities. Phospholipase A_1 and A_2 activity was found in both the acidic (pH = 3–5) and alkaline (pH = 9) ranges. An active phospholipase C was also detected in the acidic pH range. Interestingly, the phospholipase A active at pH = 9 degraded only phosphatidylethanolamine, whereas that active at pH = 3–5 de-

Table 4-2. Subcellular Distribution of Ca^{2+}-Dependent Alkaline Phospholipase Activity in Tetrahymena[a]

Subcellular fractions	Specific activity (released FFA[b]) (nmoles/min/mg)	Total protein (mg)	Total activity (nmoles/min)	Percentage of total
Pellicle	20.09	47.74	959.10	42
Mitochondria + lysosomes	5.46	74.58	407.21	18
Microsomes	1.19	37.43	44.54	2
Cilia	4.19	7.73	32.39	1
Ciliary sup	2.98	77.60	231.25	10
Postmicrosomal sup	7.62	77.80	592.84	26

[a] From Arai *et al.* (1985). An incubation mixture containing 100 nmoles of 1-[1-^{14}C]palmitoyl-2-[1-^{14}C]linoleoyl-glycerophosphoethanolamine, 100 mM glycine–NaOH (pH 9.5), and 5 mM CaC$_2$ in a total volume of 250 μl was incubated with each fraction for 30 min at 37°C. All data in the table have been corrected for the activity in the presence of 5 mM EDTA which was less than 10% of the activity in the presence of 5 mM CaC$_2$ in each fraction.

[b] FFA, free fatty acid.

graded phosphatidylcholine as well. Table 4-2 shows the subcellular distribution of these Ca^{2+}-dependent phospholipase(s) active at pH = 9.5. Even though the substrate contained radiolabeled fatty acids in positions 1 and 2 of the glycerol, no quantitation of phospholipases A_1 and A_2 was made although mention was made of high phospholipase A_1 activity in the pellicle fraction.

Arai *et al.* (1985) point out a number of at least superficial similarities between the phospholipases of this single-cell eukaryote and liver that warrant more extensive investigation. Also, the regulation of these enzymes under temperature stress could lead to a much needed understanding of the regulation of phospholipases involved in the deacylation–reacylation pathway.

Plant Phospholipases

This chapter begins the description of phospholipases of multicellular organisms. While similar problems exist between studies on phospholipases in single and multicellular organisms, their functions more than likely are different. On the other hand, undoubtedly some common themes are to be noted such as the functioning of phospholipases in the turnover of membrane phospholipid. As pointed out in the introduction to Chapter 4, single-cell eukaryotes do have some types of phospholipase that have counterparts in multicellular organisms. An understanding of the relation between the phospholipases of these two broad classes of organism is just now emerging. Comparative biochemistry of phospholipases should become an important and fascinating study for the future.

5.1. General Considerations

Historically, probably the first report of a plant phospholipase came from Contardi and Ercoli (1932) who reported that phosphatidylcholine was degraded by rice bran only after one acyl chain was removed. This would imply that lysophospholipases, but not phospholipase A, might be abundant in plants. That observation raises the question as to whether or not direct deacylation is a significant means of phospholipid degradation in plants. On the other hand, plants of many families traditionally have been the sources of phospholipase D. While phospholipase D is the primary phospholipase studied in plants, the study of this enzyme has not advanced as far as the study of other phospholipases despite the activity of a few groups such as that of Heller, Dawson, Kates, Long, Hanahan, and Shapiro. At present, few groups are actively studying plant phospholipases. Some work done on the plant phospholipase D has emphasized its function as a membrane probe and biochemical tool in lipid synthesis because of its capacity to transphosphatidylate. This has proved to be quite valuable and further studies on the mechanism of action of the enzyme(s) will add new insight into membrane structure and lipid–protein interaction. On the other hand, some investigators have emphasized more the role of phospholipase D in plant physiology by studying its tissue and subcellular distribution and its disappearance during maturation of tissues. In that regard reading this literature is somewhat reminiscent of perusing a cookbook, what stems or leaves to use, how long to blanch the carrots, and so on. Unfortunately, nothing was found on the phospholipase D in marinated artichoke hearts.

Since the vast majority of work has centered on phospholipase D, only the phospholipase D will be covered. While there is evidence that other phospholipases exist in plant tissues such as phospholipase A and/or B and phospholipase C, most work using crude tissue homogenates or cell fractions has demonstrated that relatively few other lipolytic enzymes compete with phospholipase D for substrate. Systems exist, however, to degrade phospholipids completely through pathways seemingly initiated by phospholipase D. In addition to phospholipase D, plants contain glycolipases and sulfolipases. Since they are not phospholipases, strictly speaking, they will not be covered. Although somewhat old at this point, two excellent reviews in this area are available (Kates, 1970; Heller, 1978). Since there are no clear-cut distinctions between the sources of phospholipase D, this chapter is organized in a historical fashion.

5.2. Initial Studies

The first direct reports of phospholipase D in plant tissue were made shortly after World War II. Using basic chemical assays similar to those of Macfarlane and Knight (1941), Hanahan and Chaikoff (1947, 1948) demonstrated that a filtrate of carrot homogenates removed choline and other nitrogenous bases from soybean phospholipid. No free fatty acid, phosphate, or glycerophosphate was liberated. Maximal activity was found between pH values of about 5.0 and 6.0 and higher activity occurred at 25°C than at 37°C in buffered incubations. The carrot phospholipase D preparation was rather heat stable, some activity surviving heating at 95°C for 15 min in the crude homogenate and 7.5 min blanching of carrot slice. Similar results were obtained with cabbage leaf homogenates, demonstrating that the phospholipase D was not unique to carrot. It was also concluded that the enzyme was active *in situ* since blanched leaves had a nitrogen:phosphorus ratio nearly three times that of control leaves, a finding that would be predicted by the activation of the phospholipase D. These authors suggested that the enzyme was activated by the homogenization treatment and extraction with organic solvents. These works substantiated the existence of the last of the four types of lipolytic activity postulated by Contardi and Ercoli (1932), even though the sequence of activities proposed was simpler in this pioneering study than that now known.

Within a few years, additional studies on plants were initiated and the basic characteristics of the phospholipase D reaction established. (Historically, there is some confusion in the terminology of the phospholipases. The term phospholipase C used by Kates in early studies is now referred to as phospholipase D.) The initial characterization of its subcellular localization showed it to be in the chloroplast of spinach, sugar beets, and cabbage leaves (Kates, 1953) although this is by no means universal. Optimal activity was observed between pH 4.7 and 5.8, depending on the source of the chloroplasts. Diethyl ether markedly stimulated the hydrolysis although it did not solubilize the enzyme. It was shown in this and subsequent work that certain organic solvents caused the chloroplasts and lysosomes to coalesce (Kates and Gorham, 1957). It appeared that other lipid-degrading enzymes were present, based on the slow release of inorganic phosphate. Since other possible products such as diglyceride and lysophosphatidic acid have not been isolated in these systems, the pathway(s) involved remain obscure. The possibility of a phospholipase D coupled with phosphatidate phosphatase could be involved since Kates (1953, 1954a) demonstrated that phosphatidate could be cleaved to liberate free phosphate. He also was able to demonstrate the accumulation of glycerophosphate and that chloroplasts could also degrade glycerophosphate but not glycerophosphorylcholine or lysophosphatidylcholine. The following scheme outlines the most probable pathways, although the action of phospholipase C was not ruled out (Kates, 1954b):

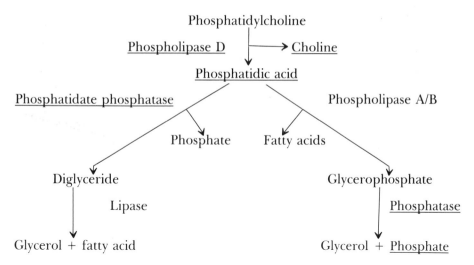

The underlined enzyme activities and compounds were detected.

It is important to recall that studies such as these were done without the benefit of isotopic tracers and therefore were relatively time consuming and complex. Indeed, the presence of endogenous precursors and products in the chloroplasts made the analysis of some compounds such as free fatty acids impossible.

A more recent study on the phosphatidate phosphatase has shown that two distinct enzymes are present in mung bean cotyledons (Herman and Chrispeels, 1980). A phosphatase active at pH 7.5 was detected in the endoplasmic reticulum, while the second active at pH 5.0 was richest in protein bodies. These protein bodies, the *lytic compartment* of the parenchymal cells, also contained the phospholipase D. These results implicate the left branch of the above scheme to be significant in plant phospholipid catabolism.

The question of phospholipase D stimulation by organic compounds is of interest since it gives some insight into the interaction of the enzyme and its aggregated substrate. Kates (1957) investigated the nature of enzyme regulation by lipid-interacting compounds by studying systematically the effect of a wide variety of organic solvents and detergents on phospholipase D activity. It was concluded that solvents with relatively symmetrically situated hydrocarbons on each side of an oxygen would promote chloroplast and liposome interaction and catalytic activity.

5.3. Purification of Phospholipase D

As with all enzyme systems, the finding of soluble phospholipase D helped immensely in its further purification and characterization. One of the earliest sources for a soluble phospholipase D was cottonseed. Tookey and Balls (1956) succeeded in purifying the enzyme some 40- to 80-fold starting with a defatting procedure that employed hexane extraction of mealed seeds. While some characteristics of the preparation were similar to those reported by Kates, unlike the finding of Kates, this preparation was not stimulated by diethyl ether. They also found an inhibitor of phospholipase D in the soluble fraction of cabbage but not cottonseed that may have accounted for Kates' inability to detect enzyme activity in the soluble fraction of cabbage leaf homogenates. Even though the preparation was far from homogeneous, the preparation hydrolyzed phosphatidylethanolamine as well as phosphatidylcholine. As shown in subsequent works with even more highly purified systems, the enzyme does have a wide substrate specificity. Table 5-1 summarizes some of the characteristics of plant phospholipases D.

The classic procedure for partial purification of the phospholipase D was described by Davidson and Long (1958). They found brussels sprout and savoy cabbage to be the best sources of the enzyme and, like many subsequent workers, selected their source according to availability at the local market. The enzyme in these sources was associated with both particulate and soluble portions of the cell and extraction with organic solvent was helpful in purification. While the relation between the enzyme in the particulate and soluble fractions is still unclear, the particulate form was tightly bound in membranes, raising doubts that the soluble form had merely washed free of the placids during homogenization. Their 40-fold purified preparations, unlike the preparation of Tookey and Balls, required ether or some other activator when diacyl but not monoacyl phospholipids underwent degradation. Likewise, they demonstrated that Ca^{2+} was required for activity and that the amount required for optimal activity was proportional to the amount of substrate present. The concentration of Ca^{2+} required for optimal activity, usually 20–40 mM, suggests that Ca^{2+} is involved in more than the catalytic event. Other divalent cations such as Ba^{2+} and Sr^{2+} appear to stimulate some but not all enzyme preparations. They extended the number of phospholipids that could be attacked by the phospholipase and found most importantly that phospholipids in the β (*sn*-2) as well as α-DL (*sn*-1 and *sn*-3) configuration could be

Table 5-1. Characteristics of Some More Common Phospholipases Purified From Plants

Source	Fold purified	pH optimum	Specific activity (μmoles/ min/mg)	Molecular weight	Phosphatidate[a] Donors	Phosphatidate[a] Acceptors	Activators[a]
Cabbage[b]	46	5.4	91.0	—	PC,PE,PS	H₂O	Ca²⁺
Cabbage[c,d,e]	—	5.4	—	—	PC	Primary alcohols, PI	Ether, Ca²⁺, anionic amphipaths
Cabbage[f]	860	6.25 7.25	8.8	112,500	PC	H₂O	Ca²⁺
Cabbage[g]	110	5.6	110.0	—	PC	H₂O, ethanol, ethanolamine, glycerol, serine	Ca²⁺, ether
Cabbage[h]	—	—	—	—	PG	PG	No Ca²⁺
Cabbage[i]	—	5.6	—	—	—	—	—
Cabbage[j]	—	9.0	(transphosphatidylation)	—	Asolectin	Choline, ethanolamine	—

(continued)

Table 5-1. (Continued)

Source	Fold purified	pH optimum	Specific activity (μmoles/min/mg)	Molecular weight	Phosphatidate[a]		
					Donors	Acceptors	Activators[a]
Carrot root[i]	—	5.5	—	—	—	—	—
Peanut seed[j,k,l,m]	1000	5.6	200.0	22,000	PC	H_2O	Ether, Ca^{2+}, CTMB, PA, PG, SDS
Fava bean[n]	10	5.7	10.0	—	PC	H_2O	SDS, Ca^{2+}
Mung bean[o]	?	5.0	—	150,000	PC	H_2O	Ca^{2+}

[a] Abbreviations: CTMB, cetyltrimethylammonium bromide; PA, phosphatidic acid; PC, phosphatidylcholine; PE, phosphatidylethanolamine; PG, phosphatidylglycerol; PI, phosphatidylinositol; PS, phosphatidylserine; SDS, sodium dodecyl sulfate.

[b] Davidson and Long (1958).

[c] Clarke et al. (1981).

[d] Dawson and Hemington (1967).

[e] Dawson (1967).

[f] Allgyer and Wells (1979).

[g] Yang et al. (1967).

[h] Stanacev et al. (1973).

[i] Saito et al. (1974).

[j] Heller et al. (1974, 1975).

[k] Heller et al. (1976).

[l] Heller and Arad (1970).

[m] Tzur and Shapiro (1972).

[n] Atwal et al. (1979).

[o] Herman and Chrispeels (1980).

degraded. This phospholipase D is therefore unlike other phospholipases in its relative lack of structural and stereochemical specificity. While acyl composition did influence the rate of activity (Kates, 1956; Davidson and Long, 1958) even distearoyl phosphatidylcholine and plasmalogen phosphatidylethanolamine were degraded. This latter observation was not confirmed in subsequent studies, although the sources of substrate were different (Lands and Hart, 1965). (This subject is covered in greater detail in Section 5.4.)

The first report of the complete purification of this enzyme from peanut seeds came from the Israeli group in the early 1970s (Tzur and Shapiro, 1972; Heller *et al.*, 1974). This amounted to roughly a 1200-fold purification of the enzyme over the initial homogenate. It was noted during the course of their work that the enzyme was inactivated by moderately low pH values, around 4.5–4.6. This precluded the use of isoelectric focusing since the isoelectric point of the enzyme is 4.65. They were successful in using preparative acrylamide disk-gel electrophoresis that yielded a pure preparation of enzyme with a specific activity of 200 units/mg protein. The enzyme apparently had a minimal molecular weight of about 20,000 daltons but it readily aggregated into a form with a molecular weight of 45,000–50,000 daltons, as well as a 200,000-dalton form.

More recently, the enzyme has been purified 600-fold over an acetone powder preparation of savoy cabbage (Allgyer and Wells, 1979). Essential in this purification scheme was rapid chromatography on Sephadex G-200 and the inclusion of 50% ethylene glycol in the solutions used during purification. Also unique to this approach was the use of a water-soluble substrate, dihexanoyl phosphatidylcholine, which permitted a rapid, semiquantitative assay. Basically, these authors developed a colorimetric assay that employed an unbuffered solution of methyl red for detection of protons liberated by the base group cleavage. The molecular weight of their preparation ranged from 112,500 (SDS–PAGE) to 116,600 daltons (sedimentation equilibrium centrifugation), perhaps representing an aggregated state similar to the peanut phospholipase D.

Lysophosphatidylcholine, not surprisingly, was also degraded; diethyl ether, however, inhibited the reaction (Davidson and Long, 1958). It was reported later that peanut seed phospholipase D could degrade lysophosphatidylcholine in the absence of both ether and Ca^{2+} but that Ca^{2+} could stimulate hydrolysis at substrate concentrations greater than 100 μM (Strauss *et al.*, 1976). Some differences were noted in the pH optimum of the cabbage system, 5.9, and the peanut system, 7.0–8.0.

5.4. Transphosphatidylation and Substrate Specificity

The plant phospholipase D catalyzes a transphosphatidylation acting through a phosphatidate–enzyme intermediate. More commonly, transphosphatidylation is termed the *base exchange reaction*. Also, the enzyme can use a wide variety of hydroxyl groups to accept the phosphatidate including the free hydroxyl on a lysophospholipid. For example, Kates (1956) found that when lysolecithin was the substrate a second product was detected in addition to the expected lysophosphatidate. In this case, the hydroxyl in position 2 of the lysophosphatidate was the acceptor and thus formed the cyclized product 1-acyl-*sn*-2,3 phosphoglycerol. At that time, the concept of transphosphatidylation was first being explored and, while Long and coworkers did not invoke these terms, they did postulate that a phosphatidate–enzyme intermediate could be formed that would lead to the cyclization process (Long *et al.*, 1967).

$$
\begin{array}{c}
\overset{\displaystyle O}{\overset{\displaystyle \|}{H_2C-OCR}} \\
| \\
HC-O \\
| \qquad\qquad P \\
H_2C-O
\end{array}
$$

1-Acyl-*sn*-2,3-phosphoglycerol

The question of substrate specificity took on new significance with the finding that the partially purified enzyme could use a number of added alcohols to transphosphatidylate. The first report of such activity, presented in 1965 by Benson's group (Yang *et al.*, 1967) at a meeting in The Netherlands, spurred a number of investigators to examine the nature of the transphosphatidylation reaction catalyzed by the phospholipase D in their partially purified preparations. This presumption, while not completely tested for all acceptor groups used, appears valid based on a number of studies comparing the hydrolytic and transphosphatidylation activities. Perhaps the first published report on the subject came from Bartels and van Deenen (1966) who concluded that phosphatidyl methanol obtained by methanol extraction of spinach leaves was the result of phospholipase D action.

In 1967, two groups reported in detail that the basic characteristic of the hydrolytic and transphosphatidylation reactions were similar (Dawson, 1967; Yang *et al.*, 1967). Perhaps one of the most convincing arguments in support of this hypothesis was presented in the initial work of Yang *et al.* (1967) who copurified the two activities through four steps that increase the specific activity 110-fold. More highly purified preparations also had a constant ratio of activities (Tzur and Shapiro, 1972). This, together with the characteristics listed in Table 5-2 that are common to the two activities, provided the basis for the acceptance that the two activities are catalyzed by a single enzyme.

Alcohols were much better phosphatidate acceptors than H_2O. Concentrations of 0.7, 0.3, 1.1, and >10% of ethanol, ethanolamine, glycerol, and serine, respectively, gave equal rates of hydrolysis and transphosphatidylation (Yang *et al.*, 1967). These data cannot be compared directly with those of

Table 5-2. Common Properties of Hydrolytic and Transphosphatidylation Activities

1. Same elution profile on DEAE–cellulose
2. Same pH for optimal activity, pH 5.5–6.0
3. Both inhibited by 10^{-4} M *p*-chloromercuribenzoate but not by iodoacetamine or *N*-ethylmaleimide (It was not clear why the latter two compounds failed to inhibit since an essential thiol appears to be present at the active site.)
4. Same inactivation kinetics at 4°C
5. Both absolutely required Ca^{2+}

Dawson (1967) who obtained slightly different results since Yang and co-workers included ether in the reaction while Dawson did not.

A scheme for the reaction sequence follows.

(1) Substrate
(2) Transphosphatidylation product
(3) Hydrolytic product
HSENZ = Enzymes active in sulfhydryl moiety

The role of a thiol intermediate is still a matter of speculation. The finding that pCMB inhibits the reaction is consistent with such a scheme (Yang *et al.*, 1967). However, that other thiol reagents do not inhibit makes the proposal less firm.

While it seems quite likely that transphosphatidylation and hydrolysis are catalyzed by a single protein, some evidence to the contrary was published based on work using a preparation of rather crude phospholipase D from brain (Saito and Kanfer, 1975). The major differences described for the two activities were the pH optimum, inhibition by hemicholinium-3, and concentration of Ca^{2+} required for optimal activity. Since the latter study was done with a brain enzyme, Heller (1978) suggested that separate enzymes might be present in the crude brain preparation used. It seems most likely that the activities now ascribed to plant phospholipase D are catalyzed by a single protein.

The list of compounds degraded by phospholipase D was extended to phosphatidylglycerol and the O-alanyl ester of phosphatidylglycerol by the Utrecht group (Bonsen *et al.*, 1965; Haverkate and van Deenen, 1965). While they were primarily interested in the structure of these compounds isolated from natural sources, they did take advantage of their chemical syntheses of the compounds to demonstrate that the *sn*-1 isomer of O-alanyl phosphatidylglycerol was degraded by phospholipase D. Furthermore, the phospholipids in natural membranes (rat liver microsomes) could be degraded, providing ether was omitted (Heller and Arad, 1970). Thus far, sphingomyelin appears to be the only phospholipid not degraded by the phospholipase(s) D.

Dawson (1967) tested nearly 20 potential acceptors for transphosphatidylation by the cabbage phospholipase D and concluded that a wide range of primary alcohols, including polyfunction alcohols, could act as acceptors. On the other hand, he concluded that secondary alcohols and hydroxy acids such as citrate were not substrates in this system. While other studies showed this not to be the case, the latter compounds are certainly not as effective in transphosphatidylation as primary alcohols. Furthermore, under the appropriate conditions, some acceptors stimulate hydrolysis as well as acting as acceptors. For example, 0.2 M glycerol gave a three-fold increase in hydrolysis while giving some 25–30% of maximal transphosphatidylation. The overall activity was increased four-fold probably due to glycerol influencing the physical interaction of the enzyme and lipid aggregate, similar to the effect of diethyl ether.

The list of *transphosphatidylation* reactions has been expanded and some rather unusual products formed. Some compounds, originally thought not to be acceptors, were incorporated if the reaction time was sufficiently long. Phosphatidylserine, in particular, could be synthesized in this manner (Comfurius and Zwaal, 1977). Perhaps the most systematic approach on transphosphatidylation acceptors was carried out by Kovatchev and Eibl (1978). Investigators interested in synthesizing a series of phospholipid analogs containing the same phosphatidyl moiety would find this paper of considerable value. Also, phospholipids that have free hydroxyls in the base moiety can serve as phosphatidyl acceptors; thus, cardiolipin can be synthesized from two molecules of phosphatidylglycerol (Stanacev *et al.*, 1973) by the cabbage phospholipase D. Similar experiments in which other phospholipids were tested failed probably because of the lack of a free primary hydroxyl or because of the rapid rate of hydrolysis with Ca^{2+}. The omission of Ca^{2+} or diethyl ether dramatically influences the relative extents of hydrolysis and cardiolipin synthesis as shown in Table 5-3 derived from the paper of Stanacev *et al.* (1973).

(1)   Hydrolysis
        Phosphatidyl-[^{3}H]glycerol $\rightarrow$ phosphatidic acid + [^{3}H]glycerol

(2)   Transphosphatidylation
        2 Phosphatidylglycerol $\rightarrow$ [^{3}H]cardiolipin + [^{3}H]glycerol

(3)   Transphosphatidylation and subsequent hydrolysis of cardiolipin
        [^{3}H]Cardiolipin $\rightarrow$ phosphatidate + [^{3}H]phosphatidylglycerol
        $\rightarrow$phosphatidic acid + [^{3}H] glycerol

Since [^{3}H]glycerol is released in both reactions (1) and (2), the liberation of [^{3}H]glycerol is a measure of total enzymatic activity. Reaction (3) was demonstrated by Heller *et al.* (1974) using the peanut enzyme, provided ether was omitted.

Even though the kinetics of this reaction are somewhat complicated by the fact that the precursor is also a product [reaction (3)], some interesting points can be made about the reaction that were brought out by Stanacev *et al.* (1973). It was concluded that the rate of transphosphatidylation is a function of the acceptor concentration and since Ca^{2+} greatly enhanced the rate of hydrolysis; the lack of Ca^{2+} keeps the acceptor levels of phosphatidylglycerol high favoring transphosphatidylation. This suggested that Ca^{2+} was not primarily involved in the formation of the phosphatidyl enzyme but in the second leading to hydrolysis of this complex. It is not clear in their postulate why Ca^{2+} should affect hydrolysis but not transphosphatidylation. They suggested that this difference might be due to charge neutralization of the product, phosphatidic acid, by Ca^{2+}. While the data available at this time do not permit a full understanding of the effect of Ca^{2+}, two factors should be

Table 5-3. Effect of Ca^{2+} and Ether on Phospholipase D Transphosphatidylation[a]

System	Incubation time (hr)	Percentage of [^{3}H]glycerol released	Percent distribution of [^{3}H] precursor and product phosphatidylglycerol	Cardiolipin
Complete	4	97.6	67.0	23.0
	24	99.7	57.4	42.6
Minus diethyl ether	4	38.9	99.1	0.9
	24	71.0	96.3	3.7
Minus CaCl$_2$	4	73.4	35.0	65.0
	24	81.2	47.2	53.8

[a] From Stanacev *et al.* (1973); measured with phosphatidyl-2-[^{3}H]glycerol.

considered. First, charge neutralization might not be the sole factor accounting for the effect of Ca^{2+} since little change in total activity was noted when ethanol was the phosphatidate acceptor (Yang *et al.*, 1967). In that case, the charge on the products, phosphatidate and phosphatidylethanol, was the same. Likewise, Dawson (1967) found that glycerol and ethanolamine that gave differently charged products were roughly equivalent as phosphatidate acceptors. Second, a number of workers have shown that negatively charged amphipaths greatly stimulate the hydrolytic reaction (Dawson and Hemington, 1967). Since diethyl ether stimulated both the rate of hydrolysis and transphosphatidylation, it influences reaction (1) by enhancing lipid–substrate interaction. As pointed out by Dawson (1967), the activation by negative amphipaths could not be correlated with the Zeta potential of the substrate aggregate. Clearly, these results point the way to some very interesting studies that will be most useful in elucidating the mechanism of action of this enzyme and its physical interaction with substrate.

Phosphatidylinositol, long thought to be inert to the attack of phospholipase D, was shown to undergo a transphosphatidylation catalyzed by an enzyme in cauliflower florets, analogous to the reaction that yields cardiolipin (Clarke *et al.*, 1981). This reaction that produces bis(phosphatidyl)inositol is somewhat surprising since other phospholipases D do not use free inositol in transphosphatidylation reactions (Allgyer and Wells, 1979). Even in the presence of diethyl ether and Ca^{2+}, bis(phosphatidyl)inositol was produced in greater quantities than phosphatidic acid, the hydrolysis product. Apparently, the enzyme did not attack a specific hydroxyl on the acceptor phosphatidylinositol molecule since more than one isomer of bis(phosphatidyl)inositol was formed. The phospholipase D from the cauliflower might very well be unique, based on the differences noted between the cauliflower and cabbage systems generally employed.

Recently, more attention has been given to the stereospecificity of the cabbage phospholipase D. Although impure enzymes have been used in these studies, they are extremely valuable in elucidating the specificities of the enzyme. Earlier work had shown that the introduction of an alkyl or alkenyl group into position 1 of the glycerol moiety drastically reduced or eliminated hydrolysis catalyzed by phospholipase D (Lands and Hart, 1965; Chen and Barton, 1971). Perhaps the highest rates on hydrolysis on ether-linked phospholipids were reported by Waku and Nakazawa (1972) who found the relative rates for *sn*-1-acyl, *sn*-1-alkyl, *sn*-1-alkenyl-2-linoleoyl phosphatidylcholines to be 100 : 16 : 10, respectively. Introduction of the alkyl group into position 2 of the glycerol likewise reduced the rate of degradation some 40-fold, but perhaps more importantly it eliminated the stereospecificity of the enzyme (Bugaut *et al.*, 1985). Figure 5-1 compares the hydrolysis of *rac*-1-lauroyl-2-oleoyl phosphatidylcholine with *rac*-1-lauroyl-2-oleyl phosphatidylcholine.

As can be seen, the diacyl substrate, while degraded at an initial rate greater than the acyl alkyl substrate, slows abruptly when one-half the substrate is degraded. Presumably, all the *sn*-1,2 enantiomer is depleted and then the *sn*-2,3 is very slowly attacked. In contrast, there is no reduction in rate when the 2-alkyl substrate is used until some 80–90% hydrolysis occurs, in-

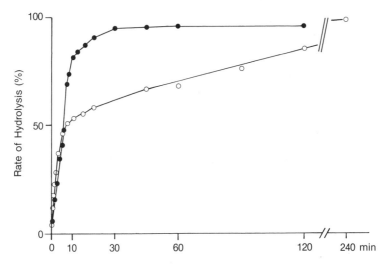

Figure 5-1. Time course of hydrolysis of *rac*-1-lauroyl-2-[1-^{14}C]oleoylglycerol-3-phosphoryl[methyl-^{3}H]choline (dashed line) and of *rac*-1-[1-^{14}C]lauroyl-2-oleylglycerol-3-phosphoryl[methyl-^{3}H]choline (solid line) by phospholipase D. Substrate, 140 and 280 nmoles of the diacyl and the acylalkyl derivatives, respectively; enzyme, 25 and 32 units, respectively, for the diacyl and the acylalkyl substrates. (From Bugaut *et al.*, 1985.)

dicating equal attack on the *sn*-2,3 and *sn*-1,2 enantiomers. The earlier results indicating that the phospholipase D is not stereospecific (Davidson and Long, 1958) probably can be explained in part by the findings of Bugaut *et al.* (1985) that phospholipase D has stereopreference but not absolute stereospecificity and that the composition of the molecule influences this stereopreference. This is indeed an unusual and provocative subject for the lipid enzymologist.

An elegant approach to the study of the stereochemistry of phospholipase D from cabbage was used by Tsai and coworkers (Bruzik and Tsai, 1984; Jiang *et al.*, 1984). In one study, they synthesized chirally labeled dipalmitoyl phosphatidylcholine and found that both hydrolytic and the transphosphatidylation reactions with ethanolamine gave overall retention of the configuration at the phosphorus. Similar conclusions were reached when a thio derivative was used, 1-2 dipalmitoyl-*sn*-glycero-3-thiophosphocholine. In this case, they were able to demonstrate that the phospholipase D is specific for the *sn*-3-S (phosphorus) isomer.

5.5. Factors Regulating Phospholipase D Action

Studies have been done to understand better the interaction of the enzyme(s) with substrates, ions, detergents, and solvents. Based on the observations of the many workers referenced above, some generalities can be made that would be useful in future experimentation. This section describes four points known to regulate activity.

1. Metal ion, Ca^{2+} in particular, is required for catalysis.
2. Charge of the lipid influences activity; negative charge is favored.
3. The enzyme works optimally at a water–lipid interface; monomers are degraded but at a low rate, relative to aggregates.
4. The lipid in the interface cannot be packed too tightly (high pressure).

As indicated earlier, the role of Ca^{2+} is complex and indeed more than one ion might be required for degradation of phospholipid. This possibility is based on the observation that 8–10 mM EDTA caused 60–70% inhibition in the presence of an optimal concentration of Ca^{2+}, 50 mM (Heller *et al.,* 1976, 1978). Since a vast excess of Ca^{2+} is still present under those conditions, it is possible that the Ca·EDTA complex is inhibitory. On the other hand, it is possible that an ion such as Zn^{2+} is required for structural integrity by analogy with the phospholipase C from *B. cereus* (Little *et al.,* 1982), although this has not been rigorously tested. A number of metal ions could activate the carrot enzyme, giving stimulation in the following order: $Ca^{2+} > Ni^{2+} > Co^{2+} > Mg^{2+} = Mn^{2+} > Zn^{2+}$ (Einset and Clark, 1958). In all cases studied thus far, extremely high concentrations of metal ions are required for full activation, which further suggests that the ions have more than a catalytic function.

Ca^{2+} was found to protect the enzyme from surface inactivation and to promote enzyme–substrate binding. Dawson and Hemington (1967) demonstrated that dodecyl sulfate and other amphipathic anions would cause binding of the enzyme to the lipid aggregate and remain active if Ca^{2+} were present. If Ca^{2+} were absent, inactivation of the enzyme occurred. Heller *et al.* (1976) made similar observations but also found that if the enzyme–substrate complex was first incubated to permit hydrolysis, the enzyme no longer remained associated with the substrate–Ca^{2+} complex. Based on the data of Dawson and Hemington, an experiment to show the separation of the effects of Ca^{2+} (e.g., enzyme stabilization and activation) would be difficult using aggregated substrate since the enzyme was inactivated by the dodecyl sulfate–phosphatidylcholine complex unless Ca^{2+} was present. It is of interest that Mg^{2+} caused binding but not catalytic activity on micelles of phosphatidylcholine and that Ca^{2+} was required for the hydrolysis of the soluble substrate dihexanoyl phosphatidylcholine (Allgyer and Wells, 1979). Another effect noted with soluble substrate was a change in the pH for optimal catalysis. With 0.5 mM Ca^{2+}, the optimum was pH 7.25 while with 50 mM Ca^{2+} the optimum was pH 6.25, closer to the value generally found with aggregates of substrate. This shift in the pH optimum was also seen if 0.5 mM Ca^{2+} and 50 mM Mg^{2+} were used (Fig. 5-2). It was concluded that a low-affinity binding of divalent metal ion caused a change in the conformation of the enzyme. A plot of velocity versus substrate concentration shows that when a critical concentration of 4.3 mM is reached, the velocity increases severalfold. This concentration is about one-half that known to be the CMC of dihexanoyl phosphatidylcholine, 10 mM. They questioned if some yet undefined lipid or lipid–enzyme aggregate is formed at 4.3 mM. This would be

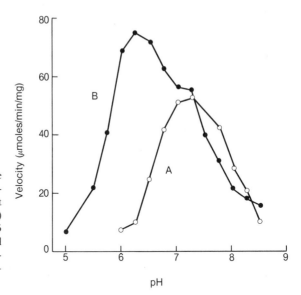

Figure 5-2. pH dependence of the phospholipase D catalyzed hydrolysis of dihexanoyl glycerol at different Ca^{2+} concentrations. (A) 1 mM dihexanoyl glycerol and 0.5 mM $CaCl_2$; (B) 2 mM dihexanoyl glycerol and 50 mM $CaCl_2$. All assays used the pH stat. (From Allgyer and Wells, 1979.)

distinct from the classical micelle since no further increase in activity was noted at the CMC. It was also found that increased ionic strength would decrease activity. This, however, was not thought to be the result of enhanced aggregation by salt or Ca^{2+} since an aggregated substrate is attacked more rapidly than monomers.

Attempts have been made to understand better the role of Ca^{2+} in surface-charge regulation of phospholipase D activity, as studied by microelectrophoretic measurements. The activities of a number of enzymes have been shown to be influenced dramatically by the surface charge of the aggregate. The cabbage phospholipase D, however, failed to show a direct relation between surface charge and hydrolysis as shown in Table 5-4 taken from Dawson and Hemington (1967).

Table 5-4. *Relation of Hydrolysis and Electrophoretic Mobility of Phosphatidylcholine[a]*

Activator	Enzyme activity (μg choline released in 15 min)	Mobility μ (sec^{-1} V^{-1} cm^{-1})	
		$-Ca^{2+}$	$+Ca^{2+}$
None	0.5	− 0.6	+ 1.98
Dicetyl phosphate	5.6	− 8.5	+ 1.58
Triphosphatidylinositol	28.3	− 7.5	+ 1.69
Monocetyl phosphate	49.2	− 8.1	+ 2.12
Phosphatidate	91.5	−11.6	+ 1.22
Dodecyl sulfate	132.0	−13.1	− 0.53

[a] From Dawson and Hemington (1967). Ca^{2+} was present in enzyme assays and the microelectrophoresis measurement, where indicated, at a concentration of 40 mM.

As can be seen, good activators such as monocetyl phosphate and phosphatidate had positive mobility not too different from neat phosphatidylcholine. The addition of Mg^{2+} caused changes in mobility similar to those caused by Ca^{2+} without stimulating activity. Furthermore, the addition of the positively charged amphipath cetyl trimethyl ammonium bromide markedly inhibited the hydrolysis of a dodecyl sulfate–phosphatidylcholine mixture without causing a noticeable change in the electrophoretic mobility. If Ca^{2+} and amphipath are regulating, the effect must be inside the double layer which cannot be detected by microelectrophoresis.

The substrate aggregation state has a considerable effect on enzyme activity, as previously indicated. Dawson and Hemington (1967) showed that small liposomes were degraded much more rapidly than large structures. Marked autocatalysis is seen with the large liposome, the result of one of the products, phosphatidic acid, stimulating the enzyme. Heller *et al.* (1978) reported that ratios of dodecyl sulfate to phosphatidylcholine normally used for substrates produced a heterogeneous mixture of particles, including mixed micelles. Ca^{2+} caused precipitation of some aggregates and those that were not precipitated were degraded by phospholipase D. They also suggested that the enzyme can undergo a reversible dissociation that leads to inactivation; this dissociation could be reversed by the phosphatidylcholine–dodecyl sulfate–Ca^{2+} complex and therefore maintain high activity.

The problem of enzyme–substrate interaction has been probed also using monomolecular films of substrate (Quarles and Dawson, 1969*a*). Films of yeast phosphatidylcholine were degraded by the cabbage phospholipase D at initial surface pressures between 12 and 25 dynes/cm. An interesting similarity between this study and that of Allgyer and Wells using soluble substrate can be noted; a low concentration of Ca^{2+} (0.2 mM) was sufficient for optimal activity and at this concentration of Ca^{2+} the pH optimum was 6.6, over a pH unit above that seen with liposomes of pure substrate (Quarles and Dawson, 1969*b*). Two phases of activity were noted, an initial slow phase followed by a more rapid secondary phase, perhaps the result of autoactivation as phosphatidate was produced. Likewise, phosphatidate but not dodecyl sulfate promoted activity at surface pressures greater than 28 dynes/cm. Although not well understood, it seems that phosphatidate facilitates the initial attack but does not influence the overall catalytic rate, somewhat analogous to lowering a K_m value without altering the V_{max}. The surface pressure dropped during the course of hydrolysis even though the phosphatidate remained in the monolayer. It could be speculated that Ca^{2+} adducts were formed that had a condensing effect and, as a consequence, phase separation occurred that favors enzyme penetration.

In summary, some partial answers can be outlined:

1. Ca^{2+} has two effects, one directly in catalysis and one in protecting or otherwise altering enzyme conformation.
2. Bulk surface charge does not regulate activity while negative amphipaths facilitate initial interaction (at least in monolayers).

3. Mixed micelles appear to be the preferred aggregate state of substrate.

5.6. Function of Plant Phospholipases D

The function of phospholipase D is far from certain but work has been done on its location within various plant tissues and among different plants. This list extends to the simplest members of the plant kingdom; most red but not green or brown algae tested have transphosphatidylation (phospholipase D) activity (Vaskovsky and Khotimchenko, 1983). The distribution of phospholipase D in plants is wide but by no means ubiquitous and the presence of phospholipase D is often dependent on the growth phase of a tissue within the plant. The list of plants studied given here is quite small and the reader is directed to the review of Heller (1978) for a more complete list.

Evidence has been obtained that phospholipase D is present in storage tissues and seeds and that its presence in those tissues could relate to rapid metabolic activity associated with growth. The work of Quarles and Dawson (1969c) shows that storage tissues such as the leaves of cabbage, kohlrabi, and lettuce, the stalks of celery and cabbage, and certain seeds, marrow, pea, soybean, and peanuts (not shown) have the highest activity. The greatest activity in many higher plants was located in the cotyledons, although the root still contained reasonable activity, especially when calculated on the basis of the nitrogen content of the tissue. Of considerable interest was the change in subcellular distribution noted; in the immature seed, most activity was associated with the particulate fractions, primarily the *mitochondria*. On germination, a shift occurred and most activity was localized in the cytosol. Such a shift in distribution demonstrates the complexity in attempting to assign a function to this enzyme. It is known that phospholipase D in cotyledons is depleted during germination; therefore, the activation of this enzyme during this or other structural changes could be responsible.

The phospholipase D activity of wheat and barley leaves was found to vary according to the growth phase of the plant and water stress, a phenomenon that produces changes in plant morphology and biochemistry (Chetal *et al.*, 1982). They found that phospholipase D activity increases with plant maturation (fillering and ear emergence stages) but decreased at the grain-filling stage. Water stress was found to decrease activity, which was thought to relate to changes in the leaf phospholipid composition. An interesting but unproven postulate was presented; a shift in the hydrolytic and transphosphatidylation activities of the phospholipase D could accompany leaf maturation. They suggested that water stress would shift activity toward transphosphatidylation, accounting for membrane composition and morphologic changes.

An interesting report on a phospholipase D from *Hevea brasiliensis* (rubber tree) was made over three decades ago (Smith, 1954). The latex of this tree is thought to be stabilized by an interfacial coating of phospholipid and protein. Upon tapping of the latex coagulation occurs, perhaps brought about by degradation of the *latex surfactant*. Indeed, a phospholipase D active on the latex phospholipid was demonstrated within the latex. While a correlation between coagulation and choline release was not made, it does raise the possibility that such an enzyme could play a specialized role in plant physiology.

An interesting question was raised by Dawson and Hemington (1974) as to how herbivores and omnivores protect themselves from the necrotic effects of dietary plant phospholipase D. To answer this, they examined the saliva from three species—cow, human, and dog. (The last sample "was obtained from a 'boxer' salivating in response to being tempted with chocolate.") They found a potent inhibitor of a grass seed phospholipase D, a 20-fold dilution of the dog saliva gave better than 50% inhibition of hydrolysis and the cow saliva, somewhat less effective, gave 90% inhibition in a five-fold dilution. The factor was purified some 25-fold by chromatography on Sephadex G-200, was distinct from salivary microprotein, and was exceptionally stable to acid and alkali treatment and to heat. At this time, its structure is still unknown.

Chapter 6

Cellular Phospholipases A_1 and Lysophospholipases of Mammals

Here we begin a series of three chapters devoted to vertebrate phospholipases of cellular origin. The extracellular, pancreatic phospholipase A_2 is not included; rather, it is covered with the venom phospholipases with similar nature and function. The functions of the cellular phospholipases in a few instances are given in these three chapters but the major description of the functions of these phospholipases is covered in Chapter 11.

Both phospholipases A_1 and lysophospholipases are covered in this chapter. In many but not all cases, lysophospholipases also have phospholipase A_1 and/or phospholipase B activity. It therefore seems reasonable to describe them together.

6.1. General Introduction to Cellular Phospholipases from Mammals

As reviewed by Zeller, it is not a simple task to fix the date of the first description of phospholipase activity in mammalian tissue. If one were to take the identification of lysophosphatidylcholine in mammalian tissue as a predictor of phospholipase A activity, then a probable year to mark as the beginning would be 1924 (reviewed by Zeller, 1951). However, if the disappearance of lecithin brought about by pancreatic juice is taken as evidence for phospholipase action, studies on mammalian phospholipases began over a century ago (reviewed by van den Bosch, 1982). The first demonstration of a precursor, phosphatidylcholine, being degraded to a lipid product, lysophosphatidylcholine, was made over six decades ago by Belfanti and Arnaudi (1932) who used pancreatic juice as the source of enzyme (phospholipase A). No matter what the date, the pancreas, known to contain a number of digestive enzymes, was the first demonstrated site of a phospholipase.

Because of the low activities of mammalian phospholipases, other than the pancreatic phospholipase, their detection and characterization was not possible until substrates with specific radiolabels were prepared in the early-to-mid-1960s, as described in Chapter 1. Metabolic tracer studies done over the past 30 years suggest that phospholipases are quite active *in situ*. Much of the impetus for the study of cellular phospholipases was the recognition that the phospholipids of membranes turned over very rapidly. Enzymes capable of degrading phospholipids must therefore be present within the cell. Table 6-1, condensed from the compilation of van den Bosch (1982) in his review on intracellular phospholipases, illustrates a number of tissues having variable and independent turnover rates of phospholipids and that even within a given tissue and class of phospholipid, the turnover rates of the individual moieties differ considerably. For example, the turnover rate of glycerol is at least 10 times higher than that of phosphate in lung phosphatidylcholine (Experiment No. 9). Furthermore, the individual molecular species of a particular class of phospholipid, labeled with a single precursor, will vary in its turnover rates. Four species of phosphatidylcholine from rat liver that differ in the number of the double bonds in their acyl chains vary 10- to 20-fold in turnover rates (Experiment No. 8). The data point out that the metabolism of mammalian phospholipids is the result of a variety of anabolic as well as catabolic events taking place, suggestive of the activity of a number of phospholipases.

All five classes of phospholipases have been described in mammalian tissues—A_1, A_2, B, C, and D. The reports of new types and sources of phospholipases are appearing at a frightening rate, at least to one surveying this

Table 6-1. Turnover of Phospholipids[a]

Experiment	Tissue	Precursor	Tissue fraction[b]	Phospholipid class[b]	Half-life (hr)
1	Rat liver	[^{32}P]phosphate	Total	Total	6–10
2	Rat liver	[^{32}P]phosphate	Total	PC	10.9
3	Rat liver	[^{32}P]phosphate	Mito	Total	38
4	Rat liver	[^{14}C]choline	Total	PC	10
5	Rat liver	[^{3}H]glycerol	Mito	Total	16
			Micro	Total	16
6	Rat liver	[^{14}C]choline	Mito	PC	15
			OM	PC	16
			IM	PC	3
			Micro	PC	8
7	Rat liver	[^{14}C]acetate	Micro	Total	48
				PC	48
				PE	48
			PM	Total	48–72
				PC	48
				PE	48
8	Rat liver	[^{3}H]ethanol	—	PC Δ0	12
				Δ1	0.7–1.8
				Δ2	2.5–4.0
				Δ4	7.8–18
9	Rabbit lung	[^{32}P]phosphate	Total	PC	88
		[^{3}H]choline		PC	13
		[^{3}H]glycerol		PC	8
		[^{14}C]palmitate		PC	10

[a] From van den Bosch (1982).
[b] Abbreviations: PC, phosphatidylcholine; PE, phosphatidylethanolamine; PI, phosphatidylinositol; Δ0, Δ1, Δ2, and Δ4, disaturated, monoenoic, dienoic, and tetraenoic species, respectively; Mito, mitochondria; OM, outer membrane; IM, inner membrane; Micro, microsomes.

vast literature. Thus far, however, few have been purified and complete purification of phospholipases is quite limited. Characterization studies on these enzymes, comparable to the characterization of the extracellular bacterial and the digestive type phospholipases A_2, have proved to be technically impractical owing to the very low yields and low specific activities. In a few instances, however, some detailed studies on the action of purified phospholipases are emerging and providing useful information on their basic mechanism of action and possible modes of regulation within the cell.

The phospholipases are widely distributed and are present in nearly all cells studied. Furthermore, they are localized in a variety of subcellular organelles within a given cell. Table 6-2 shows the distribution of the phospholipases A described in the liver, the most intensely studied tissue.

Other activities such as lysophospholipases and phospholipase C are present in liver, and in all probability other phospholipases will be described when further dissectioning of liver phospholipases is undertaken. It has usually been assumed, perhaps glibly, that these enzymes are localized in the parenchymal cell. It is quite possible that some will be found in other cells of the liver such as the Kupffer and endothelial cells; this is particularly germane in the case of the lysosomal enzymes that have similar counterparts in macrophages. Thus far, no common characteristics of the mammalian phospholipases have appeared within a given cell type. The size and charge distribution described ranges widely, as do the effects of metal ions (Ca^{2+}), pH, and detergent. However, some commonality does exist between tissues; for example, the phospholipase A_2 activity in the mitochondria from a number of tissues has some characteristics in common, as do lysosomal phospholipases. That they are the same enzyme remains to be verified. The substrate specificity is generally broad for a number of these enzymes, although some such as most phospholipases C appear to be specific in degrading phosphatidylinositol or its phosphorylated derivatives.

6.2. Phospholipase A₁: General Considerations

A number of sources for mammalian phospholipase A_1 have been described. In the partial listing (Table 6-3), some interesting patterns can be seen that reflect certain characteristics of these enzymes.

Table 6-2. Phospholipase A Distribution in Rat Liver

Subcellular site	Main activity	pH optimum	Ca^{2+} effect[a]
Plasma membrane	$A_1 + A_2$	8.0–9.5	+
Microsomes	A_1	8.0–9.5	+
Golgi membrane	$A_1 + A_2$	8.0	+
Mitochondria	A_2	8.0–9.0	+
Lysosomes	$A_1 + A_2$	4.0–4.5	−
Cytosol	$A_1 + A_2$	7.4	N.I.
	A_1	3.6	0

[a] Abbreviations: +, stimulated; −, inhibited; 0, no effect; N.I., not investigated.

Table 6-3. A Partial List of Mammalian Phospholipases A₁

Source	Localization[a]	pH optimum[b]	Ca^{2+} effect[a]	Substrate specificity[a]	Molecular weight[a]	pI[a]	Reference
Rat brain	Sol + Micro	8.0	0	PI	N.D.	N.D	Hirasawa et al. (1981a)
Calf and rat brain	N.D. (Lyso?)	4.0	0	PC,PI	N.D.	N.D.	Gatt (1968)
Human brain (white and gray matter, peripheral nerve)	N.D.	4.6	0	PC,PE,PS	N.D.	N.D.	Copper and Webster (1970)
Rat brain	Sol	9.5	+	PE	N.D.	N.D.	Doherty and Rowe (1980)
Rabbit cortex	Neuronal PM	8.0–9.0	+	PC,PE,PS,PI	N.D.	N.D.	Woelk et al. (1982)
Human platelet	Lyso	4.8	0	PC(?)	N.D.	N.D.	Smith and Silver (1973)
FL (amnionic membrane)	Lyso	3.6	N.D.	PC(?)	N.D.	N.D.	Suzuki and Matsumoto (1978)
Rat ovary	Lyso	5.0	–	PC(?)	–	–	Okazaki et al. (1977)
Sheep pancreas (pancreatic juice)	Sol	5.3	–	PI,LPI,(PE)	N.D.	N.D.	Dawson et al. (1982)
		6.2	–	PA	–	–	
Guinea pig pancreas	Sol	8.5	–	PC,(TG)	37,000	9.3	Fauvel et al. (1981a)
				PC,(TG)	42,000	9.3	
Beef pancreas	N.D.	7.5	0	PC,LPC,PE	–	–	van den Bosch et al. (1974)
Rat heart	Micro	7.5	0	PE,PC	N.D.	N.D.	Weglicki et al. (1971)
Rat liver	Sol	9.5	+	PE,LPE	N.D.	7.15	Dawson et al. (1983b)
Rat liver	PM	8.5–9.0	+	PE,PC,PS	N.D.	N.D.	Newkirk and Waite (1971)
Rat liver	PM	9.0	+	PE	N.D.	N.D.	Nachbaur et al. (1972)
Rat liver	PM	7.4	+	PE	N.D.	N.D.	Colard-Torquebiau et al. (1975)

Source	Fraction		pH	Substrate	M.W.		Reference
Rat liver	Micro	±	7.5	PE,PC	N.D.	N.D.	Waite and van Deenen (1967)
Rat liver	Micro	+	8.3	PE	N.D.	N.D.	Nachbaur et al. (1972)
Rat liver	Micro	0	7.5	PE	N.D.	N.D.	Lumb and Allen (1976)
Rat liver	Micro	+	9.5	PE	N.D.	N.D.	van Golde et al. (1971)
Rat liver	Golgi	N.D.	8.0	PE	N.D.	N.D.	Franson et al. (1971)
Rat liver	Lyso	−	4.0	PG,CL,PC	15–90,000	4.0–5.22	Hostetler et al. (1982)
Rat liver	Lyso	−	4.0	PE,PC,PI,PG	56,000	3.8–4.8	Robinson and Waite (1983)
Human plasma	Lyso	N.D.	9.0	TG,PC	75,000	4.1	Ehnholm et al. (1975)
Rat liver	PM	N.D.	9.0	TG,PL,MG	180,000	—	Kuusi et al. (1979a)
Rat liver	PM	+	9.0	TG,PL	61,700	9.7–10.0 (urea)	Jensen and Bensadoun (1981)
Rat liver	PM	+	8.0–9.0	MG,PE,TG	72,000	4.5–5.5	Waite et al. (1978)
Rabbit lung	Lyso	−	4.0	PE	N.D.	N.D.	Heath and Jacobson (1976)
Hamster heart	PM	+	6.0–9.0	N.D.	N.D.	N.D.	Franson et al. (1978a)
Rat heart	PM	±	8.0	PE(?)	N.D.	N.D.	Weglicki et al. (1972)
Rat; dog heart	Micro	±	7.5	PE	N.D.	N.D.	Weglicki et al. (1971)
Rat heart	Lyso	−	4.0	N.D.	N.D.	N.D.	Franson et al. (1972)
Rat heart	Lyso	+	5.0	N.D.	N.D.	N.D.	Franson et al. (1972)
Rat heart	Sol	±	8.4	N.D.	N.D.	N.D.	Nalbone and Hostetler (1985)

[a] Abbreviations: Sol, cytosol; Micro, microsomes; PM, plasma membrane; N.D., not determined; Lyso, lysosomes; CL, cardiolipin; LPC, lysophosphatidylcholine; LPE, lysophosphatidylethanolamine; LPI, lysophosphatidylinositol; MG, monoacylglycerol; PA, phosphatidic acid; PC, phosphatidylcholine; PE, phosphatidylethanolamine; PG, phosphatidylglycerol; PI, phosphatidylinositol; PL, phospholipid; PS, phosphatidylserine; TG, triacylglycerol.

[b] In certain cases only one substrate or pH was used; that is the one given.

A great number of these phospholipases A_1 listed in Table 6-3 probably are of lysosomal origin, in particular those that have an acidic pH optimum. Those with alkaline pH optima are often in the microsomes and/or plasma membrane. A few, however, are localized in the cytosol. The relation between the microsomal, plasma membrane, and soluble enzyme is not at all clear at this time. The phospholipase A_1 localized on the plasma membrane of the liver is also found in postheparin plasma or heparin perfusates of the liver (Newkirk and Waite, 1971; Kuusi *et al.*, 1979a; Jensen and Bensadoun, 1981). While this and other enzymes listed have broad specificity and are active on neutral glycerides, their phospholipase activity is of significance and physiologically could be a key role for the enzymes. In a great number of studies, phosphatidylethanolamine appeared to be the preferred substrate when pure phospholipids are used.

This section emphasizes those cases that have been most thoroughly characterized and probable analogous enzymes are pointed out. In certain cases, some postulates can be made about the function(s) of a phospholipase A_1. The phospholipases A_1 that are covered in greatest detail are those from the pancreas, those from rat liver lysosomes and cytosol, and heparin-releasable enzyme from rat liver. While some have broad specificity and could be classified differently, they are included here because of the emphasis placed on the work by the investigator involved.

6.2.1. Pancreatic Phospholipases A_1

One of the major jobs in detecting and isolating the phospholipase A_1 from beef pancreas was to have a means by which it can be distinguished from the phospholipase A_2 and the lipase that is known to have phospholipase A_1 activity (deHaas *et al.*, 1965). By comparing the known chromatographic characteristics of the pancreatic lipase to that of the phospholipase A_1, van den Bosch *et al.* (1974) purified the phospholipase A_1 154-fold over the crude homogenate to a state of near homogeneity. Similar to the phospholipase B of *P. notatum*, the pancreatic phospholipase A_1 had lysophospholipase activity on both the 1- and 2-acyl isomers of lysophosphatidylcholine and therefore can be classified also as a phospholipase B; however, no activity was found at position 2 of phosphatidylcholine. The enzyme was not specific for the polar head group; phosphatidylethanolamine was readily deacylated. To demonstrate hydrolysis of the 2-acyl lysophosphatidylcholine, the problems of acyl migration had to be circumvented. This was achieved when they demonstrated that the enzyme could deacylate glycophosphatidylcholine that cannot undergo acyl migration.

$$
\begin{array}{l}
\quad\ \ \text{O} \\
\quad\ \ \| \\
\text{RCO—CH}_2 \quad\ \ \text{O} \\
\qquad\quad | \qquad\quad \| \\
\qquad \text{H}_2\text{C—O—P—OCH}_2\text{CH}_2\text{—N}^+(\text{CH}_3)_3 \\
\qquad\qquad\qquad | \\
\qquad\qquad\qquad \text{O}^-
\end{array}
$$

The kinetic parameters for the deacylation of the substrates glycophosphatidylcholine and 2-acyl lysophosphatidylcholine were similar, whereas the K_m for the substrate 1-acyl phosphatidylcholine was two- to threefold higher and the V_{max} was ten-fold higher than that for the 2-acyl compounds. It is not clear if this was a reflection of differences in the aggregated state of the substrate or a property of the enzyme.

The effect of deoxycholate on the degradation of the diacyl and monoacyl was similar to that reported for the *P. notatum* phospholipase B; that is, 2.5 mM deoxycholate blocked nearly all lysophospholipase activity while the deacylation of diacylphospholipid was optimized. Without deoxycholate, very low deacylation of phosphatidylcholine was detected, giving a ratio of lysophospholipase : phospholipase A₁ activity of about 200 : 1. Although direct comparison with the *P. notatum* enzyme cannot be made owing to differences in the substrate used, it would appear that the pancreatic enzyme required a much higher detergent : substrate ratio (2.5 mM:0.4 mM) than the *P. notatum* enzyme (1.25 mM:4.0 mM) to achieve optimal phospholipase A₁ activity. Another distinguishing characteristic is the pH for optimal activity; the pancreatic enzyme was assayed at pH 7.4 while the *P. notatum* enzyme was optimally active at pH 4.0. In addition to its activity on phospholipids, the pancreatic enzyme showed a general esterase activity, capable of degrading *p*-nitrophenyl-acetate. Since the *P. notatum* enzyme did not have this esterase activity (Sugatani *et al.*, 1978), the two enzymes are clearly different.

An interesting phospholipase A₁ acting on phosphatidylinositol and phosphatidate has been detected in secretion fluids and the cytosolic fraction of sheep pancreas (Dawson *et al.*, 1982). The activity of this phospholipase A₁ had a rather low pH optimum (5.3) that was shifted even lower (pH 4.0) if an amount of Ca^{2+} or Mg^{2+} equal to the amount of substrate were added. Optimal activity on phosphatidate was at a somewhat higher pH than that for phosphatidylinositol, pH 6.2. Very low activity was noted on phosphatidylcholine and phosphatidylethanolamine. The differences in the pH for optimal activity on the two substrates and for the effect of Ca^{2+} and Mg^{2+} were ascribed to surface-charge (pH) effects. Since the surface pH of the highly acidic phospholipids is considerably lower than that of the bulk pH, neutralization of the surface pH by Ca^{2+} would require that the bulk pH be lowered to maintain the surface pH optimal for enzyme attack.

While Dawson and coworkers did not speculate on the possible function of a secretory phospholipase acting on phosphatidylinositol, they did make an interesting observation on the specificity of secretion. A phospholipase C found during the course of the investigations of the cytosolic fraction of the pancreas was not found in the secretory juices. This showed that the processing of the two enzymes at the level of the endoplasmic reticulum, Golgi apparatus, and zymogen granules must be different. Further investigations on this aspect of the problem could be quite interesting both from the point of view of enzyme function as well as models for enzyme synthesis and processing. Likewise, it will be of interest to determine if the secreted phospholipase A₁ has a zymogen form.

Distinct phospholipases A₁ that are highly cationic (pI = 9.4) have been purified from guinea pig pancreas (Fauvel *et al.*, 1981*a*). The activity against phosphatidylcholine copurified in a 1:1 ratio with a triacylglycerol lipase activity to apparent homogeneity, some 40-fold over a soluble extract. Based on the cell disruption procedure used, true subcellular localization cannot be deduced. That only a 40-fold increase in specific activity yields pure enzyme suggests that this enzyme or enzymes must be present in high concentrations within the pancreas. The two forms isolated differed in their molecular weights (37,000 vs. 42,000 daltons) and, to a minor degree, their charge, even though this was not detected by electrofocusing under the conditions employed. Similar to the findings of van den Bosch and coworkers, the activity against diacylphospholipid was markedly stimulated by deoxycholate. Another similarity between these two systems is the lack of a requirement for Ca^{2+}. That the enzyme studied by van den Bosch *et al.* (1974) was distinct from that isolated by Fauvel *et al.*, (1981*a*) was shown by the difference in sensitivity to diisopropylfluorophosphate; only the beef enzyme was inhibited.

Clearly, the pancreas is a tissue brimming with a number of phospholipases A_1/lipases. The classical pancreatic lipase with low phospholipase A_1 activity functions in digestion of dietary lipid, as might the enzyme described in sheep pancreatic secretions (Dawson *et al.*, 1982). Others described thus far have no suggested function or, for that matter, description of their subcellular localization. Further work on this gland should be of interest.

6.2.2. Liver

The liver contains a number of phospholipases A_1 that differ markedly in their distribution, pH optimum, response to Ca^{2+}, and function. Some have been purified, opening the way for detailed studies on their mode of action and function.

6.2.2.1. Cytosol

The phospholipase A_1 activity in rat liver cytosol has been known for some time (Waite and van Deenen, 1967). Its purification, however, has only recently been undertaken. In general, those enzymes found in the cytosol are more easily purified than those associated by membranes in organelles. While only a 10-fold purification of the cytosolic enzyme has been achieved thus far, it was sufficient to remove interfering phospholipase B activity (Dawson *et al.*, 1983*b*). Even though the enzyme was found in the cytosol, it might be associated with lipid, based on its elution in the void volume of a Sephadex G-150 column. As with many of the liver phospholipases, phosphatidylethanolamine was hydrolyzed in preference to other phospholipids. Almost no activity was detected in the absence of metallic cations although there did not seem to be any specificity for the cation used and even sodium EDTA stimulated. Cationic detergents that lower the surface charge such as cetyltrimethylammonium bromide activated the enzyme when present in an activator : substrate ratio of approximately 1 : 2; higher ratios inhibited. Cationic proteins produced the same stimulatory effect. Since all these additions lowered the highly acidic charge that phosphatidylethanolamine has at pH 9.5, it seems that these agents might serve to reduce charge repulsion between the anionic enzyme (pI 7.15) and the substrate. Specific bridging of enzyme and substrate by cations does not appear likely, based on the wide range of compounds that stimulate activity.

Another characteristic of the cytosolic phospholipase A_1 activity was its requirement for the form of the substrate aggregate. This enzyme, like the lysosomal phospholipase A_1, appears to act preferentially on lipids in the nonbilayer-hexagonal II phase. This is borne out in their study of lipid mixtures (Dawson *et al.*, 1983*b*). For example, addition of phosphatidylethanolamine to phosphatidylcholine enhances the activity on the latter when its ratio to phosphatidylethanolamine exceeded 1:1. As predicted, lipids that would disrupt the hexagonal organization, such as phosphatidylcholine, decreased activity more than would be predicted by surface dilution (Figs. 6-1 and 6-2).

It is difficult in these experiments to separate the effect of the state of lipid aggregation on enzyme binding from the binding of the lipid molecule in the active site without knowledge of the aggregate structure(s) present in the reaction mixture. If only a structural change were occurring, it might be predicted from these data that a 30 mole percent of phosphatidylcholine in phosphatidylethanolamine would not inhibit phosphatidylethanolamine hydrolysis so dramatically while leading to the marked increase of phosphatidylcholine degradation.

The relation between this enzyme and similar enzymes in other tissues

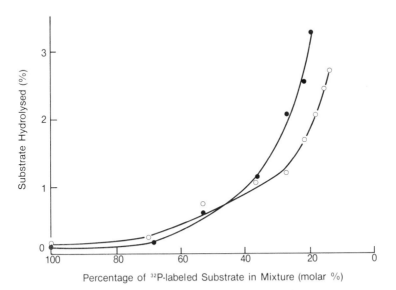

Figure 6-1. Hydrolysis of other ^{32}P-labeled phospholipids stimulated by admixture with phosphatidylethanolamine. [^{32}P]Phosphatidylcholine (0.11 μmole, ○) or [^{32}P]phosphatidylinositol (0.054 μmole, ●) was mixed with various amounts of phosphatidylethanolamine in chloroform solution, the solvent removed, and the mixed phospholipids incubated using 0.08 M KCl plus 1.5 mM MgCl₂ as activator. The ^{32}P-labeled lysophospholipids produced were separated by thin-layer chromatography and their radioactivity was determined. (From Dawson *et al.*, 1983.)

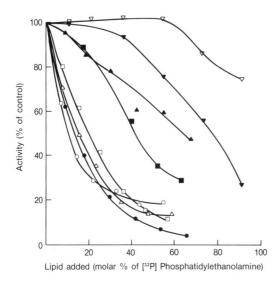

Figure 6-2. Effect of other phospholipids and fatty acids on the activity of phospholipids A_1 toward [^{32}P]phosphatidylethanolamine. The other lipids were mixed with [^{32}P]phosphatidylethanolamine in organic solvent solution, and after removal of solvent the phospholipase A_1 activity was determined using 80 mM KCl plus 1.5 mM MgCl$_2$ as activator. Symbols: phosphatidylcholine (●); sphingomyelin (○); lysophosphatidylcholine (△); phosphatidylinositol (□); phosphatidylserine (▲); phosphatidate (■); palmitate (▽); palmitolate (▼). (From Dawson *et al.*, 1983.)

is not known, similar to our lack of understanding of the relation between the cytosolic and membrane-associated phospholipases A_1 of the liver. The possibility that the enzyme preparation is lipid associated poses the question of whether it is loosely or reversibly associated with a membrane and during cell disruption is *solubilized,* hence having some role within a membrane. On the other hand, it might be a soluble enzyme that requires some activation for interaction with one or more membranes. A third possibility is that the enzyme, as it is isolated, is fully activated but requires a recognition factor within the membrane such as a particular lipid domain (hexagonal patch?). Obviously, for the assignment of function, the interaction of the enzyme with a membrane(s) must be determined.

6.2.2.2. Lysosomes

Unlike the soluble phospholipase A_1, the existence of lysosomal enzyme has been known for some time and has been under active investigation for over a decade. Its purification, like that of the cytosolic enzyme, is facilitated by its soluble nature. Two major problems have been encountered in this work, however: (1) enzyme instability and (2) heterogeneity. While the first problem has been resolved reasonably well, the second problem remains. Since there is now a fair amount known about the enzyme and its regulation by substrate structure, it will be covered in some detail.

Two groups have published methods for its purification (Waite *et al.,* 1981; Hostetler *et al.,* 1982; Robinson and Waite, 1983) and there is good agreement on a number of the properties of the enzyme. Since it binds concanavalin A it is thought to be a glycoprotein containing α-D-glucosyl- or α-D-mannopyranosyl terminus. Hostetler *et al.* (1982) isolated five forms (isoenzymes?) and demonstrated the characteristics outlined in Table 6-4. The mo-

Table 6-4. Phospholipases A₁ from Rat Liver Lysosomesᵃ

pI	Molecular weight	K_m	V_{max}	$\dfrac{\text{PLA}}{\text{LPL}}1^{b,c}$	Yieldᵈ
<4.0	24,000	1.2×10^{-4}	0.27	6.1	0.4
<4.0	44,000	3.5×10^{-4}	1.3	6.4	2.0
5.2	15,000	1.8×10^{-4}	0.44	11.4	0.5
5.2	34,000	4.3×10^{-4}	16.2	9.1	5.0
5.2	90,000	6.5×10^{-5}	0.01	6.1	0.2

ᵃ From Hostetler *et al.* (1982).
ᵇ Ratio of phospholipase A₁ to lysophospholipase activity (1-acyl lysophosphatidylcholine; no activity was found at position 2).
ᶜ Abbreviations: PLA₁, phospholipase A₁; LPL, lysophospholipase.
ᵈ Percentage of starting activity.

lecular weights were determined by gel filtration and the pI by chromatofocusing (Table 6-4).

The various characteristics described do not fit any particular pattern that might help in understanding the significance of these forms. Likewise, these forms do not vary much in their response to the addition of salts (NaCl, $CaCl_2$, $HgCl_2$, EDTA), Triton X-100, or bromophenylacyl bromide which caused only minimal inhibition. Likewise, there did not appear to be any major differences in substrate specificity.

These general characteristics of the multiple forms were found in our laboratory as well and were viewed with some annoyance since this heterogeneity complicated purification. Hostetler suggested that these different forms could represent different stages of processing and maturation of the enzyme which could have some interesting implications. He also raised the possibility that there is specificity of the enzyme according to the cell of origin (hepatocyte, Kupffer, endothelial). On the other hand, these differences in molecular properties could be artifactual, arising during cell disruption or processing. Despite differences in lysosome isolation, no major differences were found between the two reports, although the range of molecular weight found by our laboratory was 45,000–60,000 daltons. Our attempts to resolve the question of enzyme degradation during isolation suggested that some degradation did occur that caused an increase in the pI of the enzyme. This, however, would not relate well to variation in molecular weights found by Hostetler. The molecular weights reported by both groups should be viewed with caution since these are based on gel filtration analysis. Unfortunately, the amount of enzyme recovered thus precludes detailed molecular analysis.

More is known about substrate specificity and enzyme–lipid interaction than about the structure of the enzyme itself. Initially, we found that phosphatidylethanolamine was the preferred substrate and that the acidic phospholipids were barely degraded. This was surprising because at pH 4.0 phosphatidylethanolamine is nearly neutral and forms aggregated hexagonal arrays that are difficult to keep in suspension. Under the same conditions, phosphatidylcholine is neutral in charge yet easily forms liposomes upon sonication. However, phosphatidylcholine liposomes are not readily attacked by the enzyme, despite their fine dispersion. Likewise, if the phosphatidylethanolamine aggregates are better dispersed with Triton X-100 in ratios as low as 1 : 1, drastic inhibition occurs. Studies in Hostetler's laboratory in which the substrate was mixed with Triton X-100 gave results quite different from those described above. As shown in Table 6-5, the acidic phospholipids were most readily attacked and phosphatidylethanolamine that we found to be the best substrate in the absence of Triton was hydrolyzed weakly.

We reinvestigated the problem and found that the apparent substrate specificity was determined to a large extent by the type of substrate aggregate used (Fig. 6-3).

These data show that while Triton inhibits the activity on phosphatidylethanolamine, it enhances activity on all other phospholipids. (Phosphatidylserine, for some reason, requires a slightly higher pH for optimal degradation, pH 4.5.) Like the cytosolic phospholipase A_1, it would appear that the hexagonal array is preferred over the bilayer liposome. A mixed micelle predicted to predominate at a Triton : phospholipid ratio of 1 : 2 or higher is most readily attacked, even at Triton : phospholipid ratios where surface dilution of substrate would be expected. In the case of the strongly acidic phospholipids, we predicted that surface-charge dilution is a significant factor in enhancing the activity with increasing Triton. This is supported by the fact that Ca^{2+} stimulates the activity on mixed micelles of acidic phospholipid but inhibits the activity of hexagonal aggregates of phosphatidylethanolamine. Indeed, we found a very definite surface-charge requirement of the enzyme using pure phosphatidylcholine or phosphatidylethanolamine (Figs. 6-4 and 6-5).

Similar to the effect on the cytosolic phospholipase A_1 (Dawson *et al.,* 1983*b*), the addition of phosphatidylcholine to phosphatidylethanolamine stimulated phosphatidylcholine degradation (Fig. 6-6). In this case, however, the inhibition of phosphatidylethanolamine degradation, while significant, was not as great as that noted with the cytosolic enzyme. While some details might be different, the notion that both phospholipases A_1 require the nonbilayer aggregates appears reasonable.

Table 6-5. Substrate Specific Activity Determined in the Presence of 0.8 mM Triton X-100[a]

	μmoles/hr/mg protein (34,000-dalton enzyme)
Phosphatidylglycerol	82
Cardiolipin	58
Phosphatidylcholine	44
Phosphatidylinositol	26
Phosphatidylserine	22
Phosphatidylethanolamine	16
Bis(monoacylglycero)phosphate	7
Triglyceride	3.7

[a] From Hostetler *et al.* (1982). With 0.2 mM substrate, a substrate : Triton ratio of 1:4.

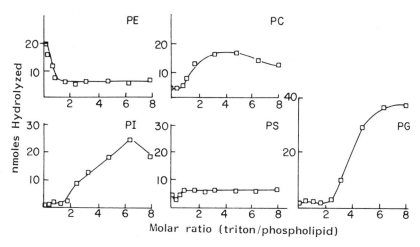

Figure 6-3. Effect of Triton WR1339 on the hydrolysis of phospholipids by phospholipase A$_1$. Each assay contained 100 nmoles of each phospholipid substrate and 0–0.8 μmoles of Triton WR1339. (From Robinson and Waite, 1983.)

The studies on surface-charge requirements are particularly useful in understanding what role lysosomal phospholipase A$_1$ might play *in situ* and in predicting ways by which the enzyme might be inhibited. There is a wide range of lipophilic cations such as local anesthetics and antimalarial agents that inhibit this enzyme. A similar compound, 4,4'-diethylaminoethoxyhexestrol (DH), produced a lipid storage disease that was initiated at the lysosomal level (Matsuzawa and Hostetler, 1980). These findings suggested to Hostetler and his colleagues that a lysosomal phospholipase could be involved. While such a prediction has not yet been fully substantiated, it would certainly appear that the enzyme functions in normal turnover of membrane lipid by *in vivo* endocytosis. Likewise, it might function in the degradation of endocytized particles such as bacterial and circulating lipoproteins, depending on the cell type. Blockage of the enzyme by changes in surface charge would then lead to a lipidosis. While this is still speculative, information available supports such a hypothesis.

A number of factors influence its activity on aggregates and pH is thought

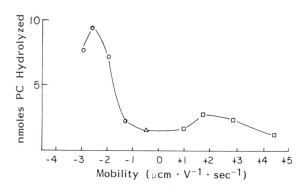

Figure 6-4. Hydrolysis of PC vesicles as a function of mobility. Mobility was altered with the addition of (right to left) 1, 2.5, 5, and 10 mole % dicetylphosphate (○), no addition (△), or (left to right) 1, 2.5, 5, and 10 mole % stearylamine (□). (From Robinson and Waite, 1983.)

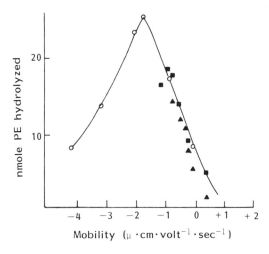

Figure 6-5. Hydrolysis of PE vesicles as a function of mobility and Ca^{2+}. Mobility was altered with the addition of (left to right) 10, 5, 2.5, and 1 mole % dicetylphosphate, no addition, and 1 mole % stearylamine. Activity and mobility were measured for each series of amphipath molecule-altered vesicles with no addition (○), 1 mM $CaCl_2$ (■), or 2 mM $CaCl_2$ (▲). (From Robinson and Waite, 1983.)

to be at least one component that will regulate its activity within the lysosome (Weglicki *et al.*, 1974). Lysosomes were relatively stable when incubated at neutral pH. When the pH was lowered to 5.5, membrane lipid was degraded and the lysosome became labile. While this would be a pathological response, two important points are to be noted: (1) The enzyme does degrade membrane phospholipid (confirmed in direct experiments), and (2) its sharp pH optimum with no detectable activity above pH 6.0 can limit activity.

6.2.2.3. *Plasma Membrane (Heparin Releasable)*

This phospholipase A_1 was first described in the early 1970s when techniques became available for the isolation of plasma membranes with sufficient purity to allow characterization of their enzyme content (Newkirk and Waite, 1971). While this enzyme does have high activity against neutral glycerides

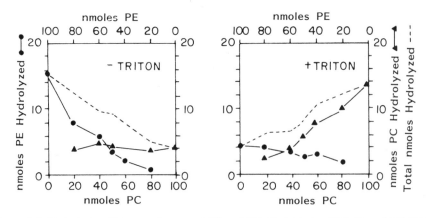

Figure 6-6. Hydrolysis of binary mixtures of PC and PE. Hydrolysis of PE (●) and PC (▲) in binary mixtures with varying mole % of each phospholipid in the absence (*left*) or presence (*right*) of 0.5 μmoles of Triton WR1339. (From Robinson and Waite, 1983.)

and is thought to be involved in lipoprotein metabolism, its activity on phosphoglycerides could be of considerable importance physiologically and is covered in some detail here. Two observations had been made suggesting that
it is an ectoenzyme with broad substrate specificity. First, a phospholipase A_1
was identified in postheparin plasma that catalyzed a transesterification in
which the fatty acid in position 1 of phosphatidylethanolamine was transferred
to an alcohol acceptor (Vogel *et al.*, 1965). Second, the phospholipase A_1 of
rat liver plasma membrane was released by heparin and catalyzed a transacylation between two molecules of monoacylglycerol (Waite and Sisson, 1973*a*,
1973*b*).

$$2 \text{ 1-acylglycerol} \rightarrow 1,3(2)\text{diacylglycerol} + \text{glycerol}$$

In this reaction, the diacylglycerol formed was a mixture of 1,2 and 1,3
diacylglycerol which probably reflects the lack of positional specificity for the
acyl acceptor, unlike the donor that was specific for a position. The diacylglycerol was then degraded and, given sufficient time, only fatty acid continues
to accumulate. As shown in Fig. 6-7, both phosphatidylethanolamine (diacyl
GPE) and monoglyceride served as acyl donors to form diglyceride.

These substrates were used either singly or in mixtures with 1-acylglycerol
as the acyl donor or acceptor. In the latter case, the substrate mixture was
prepared such that aggregates were comprised of both lipids, thus maximizing

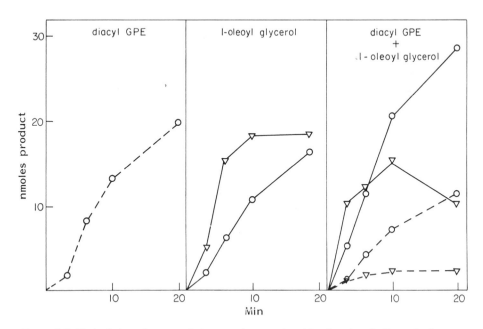

Figure 6-7. Hydrolysis and transacylation reactions catalyzed by the phospholipase A_1 from rat
liver plasma membranes. The enzyme was incubated with diacylglycerophosphorylethanolamine
(diacyl GPE) (---) and oleoylglycerol (———) singly or together. The products formed were free
fatty acid (○) and diglyceride (▽).

the potential for transacylation. The availability of a given hydroxyl, an alcohol or water, determined what product would be formed, an ester or the free acid. Since monoacylglycerol was most rapidly attacked and since a general acyl transfer was catalyzed in those studies, the name monoacylglycerol acyl transferase (MGAT) was suggested (Waite and Sisson, 1976*b*). This enzyme is referred to in the literature most often as hepatic triglyceride lipase since triglyceride is a commonly used substrate. However, in this chapter it is referred to as the plasma membrane phospholipase A_1 and the work covering its activity is emphasized.

The purification of this enzyme has been greatly facilitated by the use of liver perfusion with heparin. This releases the enzyme with few contaminating proteins. Most purification procedures employ affinity chromatography on heparin–Sepharose columns; the phospholipase A_1 elutes with 0.7 M NaCl, regardless of its source. A recent procedure that is attractive involves four steps with a 421-fold purification and 31% yield starting with the liver perfusate (Jensen and Bensadoun, 1981). Important in this procedure was the stabilization of the enzyme with Triton N-101. The pure enzyme had an apparent molecular weight of 180,000 daltons as determined by gel filtration and SDS–PAGE (Kuusi *et al.*, 1979*a*). This value might represent an aggregate since others have reported lower values in the 60,000–70,000-dalton range (Colard-Torquebiau *et al.*, 1975; Jensen and Bensadoun, 1981; Belcher *et al.*, 1985). Some uncertainty remains concerning the molecular weight values reported at this time which could be the result of enzyme proteolysis (Kuusi *et al.*, 1979*b;* Jensen *et al.*, 1982; Tsujita *et al.*, 1984).

The substrate acyl donors and acceptors tested with the phospholipases A_1 from plasma membranes are listed below (Waite and Sisson, 1973*a*, 1973*b*, 1974) (Table 6-6).

Detailed studies that have been carried out to define better the degradation of several structures made up of different potential substrates include the purported natural substrates, the lipoproteins (Table 6-7).

Table 6-6. Substrate Phospholipase A_1[a]

Acyl donors	Acyl acceptors
1-Acylglycerol	1- and 2-Acylglycerol
1,2- and 1,3-Diacylglycerol	1-Acyl propanediols
1-Acyl-GPE[b]	Acylglycol
Diacyl-GPE	1- and 2-Alkylglycerols
Triacylglycerol	H_2O
	Acetyl alcohol
Not acyl donors	*Not acyl acceptors*
2-Acylglycerol	Diacylglycerol
2-Acyl-GPE	1-Acyl- or 2-acyl-GPE
	Cholesterol

[a] From Waite and Sisson (1974).
[b] Abbreviation: GPE, glycerophosphorylethanolamine.

Table 6-7. Activity of Phospholipase A_1^a

Substrate[b]	Specific activity (μmoles/min/mg)
Monooleoylglycerol	429.90
Chylomicron TG	42.15
Remnant TG	24.25
Gum arabic/TG emulsion	108.50
PC vesicles	8.40
Chylomicron PC	3.65
LDL/HDL-1 PC	0.44
HDL-2 PC	0.14

[a] From Belcher *et al.* (1985). Lipolytic measurements were carried out in the absence of Triton X-100 under optimal conditons with $n = 2$; for protein determinations, $n = 1$. All assays were performed on the same enzyme preparation on the same day. Results are means ± SD.

[b] Abbreviations: TG, triacylglycerol; PC, phosphatidylcholine; LDL, low-density lipoprotein; HDL, high-density lipoprotein.

The phospholipase A_1 activity toward phosphatidylcholine shows the latter not to be the preferred substrate. As shown in Fig. 6-8, phosphatidylethanolamine is degraded much more rapidly than phosphatidylcholine when present with Triton X-100. Also, it can be seen that Triton X-100 increases hydrolysis severalfold.

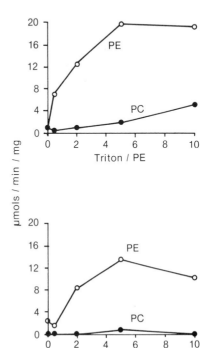

Figure 6-8. Hydrolysis of phosphatidylethanolamine (PE) and phosphatidylcholine (PC) in mixed micelles with Triton X-100. The top panel shows hydrolysis when the two substrates were hydrolyzed separately, while the bottom panel shows hydrolysis of the two in the same reaction mixture. The abscissa indicates the molar ratio of Triton to phospholipid.

There is, however, a great deal of uncertainty about substrate specificity of this enzyme since the activity on triacylglycerol is measured in different types of emulsion and in liproproteins, on monoacylglycerol with Triton X-100 that inhibits its degradation and on various phosphoglycerides. Even the acyl group structure of monoacylglycerol will influence activity, adding to the complexity of the system (Hulsmann *et al.*, 1980). For example, stearoyl glycerol is degraded at 5–10% the rate of oleoyl glycerol and linoleoyl glycerol (Miller *et al.*, 1981). In mixtures of saturated and unsaturated acyl glycerols, the enzyme still degrades the unsaturated molecules more rapidly than the saturated molecules. This difference was attributed to the recognition of different phases of substrate in the aggregate rather than there being some specificity at the catalytic site. Likewise, unsaturated phosphatidylcholine degradation was greater than when disaturates were used (Hulsmann *et al.*, 1980). A model was proposed that included an acyl–enzyme intermediate, although direct evidence for this is lacking owing to the rapid hydrolysis that would occur if such an intermediate were formed (Waite and Sisson, 1973*a*).

The argument for an acyl intermediate is based on the observation that there is a bulk limit for the acceptor molecule. For example, cholesterol, diacylglycerol, and lysophosphatidylethanolamine are not acceptors yet the latter two can function as donors. It would appear then that the equivalent of two acyl or alkyl chains can fit into the active site at a given moment, providing the polar region is not too large. While not totally satisfying, all observations made thus far indicate such a model for the active site warrants further investigation.

An important observation was made by Kuusi *et al.* (1979*b*), who demonstrated that a protease decreased the capacity of purified phospholipase A_1 to degrade triglyceride. They detected a change in the behavior of the enzyme when chromatographed on Ultragel Aca34, indicating a decrease in molecular weight. Since collagenase and proteolytic enzymes are known to be present in liver, these workers suggested that enzyme modification might have occurred in procedures of cell isolation or disruption. It was speculated that the peptide fragment lost was a recognition site or component specific for triglyceride. While this certainly is a likely possibility, a change in the structure of the active site could occur such that a long acyl chain triglyceride could no longer be accommodated.

These results were subsequently confirmed and extended. As shown by Jensen *et al.* (1982), the enzyme's capacity to hydrolyze triglyceride was more susceptible to pepsin cleavage than its capacity to degrade monoglyceride or phospholipid (Fig. 6-9).

This susceptibility to proteolysis was shown by Tsujita *et al.* (1984), most likely to be the destruction of a hydrophobic binding region since the enzyme's activity on soluble tributyrin was unaffected by trypsin proteolysis whereas the activity on triolein was abolished.

The physiologic role of this enzyme is of considerable interest, in particular to those studying lipoprotein metabolism. Perhaps one of the most important sets of observations demonstrating that the phospholipase A_1 is involved in lipoprotein catabolism was made by the injection of antibody to the

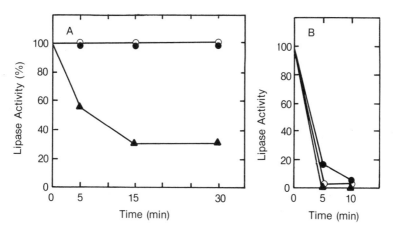

Figure 6-9. Proteolytic digestion of phospholipase A_1 with pepsin. Highly purified phospholipase A_1 from gel filtration was diluted five-fold in 50 mM Tris-HCl, pH 7.2, that contained pepsin to yield net protease concentrations of 0.89% (w/v) (A) or 8.9% (w/v) (B). These preparations were maintained at room temperature and assayed at the indicated time points. Lipase activities are expressed as percentage of simultaneous controls that were diluted in Tris-HCl buffer lacking pepsin and assayed at each time point. Initial activities were 69.6, 22.7, and 2.08 µmoles fatty acid/ml/hr for the triacylglycerol lipase (▲), monoacylglycerol lipase (○), and phospholipase (●) activities, respectively. The relative sensitivity of the triacylglycerol lipase activity to proteolytic digestion was also found in proteolytic digestions with collagenase, trypsin, chymotrypsin, or thermolysin. (From Jensen *et al.*,1982.)

enzyme into rats (Kuusi *et al.*, 1979*c*; Jansen *et al.*, 1980). Both low-density lipoprotein (LDL) and high-density lipoprotein (HDL) phospholipids increased, which led those investigators to propose that *in situ* the plasma membrane phospholipase A_1 functions as a phospholipase. Furthermore, cholesterol in HDL also accumulated, suggesting that cholesterol uptake by the liver required phospholipid hydrolysis catalyzed by the phospholipase A_1. If this is the case, then this enzyme might regulate cholesterol flux and factors such as dietary polyunsaturated fat could promote their cholesterol lowering effects through their stimulation of phospholipase A_1 (Hulsmann *et al.*, 1980).

Specifics of this proposal have been further refined. van't Hooft *et al.* (1981) presented a model in which the phospholipid of HDL_2 is degraded by the plasma membrane phospholipase A_1, reducing the surface coat of the particle. This leads to a "squeezing out" of the central core of cholesterol ester into the liver cell. This results in the formation of HDL_3, which is smaller and poorer in phospholipid and cholesterol ester than the HDL_2. Direct evidence for this mechanism was provided by Bamberger *et al.* (1985), who demonstrated a marked increase in cholesterol uptake from phospholipase A_1-treated HDL by cultured rat hepatoma cells, as compared with untreated HDL. It is curious, however, that the major phospholipid in HDL is phosphatidylcholine, whereas the plasma membrane phospholipase A_1 preferentially degrades phosphatidylethanolamine, at least in model systems (Fig. 6-7). Perhaps an interaction of phosphatidylcholine with another component(s) of the lipoprotein will enhance its hydrolysis.

Evidence has been accumulated that the phospholipase A_1 is synthesized in the parenchymal cell and can be released from cultured hepatocytes (Sundaram *et al.,* 1978). The enzyme did not accumulate in the cells but was released into the medium (Jansen *et al.,* 1979). At present, the mechanism of enzyme synthesis and release is somewhat obscure since cycloheximide but not puromycin blocked its appearance in the medium. It appeared that the cytoskeletal system is involved in its release since colchicine also blocked release. If, however, monoacylglycerol acyl transferase activity is assayed on cultured hepatocytes, a rapid degradation of monoacylglycerol but not triacylglycerol was observed (El-Maghrabi *et al.,* 1977). It is possible that the enzyme associated with the parenchymal cell lacks the capacity to act on triacylglycerol because of the cell isolation procedure used (Kuusi *et al.,* 1979*b*). The physiologic significance of this is totally unknown at this time and indeed could be the result of some enzyme modification that occurred during cell isolation, such as proteolysis. The latter seems unlikely though since we have found that long-term culture of the parenchymal cells does not lead to increased triacylglycerol degradation and Jansen *et al.* (1979) showed that activity against triacylglycerol did appear in the medium.

It has been suggested that the phospholipase A_1, once released from the parenchymal cell, binds to the endothelial cell where it functions (Jansen *et al.,* 1978*b*). Furthermore, the phospholipase A_1 bound to the endothelial cell was released by heparin. The conclusion of Jansen and coworkers was supported by Kuusi *et al.* (1979*d*) who found the phospholipase A_1 to be localized exclusively on endothelial cells using immunoelectron microscopy. This enzyme must be lost from the endothelial cells during their isolation since our laboratory found little activity of the lipase or phospholipase activities *in vitro* or *in vivo* (Lippiello *et al.,* 1981, 1985). We found that both Kupffer and parenchymal cells take up phospholipid, neutral glyceride, and cholesterol (ester) from preparations of remnant lipoproteins. While this did not directly demonstrate the activity of the phospholipase A_1, several lines of evidence suggested it was present.

A number of points are very difficult to reconcile at this time. The organization of the system as proposed by Jansen and Kuusi has precedent with the similar lipoprotein lipase (Nilson-Ehle *et al.,* 1976). Although considerable evidence exists that the phospholipase A_1 and hepatic lipase are one and the same, it is possible that different or additional enzymes are present on the cell surface. While it is generally thought that the conversion of the phospholipase A_1 to a form not active on triacylglycerol is artifactual, it is possible that this is of physiological importance worthy of further investigation. Obviously, there are numerous opportunities remaining in this area of research.

6.2.3. Heart

Like the liver, the heart contains quite a number of phospholipases, including phospholipases A_1. Also like the liver, there is uncertainty as to the cellular origin of these phospholipases A_1. Nonetheless, their presence implies that the turnover of phospholipids in the heart is significant and potentially the activity of phospholipases in the heart might be related to pathophysiologic conditions such as infarctions or ischemia (Franson *et al.*, 1979). Thus far, most work on the phospholipases A_1 in the heart have been limited to the description of their subcellular localization and the establishment of the assay conditions.

6.2.3.1. Heart Microsomal and Sarcolemmal Phospholipase A_1

These two membrane preparations are treated together since there is evidence that the enzymes in the microsomal preparations studied have been clearly differentiated from those in the sarcolemmal preparations studied. The first report of phospholipase A_1 activity in a microsomal preparation from rats and dogs was provided by Weglicki *et al.* (1971). The hydrolysis of phosphatidylethanolamine was much greater than that of phosphatidylcholine and little effect was noted when EDTA replaced Ca^{2+}. Subsequent work showed that an eightfold purified sarcolemmal fraction from guinea pig heart had enriched phospholipase A_1 activity with two pH optima, 6.0 and 9.0 (Franson *et al.*, 1978a). Based on the difference between these observed pH optima and that reported earlier for the microsomal preparation, pH 8.0, these workers suggested that these two membrane preparations contained distinct phospholipases. While this is supported by differences in Ca^{2+} response, it is unclear whether the differences noted are the result of species differences. Comparison of a dog heart sarcolemmal preparation comparable to that of the guinea pig had optimal pH activity at pH 7.5. Interestingly, Franson *et al.* (1979) found a threefold increase in phospholipase activity when 1×10^{-5} M isoproterenol was present and suggested that this might be a cause of necrosis of the heart observed when animals are chronically treated with this class of drugs.

6.2.3.2. Heart Lysosomal Phospholipase A_1

There is considerable difficulty in obtaining sufficient purification of rat heart lysosomes for adequate characterization. The use of Triton WR1339 as done to alter the density of the liver lysosome alters the density of only a portion of the heart lysosomes (Franson *et al.*, 1972). Presumably, these lysosomes originated from two different cell populations within the heart. Assuming that the lysosomes that had altered density upon Triton injection into the rats were from macrophages whereas those whose density did not shift originated from muscle cells, some differences in the cell distribution of lysosomal phospholipases were suggested. Phospholipase A_1 with optimal activity at pH 4.0 in the presence of EDTA was found in both populations of lysosomes while that with optimal activity at pH 5.0 and required Ca^{2+} was found primarily in the lysosomes of purported macrophage origin.

6.2.3.3. Heart Cytosolic Phospholipase A_1

The cytosolic fraction of heart has been studied only recently. Nalbone and Hostetler (1985) did a rather detailed fractionation of rat heart. This paper describes the distribution of three phospholipases—one active at acid pH, one active at neutral pH, and a lysophospholipase (Table 6-8). As seen in Table 6-8, nearly half of neutral-active (pH 8.4) phospholipase A and lysophospholipase activities were recovered in the cytosolic fraction. This enzyme was stimulated only slightly by Ca^{2+}, similar to the microsomal enzyme. The cytosolic and the light mitochondrial (lysosomal) fractions had phospholipase A_1 primarily. In this study, phosphatidylcholine was used as the substrate which yielded lower specific activities than those reported earlier in which phosphatidylethanolamine was used (Weglicki et al., 1971).

6.2.3.4. Heart Mitochondrial Phospholipase A_1

In most tissues, the phospholipase A_2 predominates in the mitochondrial fraction. Based on the composition of fatty acids released from mitochondrial membranes, evidence for the presumed action of a phospholipase A_1 was presented by Palmer et al. (1981). Similar to the findings of Nalbone and Hostetler (1985) given in Table 6-8, the subsarcolemmal and interfibrillar mitochondria had phospholipase A_1 and phospholipase A_2 activity. Presumptive evidence was given that the phospholipase A_1 is present on the outer membrane of both classes of mitochondria. Using exogenous substrates, Weglicki et al. (1971) demonstrated that dog and rat heart mitochondria had predominately phospholipase A_2 activity.

6.3. Lysophospholipases: General Considerations

This group of enzymes probably is distributed as widely as the phospholipases A and many of the references given in Tables 6-1 and 7-1 include studies on the distribution of lysophospholipases. Often where the distribution of phospholipases in a tissue has been investigated, lysophospholipase activity is implied to explain the lack of stoichiometry between the recovery of free fatty acid and lysophospholipid, a conclusion that may not always be valid. A true definition of lysophospholipase is difficult, as pointed out by van den Bosch (1982). A number of enzymes that are termed lysophospholipase have activity on diacyl phospholipids and therefore can be termed phospholipase

Table 6-8. Percentage of Enzyme Activities Expressed as the Percentage of the Total Recovered Activity[a]

Fraction	Acid phospholipase A	Neutral phospholipase A	Lyso-phospholipase
Nuclei and debris	15.2	8.7	12.8
Subsarcolemmal mitochondria	9.7	12.2	6.9
Interfibrillar mitochondria	4.0	2.4	6.3
Light mitochondria	12.8	8.5	2.7
Microsomes	11.6	17.4	17.2
Cytosol	11.3	45.8	45.7

[a] From Nalbone and Hostetler (1985). Each value represents the average of four separate preparations ± standard deviation.

B or A_1 (van den Bosch *et al.*, 1974). Likewise, a number of enzymes termed lysophospholipase have broad esterase activity and do not necessarily require the presence of a phosphate group. The type of activity observed is often dependent on the conditions used. Almost without exception, lysophospholipase activity is inhibited by detergents suggesting that mixed micelles are not suitable substrates. The activity toward diacyl phospholipid catalyzed by the same enzyme is often enhanced by detergent. In this part of the chapter, those enzymes having acyl hydrolase activity, primarily on the lysophospholipid, are covered although in some cases the full range of activities has not been explored. Emphasis is placed on those from liver, lung, and placenta that have been studied in some detail.

Lysophospholipases undoubtedly function to stabilize membranes by controlling the levels of lytic lysophospholipids. While such a generalized statement appears as a simplistic truism, the activity of this group of enzymes undoubtedly is coordinated in some fashion with other enzymes which metabolize its substrate, lysophospholipid. This is particularly true of its potential competition with the enzymes that reacylate lysophospholipid and introduce the long-chain polyunsaturated acyl groups in position 2 of the glycerol. Indeed, were the phospholipases not in balance with reacylation enzyme, both the quantity and quality of membranous phospholipid would change. This has been postulated to occur in hepatoma cells that have a low phospholipid : protein ratio as well as a low percentage of polyunsaturated acyl groups (Waite *et al.*, 1977). This group of enzymes serves a key role in phospholipid metabolism and further studies aimed at substrate specificity and its relation to turnover of phospholipids and to the other phospholipid-metabolizing enzymes are essential to understanding membrane structure and function.

6.3.1. Early Observations

One of the earliest distinctions made in mammalian tissue to demonstrate that lysophospholipase could be different from phospholipase B was presented by Shapiro (1953). In this work, a lysophospholipase was purified 25-fold from the soluble fraction of bovine pancreas and was shown to be devoid of phospholipase activity under the conditions that favored lysophosphatidylcholine hydrolysis. Subsequently, it was shown that a lysophospholipase activity in a particulate fraction from intestinal mucosa was autocatalytic and that free fatty acid stimulated the hydrolysis of lysophosphatidylcholine (Epstein and Shapiro, 1958). A similar enzyme capable of degrading both lysophosphatidylethanolamine and lysophosphatidylcholine was described in the soluble fraction of liver, although no investigation of hydrolysis of diacyl phospholipid was carried out (Dawson, 1956). In both cases, the enzyme was in a solubilized form.

The interest in this enzyme was quite high at that time since it was felt that lysophospholipases could protect the cell against the lytic action of lysophospholipids. The widespread distribution of this enzyme, as shown in Table 6-9, shows that it might serve such a generalized protective function (Marples and Thompson, 1960).

Table 6-9. *Degradation of Lysophosphatidylcholine by Various Rat Tissues*[a]

Tissue	Activity (μmoles/g/hr)			
	Fatty acid	Total choline	Free choline	GPC[b]
Lung	57.2 (2)	51.6 (2)	7.1 (2)	44.5
Ileum	71.2 (3)	72.3 (2)	2.6 (2)	69.7
Spleen	52.3 (2)	46.7 (4)	3.2 (4)	43.5
Liver	29.4 (5)	31.5 (4)	0 (3)	31.5
Pancreas	27.2 (3)	—	—	—
Kidney	9.4 (2)	9.6 (2)	4.4 (2)	5.2
Testis	6.7 (2)	5.3 (2)	0 (2)	5.3
Skeletal muscle	4.4 (2)	5.6 (2)	0.1 (2)	5.5
Heart	—	2.1 (2)	0 (2)	2.1
Brain	—	1.7 (3)	0 (3)	1.7
Spinal cord	—	0.8 (2)	0 (2)	0.8
Adipose tissue	0 (2)	—	—	—

[a] From Marples and Thompson (1960).
[b] GPC, glycerophosphatidylcholine.

The solubilized lysophospholipase studied had a common pH optimum, pH 6.0, and did not require Ca^{2+}.

Further studies on the pathways of lysophosphatidylcholine metabolism revealed a novel anabolic reaction involving a transacylation (Erbland and Marinetti, 1965a). Three metabolic fates of lysophosphatidylcholine were then known.

(1) Lysophosphatidylcholine + H_2O → fatty acid + glycerophosphorylcholine

(2) 2 Lysophosphatidylcholine → phosphatidylcholine + glycerophosphorylcholine

(3) Lysophosphatidylcholine + acyl CoA → phosphatidylcholine + CoA

Reactions (1) and (2) were found primarily in the cytosolic fraction, whereas reaction (3) was predominantly in the microsomes (Erbland and Marinetti, 1965b). Diacyl phospholipids did not function in this transacylation process and hydrolysis is the primary reaction catalyzed by the lysophospholipase at low (physiologic) concentrations of lysophosphatidylcholine. Therefore, reaction (2) might not be of physiologic significance. On the other hand, the presence of phospholipids with identical acyl chains in positions 1 and 2 could be the result of such a reaction (i.e., lung).

The lysophospholipase activity in rat liver cytosol was found to be rather nonspecific in that both 1- and 2-acyl lysophosphatidylcholine, stearoyl ethylene glycol phosphorylcholine, and palmitoyl propane diol-phosphorylcholine were degraded (van den Bosch et al., 1968). Likewise, no stereospecificity was observed. A key observation was the finding that lysophosphatidylcholine present in a bilayer system could not be degraded, emphasizing that the physical structure of the substrate is crucial. This was further brought out by the finding that the saturation kinetics of the reaction were dependent on the protein concentration (Fig. 6-10).

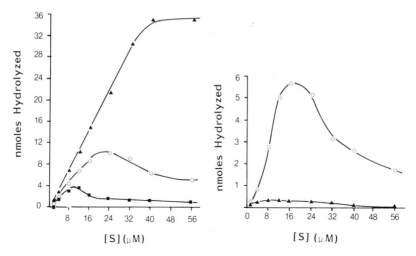

Figure 6-10. Effect of substrate concentration on various levels of lysophospholipase. The incubation mixture consisted of varying amounts of 1-acyl-3-*sn*-glycerophosphatidylcholine, 20 μmoles of Tris-HCl buffer (pH 7.2), and either 500 μg (▲———▲), 167 μg (○————○), or 56 μg (■———■) of supernatant protein in a final volume of 1.0 ml. From van den Bosch *et al.* (1968).

Presumably, the enzymatic activity is regulated by the dispersion of the lipid substrate by the protein. This plus the finding that albumin enhanced the lysophospholipase activity in some but not all cell preparations led to a reevaluation of the distribution of the lysophospholipase (lysolecithinase) (Leibovitz-Ben Gershon and Gatt, 1976). Using what was considered to be optimal conditions, the particulate fractions of five tissues were found to be more active than the cytosol (Table 6-10).

Similar but more detailed studies led the same group of workers to conclude that lysophospholipase of rat brain acted on monomers rather than micelles of substrate. This conclusion was challenged since the CMC values used in the study with the brain enzyme have been found to be about 20-fold lower than that estimated in the study with the brain enzyme (Aarsman and van den Bosch, 1979). However, the detailed analysis of the distribution of

Table 6-10. Lysolecithinase Activity in Subcellular Fractions of Rat Organs[a]

Tissue	Total activity (units/g)	Heavy particles	Light particles	Microsomes	Supernatant
		\multicolumn{4}{c}{Specific activity of subcellular fraction (units/mg protein)}			
Liver	23,780	120	460	195	70
Kidney	41,976	300	800	750	40
Lung	37,800	90	340	700	740
Intestine	412,086		15,880	20,300	550
Brain	17,147		776	2,117	66

[a] From Leibovitz-Ben Gershon and Gatt (1976).

the enzyme activity in the various fractions is probably an accurate reflection of activity.

6.3.2. Adrenal Medulla Lysophospholipase

The lysophospholipase activity in the adrenal medulla has been of considerable interest since it is known that the chromaffin granules of these glands that store high levels of medullary hormones contain a high level of lysophosphatidylcholine (Blaschko *et al.*, 1968). Phospholipases have been demonstrated in the lysosomal fraction that have the usual characteristics of acidic pH optima (Smith and Winkler, 1968). Also, a lysophospholipase is known to be present in microsomal fraction that is active at pH 6.6 in the absence of Ca^{2+} (Hortnagl *et al.*, 1969). These authors found comparable lysophospholipase activity in dialyzed and undialyzed crude homogenates, a point raised in subsequent work. The assay used by these workers involved the titration of released fatty acid and therefore high levels of substrate, 6 mM. Subsequently, it was shown that the lysophospholipid composition of granules changes significantly upon dialysis, suggesting that the phospholipase(s) was active during dialysis at 4°C (Franson and van den Bosch, 1982). They found that the percentage of lysophosphatidylcholine decreased from 14.5 to 6.8% of the total phospholipid, while the percentage of lysophosphatidylethanolamine increased from 0.5 to 5.9%. Also, the phospholipid : protein ratio decreased some 10%, showing that this organelle could catalyze the complete degradation of phospholipid. Upon further study, it was found that fresh but not dialyzed granules could degrade exogenous radiolabeled lysophosphatidylcholine at concentrations of substrate below the maximal amount that could be absorbed by the granule preparation (Fig. 6-11).

It was concluded that the enzyme was inhibited by excess lysophosphatidylcholine when no more could be absorbed by the granule fraction, about 10 μM. It likewise was concluded that optimal enzyme activity occurred when lysis of the granule occurred. The microsomal lysophospholipase was not influenced to the same degree as granule enzyme by dialysis and high concentrations of lysophosphatidylcholine. At this point, it is not clear if the differences between the phospholipase activity observed in the two organelles are the result of distinct enzymes. These findings have some interesting possible implications. Microsomes, primarily endoplasmic reticulum, probably are not involved in fusion events and remain sealed. Secretory organelles such as the chromaffin granules fuse with the plasma membrane and open to the cell exterior, perhaps requiring lysophospholipid as a membrane-destabilizing agent. These high levels of lysophospholipid must be carefully regulated since a lysophosphatidylcholine : protein ratio of 160 nmoles : 1 mg initiated granule disruption, an amount equal to 20% of the total phospholipid present in the granule. If indeed the physiologic function of the organelle is related to the lysophospholipid level, fine regulation must exist, balancing the source of the lysophosphatidylcholine, the interaction of the enzyme and substrate, and membrane destabilization.

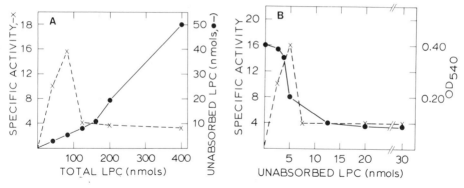

Figure 6-11. Lysophospholipase activity as a function of unabsorbed lysophosphatidylcholine (LPC) and as a function of granule integrity. Standard reaction mixtures in a total volume of 1.0 ml were prepared in duplicate and contained 100 mM, Tris-HCl, pH 7.5, 0.50 mg granule protein, and the indicated concentrations of substrate [^{14}C]LPC, which in this case was 5000 cpm/nmoles LPC. One set of reaction mixtures was incubated at 37°C for 10 min and analyzed to obtain specific activities (x). To measure unabsorbed LPC, the duplicate set of reaction mixtures was vortexed and maintained at 4°C for 5 min. The experimental values are corrected for sedimentable [^{14}C]LPC (in the absence of granules) which was constant (42 = 2%) over the entire concentration range of the substrate. These mixtures were then centrifuged at 30,000g for 20 min and the radioactivity in an aliquot (0.5 ml) of the supernatant fraction was determined (●, A). Lysis of granules was monitored by absorbance changes at 540 nm (■, B). (From Franson and van den Bosch, 1982.)

6.3.3. Beef Liver Lysophospholipases

The lysophospholipases of beef liver have been purified and thoroughly characterized. Lysophospholipase I was found to have a dual localization being recovered from both the cytosol and the matrix space of the mitochondria, whereas lysophospholipase II was localized in the microsomes (van den Bosch and deJong, 1975). Table 6-11 lists the properties of the two preparations (van den Bosch, 1982). The comparisons indicate that while certain properties are shared, they appear to be distinct proteins. This conclusion is strengthened by the finding that they do not cross-react immunochemically (deJong *et al.*, 1976).

A series of lysophosphatidylcholine analogs have been compared as substrates, in part to demonstrate the relative activities on oxyacyl and thioacyl esters and in part to develop a rapid and sensitive spectrophotometric assay. The latter is based on the coupling of the released thiol with DTNB (5,5'-dithiobis-2-nitrobenzoate). Table 6-12 shows the relative kinetic parameters using five substrates (Aarsman and van den Bosch, 1979).

While the absolute kinetic values cannot be compared because of possible differences in the aggregated nature of the substrates, several interesting features were deduced from the relative values obtained. The enzymes bound the thiol compound more tightly, which was interpreted to be the result of the greater hydrophobicity of thio ester when compared with the oxy ester. The change in hydrophobicity was reflected in the thiol phospholipids having

Table 6-11. Lysophospholipase[a]

Characteristic	I — Mitochondria cytosol	II — Microsomes
Fold purified	3,590	770
Molecular size	25,000	60,000
pH optimum	6.0–8.0	8.5
Isoelectric point	5.2	4.5
Effect of Ca^{2+}	0	0
Nonspecific esterase[b]	+	+
Specific activity (units/mg)	1.4	1.4
Sensitivity to sulfhydryl agents	+ +	+
Sensitivity to detergents	+	+ +
Sensitivity to bis[p-nitrophenyl]phosphate[c]	±	+ +
Percentage activity on long-chain diacyl phospholipid	2%	2%
Sensitivity to diisopropylfluorophosphate	?	+

[a] From van den Bosch (1982).
[b] Characterized by hydrolysis of tributyrylglycerol and p-nitrophenyl-acetate.
[c] Inhibitor of the carboxyesterase activity (deJong *et al.*, 1974; van den Bosch and deJong, 1975; Aarsman *et al.*, 1977; Aarsman and van den Bosch, 1979; van den Bosch, 1982).

CMC values roughly 40% lower than the oxy phospholipid. The differences in kinetic properties were not simply related to the CMC since there were differences in the CMC of the oxy phospholipids used.

Based on the various properties described for these enzymes, a mechanism of action was proposed involving an acyl–ester intermediate.

$$EOH + R_1COCOR_2 \underset{k_{-1}}{\overset{k_1}{\rightleftharpoons}} EOH \cdot R_1COCOR_2 \xrightarrow{k_2} EOCOR_2 \xrightarrow{k_3} EOH$$
$$+ HOCOR_2 \quad\quad R_1\overset{\downarrow}{C}OH \quad\quad \overset{\nearrow}{H_2O}$$

EOH = serine enzyme (based on sensitivity to diisopropylfluorophosphate)

R_1CO = alcohol moiety of substrate

R_2OC = acyl moiety

Table 6-12. Lysophospholipase

	I — V (μmoles/min/mg)	II — V (μmoles/min/mg)	K_m
Thioglycolphosphatidylcholine	3.50	1.34	6.8
Oxyglycolphosphatidylcholine	0.82	0.43	30.0
Thiodeoxylysophosphatidylcholine	1.30	1.29	7.2
Oxydeoxylysophosphatidylcholine	0.27	0.60	50.0
1-Acyllysophosphatidylcholine	0.70	0.60	50.0

[a] From Aarsman and van den Bosch (1979).

This model, in which the ester oxygen or sulfur remains with the alcohol rather than the acyl moiety, was supported by the detection of the thiol group in water-soluble 1-thiopropylphosphorylcholine when thiodeoxylysophos-phatidylcholine was used as substrate. Although attempts were made to elucidate kinetic constants involved in the partial reactions, inconclusive results were obtained despite the kinetic advantages of the spectrophotometric assay. It was suggested, however, that a step prior to enzyme deacylation was rate limiting; that is, either k_1 or k_2 is slower than k_3.

The experiments using the five different analogs suggested that hydrolysis is minimally influenced by the CMC of the substrate. This suggestion was verified by experiments using 1-decanoyl and 1-dodecanoyl lysophosphatidylcholine (van den Bosch, 1982) (Fig. 6-12).

These studies show that the lysophospholipases II attack both micellar and monomeric substrates. If only the monomers were attacked, one would expect that no increase in activity would be found above the CMC. That micelles are degraded was further substantiated by the finding that lysophospholipase II could degrade short-chain diacyl phosphatidylcholines that form micelles rather than liposomes and, in this case, the enzyme exhibited phospholipase A₁ activity, similar to that exhibited by the pancreatic phospholipases A₁. It is not clear at present, however, if the liver enzymes have phospholipase B activity.

6.3.4. Rat Lung Lysophospholipase

In a search for potential mechanisms for lung surfactant synthesis, Brumley and van den Bosch (1977) purified an enzyme from lung supernatant having both hydrolytic activity and transacylation activity on lysolecithin. It was reasoned that a lysophospholipase potentially could convert two molecules of 1-palmitoyl phosphatidylcholine to dipalmitoyl phosphatidylcholine, an essential component of lung surfactant. Like a number of lipolytic enzymes from mammals, glycerol was found to stabilize the preparation. Its molecular properties were somewhat different from the liver enzymes; its molecular

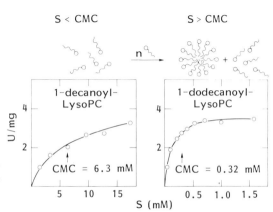

Figure 6-12. Activity of bovine liver lysophospholipase II on the monomeric and micellar lysophosphatidylcholine. (From van den Bosch, 1982.)

weight was 50,000 daltons and its pI was 4.0–4.5. Likewise, under the same experimental conditions, lysophospholipases from beef liver and pancreas did not catalyze transacylation.

The reaction kinetics were rather interesting; while the two activities were linear with respect to time and protein concentration, the dependence on substrate concentration was complex (Brumley and van den Bosch, 1977) (Fig. 6-13).

The rapid rate of hydrolysis increased linearly until saturation was reached at 100 μM. On the other hand, phosphatidylcholine formation, the result of transacylation, showed three distinct rates. Since the break in the rate of phosphatidylcholine formation occurred well above the CMC, 7 μM, transacylation presumably occurred with micellar substrate and not with monomers. The product of transacylation, phosphatidylcholine, almost completely inhibited transacylation while decreasing hydrolysis some 60%. On the other hand, the product of hydrolysis inhibited transacylation about 60% and hydrolysis 80%. It would therefore appear that the reactions are regulated by their products, mediated in all probability by the physical nature of the lipid aggregate.

6.3.5. Amnionic Lysophospholipase

A neutral active, Ca^{2+}-independent, lysophospholipase was found to be enriched in amnionic membranes (Jarvis *et al.*, 1984). Since this membrane is of fetal origin and is implicated in the initiation of labor, purification and characterization of this enzyme were of considerable physiologic interest. This is particularly true since phospholipid deacylation and prostaglandin synthesis are associated with labor.

This enzyme was purified about 1200-fold over a 10,000 *g* supernatant of human amnionic membranes. Considerable care was taken in an initial step to dissociate the enzyme into monomers with urea and to protect the enzyme against inactivation with Triton X-100. Triton X-100 was found to be inhibitory, however, when the substrate, lysophosphatidylcholine, was in concentrations above its CMC, 7 μM. On the other hand, when the substrate concentration was below the CMC, Triton X-100 below its CMC, 0.33 mM, was stimulatory (Fig. 6-14).

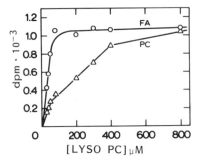

Figure 6-13. Lysophospholipase–transacylase activity versus increasing substrate concentration. 1-[^{14}C]Palmitoyl-*sn*-glycero-3-phosphocholine (specific activity 400 dpm/nmoles) was incubated with 0.2 μg of enzyme protein. (From Brumley and van den Bosch, 1977.)

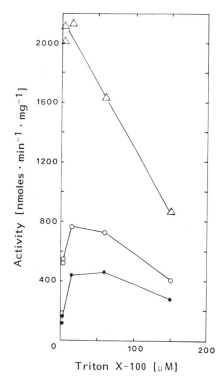

Figure 6-14. Effect of Triton X-100 on lysophospholipase activity at low substrate concentrations. Triton X-100 was varied between 3.3 and 150 μM; the lysophosphatidylcholine concentrations used were 2.7 μM (●), 5.3 μM (○), and 26.7 μM (△). The 3.3 μM Triton assays were performed in duplicate and both data points are shown; 0.075 μg of enzyme was used in each assay. (From Jarvis *et al.*, 1984.)

Table 6-13. Lysophospholipase Activity toward Various Lipids[a]

Substrate[b]	Concentration (μM)	Physical form	Specific activity (nmoles/min/mg)	Assay method[c]
Palmitoyllyso-PC	150	Micelle	2680	D
Palmitoyllyso-PC	188	Micelle	2770	A
Palmitoyllyso-PC	16	Micelle	1870	A
Myristoyllyso-PC	187	Micelle	3920	D
Myristoyllyso-PC	14	Monomers	850	D
Palmitoyllyso-PE	290	Dispersion	860	D
Bovine brain lyso-PS	360	Micelle	990	D
Palmitoyllyso-PG	189	Micelle	730	D
Dipalmitoyl-PC	100	Sonicated vesicles	ND[d]	B
Monooleoylglyceride	120	Dispersion	ND	C
Dioleoylglyceride	120	Dispersion	ND	C
Trioleoylglyceride	120	Dispersion	ND	C

[a] From Jarvis *et al.* (1984).

[b] PC, phosphatidylcholine; PE, phosphatidylethanolamine; PS, phosphatidylserine; PG, phosphatidylglycerol.

[c] A, radioactive lyso-PC assay, detection limit 2 nmoles/min/mg; B, radioactive PC assay, detection limit 20 nmoles/min/mg; C, radioactive acylglyceride assay, detection limit 2 nmoles/min/mg; D, phosphate assay, detection limit 200 nmoles/min/mg.

[d] N.D., not detectable.

Table 6-13 shows that the purified lysophospholipase has a wide substrate specificity on lysophospholipids but is inactive on vesicles of dipalmitoyl phosphatidylcholine. It also shows that it is roughly one-half as active on monomers as micelles. It was tentatively concluded therefore that, because of the enhanced activity found when Triton X-100 was present below the CMC of lysophosphatidylcholine, the comicelles formed even though the Triton X-100 also was below its CMC. The inhibition found above the CMC of the substrate was attributed to dilution of the substrate in the micelle by Triton X-100.

Chapter 7

Phospholipase A₂ of Mammalian Cells

This chapter continues with the theme of mammalian phospholipases, also the subject of Chapters 6 and 8. The work described here is most closely related to that in the previous chapter on the other classes of mammalian acyl hydrolases, the phospholipases A_1 and lysophospholipases. The literature on mammalian phospholipases A_2 is larger than that on the phospholipases A_1 and lysophospholipases, perhaps because of interest in the metabolism of essential polyunsaturated fatty acids.

7.1. General Considerations

Considerable interest in phospholipases A_2 recently is the result of their agonist-mediated response activity such as in the *arachidonate cascade*. This cascade leads to the formation of bioactive molecules, the eicosanoids, a term used for all the oxidative bioactive products of arachidonate (prostaglandins, leukotrienes, thromboxanes, etc.). Another factor that influenced the nature of research in the area was the relative ease with which the phospholipase A_2 could be assayed, relative to the phospholipase A_1. Until a number of radiolabeled substrates became commercially available, the preparation of phospholipids labeled at position 2 followed by the measurement of radiolabeled fatty acid released by enzyme preparation provided the simplest and quickest assay for phospholipases. Now it is possible to obtain substrates that permit assay of other phospholipases with equal ease. This family of enzymes is widely distributed and undoubtedly has a variety of functions, only some of which are known at this time. Table 7-1 lists a number of sources of phospholipases A_2 along with certain of their characteristics.

With the exception of phospholipase A_2 of lysosomal origin, most, if not all, are either stimulated by or have an absolute requirement for Ca^{2+}. Likewise, most are optimally active at neutral to high pH values, again with the exception of that in the lysosomes. In the cases where detailed substrate specificity studies have been carried out, phosphatidylethanolamine most often is the preferred substrate, although most can degrade a number of phosphoglycerides under the appropriate conditions. They are rather broadly distributed within the cell with the exception of the microsomal fraction.

Table 7-1. A Partial List of Mammalian Phospholipases A₁ (Excluding the Pancreatic Secretory Enzyme)

Source	Localization[a]	pH optimum[b]	Ca²⁺ effect	Substrate[a,c] specificity	Molecular weight	pI	References
Rat spleen	Sol (?)	7.0	+	—	15,000	7.4	Rahman et al. (1973)
Bovine brain (gray matter)	Micro	—	+	PI	—	7.5	Gray and Strickland (1982a)
Bovine brain	Syn Ves	9.0	+	PC	N.D.	N.D.	Moskowitz et al. (1982)
Rat aorta smooth muscle	M	7.5	+	PE	N.D.	—	Thakkar et al. (1983)
Human lung secretion	(?)	9.0	+	Egg yolk	75,000	N.D.	Sahu and Lynn (1977)
Mammary tumors	Soluble and M	7.7	+(?)	PC	N.D.	N.D.	Rillema et al. (1980)
Rat ascites hepatoma	PM	—		PE,LPE,PG	—	—	Natori et al. (1980)
Bovine adrenal medulla	Lyso	4.2–6.5	±	PE,PC	N.D.	N.D.	Smith and Winkler (1968)
Cat adrenocortical cells	Lyso	4.5	—	PE	N.D.	N.D.	Laychock et al. (1977)
	M (?)	7.4	+	PE	N.D.	N.D.	Laychock et al. (1977)
Mouse peritoneal macrophages	Lyso	4.5	—	PE,PC	N.D.	N.D.	Wightman et al. (1981a)
		8.5	+	—		—	Wightman et al. (1981a)
Rabbit alveolar macrophages	Lyso	4.5	—	PE,PC	—	—	Franson and Waite (1973)
		8.0	+	PE,PC		—	Franson et al. (1973)
		7.0	+	—			Lanni and Franson (1981)
Rat liver	Mito	9.0–9.5	+	PE,PC	15,000		Waite and van Deenen (1967) de Winter et al. (1982) Natori et al. (1983) Nachbaur et al. (1972)

Tissue/cell	Fraction	pH optimum	Ca^{2+}[b]	Substrate	MW	pI	Reference
Rat liver	PM	7.0–9.5	+	PG,PE	N.D.	N.D.	Colard-Torquebiau et al. (1975); Victoria et al. (1971)
Rabbit lung	Lam Bod	4.0	–	PE	N.D.	N.D.	Heath and Jacobson (1976)
Rabbit lung	M	7.0	+	PE	N.D.	N.D.	Franson and Weir (1982); Longmore et al. (1979); Garcia et al. (1975)
Rabbit neutrophils	Granules	7.5	+	PE	14,000	9.5–10.0	Elsbach et al. (1979)
Human neutrophils	PM	7.5	+	PE	N.D.	N.D.	Victor et al. (1981)
Human platelets	PM (and others)	9.5	+	PC	44,000		Trugman et al. (1979); Apitz-Castro et al. (1979)
Rabbit platelets	—	—		—	12,000		Kannagi and Koizumi (1979)
Rat heart	Mito	9.0	+	PE,PC	N.D.	N.D.	Franson et al. (1972)
	Lyso	4.0	–	PE	N.D.	N.D.	Palmer et al. (1981)
Rat and pig intestine (distal)	Crystal cells	8.0	+	PG	14,100	9.6	Verger et al. (1982)
Macrophage line (P388D$_1$)	Fcγ2b frag	9.0	+	PC	30,000	5.8	Suzuki et al. (1982)
Rat adipose	PM	8.5	+	PE	N.D.	N.D.	Bereziat et al. (1978)
Sheep erythrocytes	ESM	8.0	+	PE,PC	18,500	—	Kramer et al. (1978)

[a] Abbreviations: Sol, cytosol; Micro, microsomes; Syn Ves, synaptic vesicles; PM, plasma membrane; M, membranes (unspecified); Lam Bod, lamellar bodies; N.D., not determined; Lyso, lysosomes; Mito, mitochondria; frag, fragment; ESM, external surface of membrane; LPE, lysophosphatidylethanolamine; PC, phosphatidylcholine; PE, phosphatidylethanolamine; PG, phosphatidylglycerol; PI, phosphatidylinositol.

[b] In certain cases only one substrate or pH was used; that is the one given.

[c] Unpublished data.

Although phospholipase A_2 activity has been noted in this cell fraction, the phospholipase A_1 seems to show the dominant activity. A number that have been purified have rather low molecular weights ranging from about 12,000 to 18,000 daltons. It is possible that the higher molecular weights reported such as 30,000, 44,000, and 75,000 daltons could be the result of polymer formations. This would not be unlikely since most are membrane associated and therefore hydrophobic. A few of the more thoroughly investigated enzymes are described in detail.

7.2. Phospholipase A_2 of Neutrophils (Polymorphonuclear Leukocytes)

Phospholipid metabolism in neutrophils has been a long-standing interest of Elsbach's group. Initially, they investigated the potential of the neutrophil to metabolize a number of lipid precursors in order to understand better the mechanisms by which membrane alterations could take place during phagocytosis. During the course of these studies, they described a phospholipase activity with optimal activity at pH 6.0, although their assay system could not distinguish the positional specificity of phospholipase being measured (Elsbach and Rizack, 1963). Subsequently, they demonstrated another phospholipase that was membrane associated and could degrade the phospholipids in *E. coli* membranes, primarily phosphatidylethanolamine (Elsbach *et al.*, 1972). Interestingly, the neutrophil incorporated the fatty acids released from the *E. coli* into the cellular membranes. The finding that the killing of *E. coli* by neutrophils is accompanied by increases in permeability of the *E. coli* envelope prompted them to examine the possibility that this permeability factor(s) in neutrophils has phospholipase A_2 activity and that the increased permeability activity is the result of phospholipid degradation. Indeed, the permeability factor copurified some 1000-fold with phospholipase A_2 and the two had a number of properties in common (Weiss *et al.*, 1975).

Since a number of other highly cationic proteins in the granules had been extracted from cellular homogenates with 0.15 N H_2SO_4, these workers successfully attempted this rather drastic procedure in the isolation of the permeability factor and phospholipase A_2. This relatively simple procedure gave nearly quantitative recovery of both activities with a 20-fold purification over the homogenates. This procedure has now been applied in the isolation and purification of phospholipases A_2 from a number of sources. Overall, the enzyme was purified over 8000-fold with a 15% recovery. The pI of the phospholipase A_2 was higher than that of lysozyme, 10.5.

When they succeeded in completely purifying the phospholipase A_2 and permeability factor, it was shown that two proteins were responsible and were quite different in their molecular properties (Elsbach *et al.*, 1979). For example, the molecular weight of the permeability factor was found to be 50,000 daltons while that of the phospholipase A_2 was 14,000 daltons. Even though only small amounts of enzyme could be obtained (less than 80 μg from 2.1×10^{10} cells), amino acid analyses have been done and found to be similar to that reported for the pancreatic and snake venom phospholipases A_2. The phospholipase A_2, once separated from the permeability factor, had no activity on

E. coli unless the permeability factor was included in the reaction mixture. This was not thought to "merely pave the way for phospholipase A₂" (Elsbach, 1980) since there was a high degree of specificity in the interaction of the permeability factor and phospholipase A₂. For example, the human permeability factor, although having biologic activities identical to that of the rabbit factor, was unable to stimulate the rabbit phospholipase A₂ (Weiss *et al.*, 1979). Likewise, the association of the rabbit enzyme and permeability factor during the initial stages of purification was not seen with the human system. The nature of this interaction has been partially elucidated by the finding that the antibiotic polymyxin B stimulates the hydrolysis of *E. coli* membranes by the phospholipase A₂ without the addition of Ca^{2+} (Elsbach *et al.*, 1985). In an elegant study, they demonstrated that membrane-bound Ca^{2+} in the *E. coli* was sufficient for hydrolysis and that the cationic polymyxin B stimulated hydrolysis by displacing Ca^{2+} from anionic binding sites, thus making Ca^{2+} available to the phospholipase A₂. An excellent review on the mechanism that these two proteins might play in bacterial degradation by phagocytes is suggested reading (Elsbach, 1980).

The molecular mechanism of this enzyme has not been studied in great detail. In most studies, its activity was measured using autoclaved *E. coli* as the substrate. It was shown, however, that the enzyme degraded aggregates of phosphatidylethanolamine more readily than liposomes of phosphatidylcholine. Likewise, there was an absolute requirement for Ca^{2+} ions but concentration of Ca^{2+} greater than 1 mM caused inhibition. In a study to understand the basis for these characteristics, Franson and Waite (1978) found that surface-charge differences between the two substrates could not account for the preference found for phosphatidylethanolamine. When the electrophoretic mobility of phosphatidylcholine liposomes was made slightly negative and equal to that of phosphatidylethanolamine by the addition of dicetylphosphate, the activity was one-half or less than that on phosphatidylethanolamine. The effect of Ca^{2+} was rather complex (Fig. 7-1).

As can be seen, optimal activity was found within a rather narrow range of surface charge (mobility). In this case, the surface charge producing optimal activity was dependent on the Ca^{2+} concentration. While the reason for this is unknown, this effect could be due to a bridging effect by Ca^{2+} on the substrate, thereby influencing molecular packing and/or the aggregated nature of the substrate. On the other hand, since the enzyme is strongly cationic, it might effectively compete with Ca^{2+} for the phosphate of phosphatidylethanolamine. The failure of dicetylphosphate to enhance activity on zwitterionic phosphatidylcholine to the level of that on phosphatidylethanolamine could be due to a difference in fit of the two substrates into the active site of the enzyme. While this is speculative, such a model can be tested by using more highly charged phospholipids and by comparing the enzymes from human and rabbit cells. Likewise, the role that the permeability factor plays in enzyme–substrate interaction should be useful in understanding regulation at both the physiologic and mechanistic levels.

Physiologically, this enzyme clearly has an important role in bactericidal function of neutrophils. While the phospholipase A₂ cannot directly degrade the lipid in intact cells, under the appropriate conditions many but not all

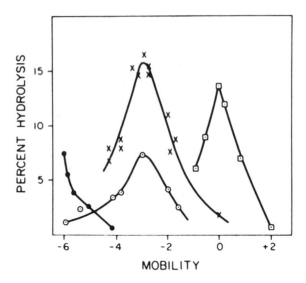

Figure 7-1. Hydrolysis of amphipath-altered phosphatidylethanolamine liposomes as a function of surface charge and Ca^{2+} concentration. Mobility and enzyme activity were measured. Mobility was altered at each concentration of Ca^{2+} (●, 0.01 mM; ⊙, 0.1 mM; x, 1.0 mM; ▣, 10.0 mM) with diacetyl phosphate or cetyltrimethylammonium bromide. (From Franson and Waite, 1978.)

membranes undergo rapid degradation. The fact that some membranes are not degraded can be ascribed in all probability to the lack of accessibility of the enzyme to the membrane (Weiss *et al.*, 1979). The enzyme might also be involved in inflammatory responses based on the observation that the purified enzyme is inhibited by the anti-inflammatory drug indomethacin and certain of its pharmacologically active analogs (Kaplan *et al.*, 1978; Jesse and Franson, 1979; Kaplan-Harris and Elsbach, 1980). In such a case, one would have to postulate that the enzyme attacks the cellular phospholipids and that the products of hydrolysis might have a direct inflammatory action. If arachidonate were released, the enhanced production of eicosanoids or related compounds could also lead to inflammation. The localization of phospholipase A_2 in the specific and azurophilic granules and in the plasma membrane is consistent with the enzyme(s) being involved in inflammatory response and the arachidonate cascade (Franson *et al.*, 1974; Victor *et al.*, 1981). Thus far, no acid-active phospholipases have been detected in neutrophils as found in macrophage lysosomes (Franson and Waite, 1973). However, a phospholipase A_2 requiring Ca^{2+} and having a broad pH optimum for activity was released from peritoneal neutrophils upon degranulation. It was therefore concluded that this phospholipase A_2 is of lysosomal origin (Traynor and Authi, 1981).

7.3. Phospholipase A_2 of Brain

The gray matter of bovine brain contains a phospholipase A_2 that acts on phosphatidylinositol (Der and Sun, 1981; Gray and Strickland, 1982*b*).

While the enzyme was found in both the microsomal and soluble fractions, its purification was initiated using microsomes since the soluble fraction had competing lipolytic enzymes, primarily phospholipase C. The phospholipase A_2 was solubilized from the membrane with 0.05% Triton X-100 and purified using rather standard techniques with the exception that detergents, asolectin, and glycerol (10%) were present throughout most steps to minimize inactivation. When run on Sephadex G-200 with asolectin, 0.01% Triton X-100, and 10% glycerol, the enzyme eluted as a small molecular weight protein, indicating that it was not aggregated under these conditions and that, at most, it interacted a few lipid molecules. The enzyme, purified some 1600-fold over the homogenate, had a specific activity of 0.74 unit/mg when assayed in the presence of 5 mM Ca^{2+}. Other metal ions, Mg^{2+}, Mn^{2+}, and most noticeably Zn^{2+}, inhibited the Ca^{2+}-stimulated activity. There was not an absolute requirement for added Ca^{2+}, however; 40% of maximal activity obtained with 5 mM Ca^{2+} was found without any added metal ion. The amino acid composition of this enzyme, which had a molecular weight of 18,521 daltons, was quite different from that of a number of other phospholipases. From a total of 173 amino acids, it had 46 Asp, 26 Glu, 21 Gly, and only 1 Cys and it was suggested that the enzyme might be a glycoprotein.

The phospholipase A_2 of brain had a rather broad specificity (Gray and Strickland, 1982*b*) (Table 7-2). Based on the observed specificity for the acyl group at position 2 of the glycerol, it was postulated that the enzyme might be involved in a deacylation–reacylation cycle that would remove saturated fatty acid in position 2 and allow for reacylation with a polyunsaturated acid, in particular arachidonate. Since it has been shown that lysophosphatidylinositol is preferentially acylated by arachidonoyl–CoA (Baker and Thompson, 1973), this deacylation–reacylation cycle would account for the high percentage of 1-stearoyl-2-arachidonoyl phosphatidylinositol present in brain and other tissues. Furthermore, the action of this phospholipase A_2 was suggested to be involved in membrane fusion and presynaptic neurotransmission (Moskowitz *et al.*, 1982). The experiments from which Table 7-2 was derived were done using a single set of assay conditions. Further studies to characterize this enzyme would be of interest, particularly since it was shown that concentrations of phosphatidylinositol greater than 0.4 mM were inhibitory.

Table 7-2. Specificity of Brain Phospholipase A_2[a]

Substrate[b]	Relative rate of hydrolysis
2-Palmitoyl phosphatidylinositol	5.0
2-Stearoyl phosphatidylinositol	1.9
2-Oleoyl phosphatidylinositol	1.0
2-Arachidonoyl phosphatidylinositol	0.4
2-Oleoyl phosphatidylcholine	0.62
2-Oleoyl phosphatidic acid	0.33
2-Oleoyl phosphatidylethanolamine	0.25
2-Oleoyl phosphatidylserine	0.22

[a] From Gray and Strickland (1982*b*).
[b] The one acyl position did not have a specified acyl group; however, this was not considered as a contributory factor.

7.4. *Phospholipase A₂ of Platelet*

The phospholipase A_2 of the platelet has become the center of focus for a number of investigators who are interested in arachidonate metabolism by lipoxygenase and cyclooxygenase. Considerable interest surrounds its regulation and how it is stimulated by factors that initiate the arachidonate cascade. This is of particular significance since the mechanism of arachidonate release is still somewhat unclear (see Chapter 11). The central questions that exist at present are the localization of the enzyme within the cell, its substrate specificity including acyl specificity, and the determination of the factors that initiate and terminate enzyme activity. Considerable progress is being made in the field but much remains to be established.

The platelet phospholipase A_2 is membrane associated primarily (Trugnan *et al.*, 1979), unlike the phospholipase C that recovered from the cytosolic fraction (Mauco *et al.*, 1979). Likewise, the platelet is also rich in phospholipase A_1 activity, which is weakly membrane associated, and lysophospholipase activity, which is localized in the cytosol. A phospholipase A_1 had previously been reported to have an acidic pH optimum and presumably was in the lysosomes (Smith and Silver, 1973). While cell fractionation studies have not shown a single localization for the enzyme, a substantial amount is localized in the plasma membrane fraction. This activity, measured by the hydrolysis of exogenously added phosphatidylcholine at pH 9.0, was equally as high in a dense granule fraction (Trugnan *et al.*, 1979). The mitochondria, unlike a number of other cells, had low phospholipase A_2 but high phospholipase A_1 activity. While it is tempting to speculate that the phospholipase A_2 that deacylates phosphatidylcholine in response to agonist stimulation is functioning in the plasma membrane, other possibilities cannot be excluded. It would be of interest to determine the substrate specificity of the phospholipases A_2 in these fractions; such determinations could add further insight into the role of various phospholipases in arachidonate release.

Various procedures can be used to solubilize the enzyme from membranes. Some evidence would suggest that the phospholipase A_2 is a peripheral protein since salt extraction frees the enzymes from the rabbit platelet membranes (Kannagi and Koizumi, 1979). Their preparation, purified some 1000-fold, had a molecular weight of 12,000 daltons, as determined by gel filtration. This preparation was not homogeneous, as shown by SDS–PAGE. This or a similar phospholipase A_2 can be solubilized also by the H_2SO_4-extraction procedure (Apitz-Castro *et al.*, 1979; Franson *et al.*, 1980; Jesse and Franson, 1979). Affinity chromatography was reported to be a very effective means for the purification of the enzyme from human cells; a 230-fold increase in specific activity was obtained using 1-(9-carboxyl)nonyl-2-hexadecyl-glycero-3-phosphocholine coupled to Sepharose. The molecular weight of this preparation was considerably higher than the 12,000 daltons reported for the human enzyme; in this case, the molecular weight was found to be 44,000 daltons as determined by SDS–PAGE (Apitz-Castro *et al.*, 1979). The reason for the difference in molecular weight is not clear at present; while this could be the result of species differences, it is possible that the method of extraction could influence properties of the enzyme. Alternatively, since the H_2SO_4-extractable phospholipase A_2 is strongly cationic, it could have been complexed with another protein(s). The two preparations were purified to roughly the same extent but the final specific activity of the salt-extracted enzyme was about 50% higher than that of the H_2SO_4-treated preparation (0.83 vs. 0.53 unit/mg protein although the assay conditions were somewhat different). In that regard, the activity on *E. coli* phosphatidylethanolamine by a 3500-fold purified preparation was considerably lower, 0.035 unit/mg protein. The activity reported for the platelet enzyme was 5–10% of that reported for the highly purified phospholipase A_2 from rabbit peritoneal fluids, 0.50 unit/mg protein (Franson *et al.*, 1978*b*).

In an approach to understand better the regulation of the platelet phospholipase A_2, Ballou and Cheung (1983) applied a membrane preparation of human platelets directly to a DEAE–cellulose column. The phospholipase A_2 was eluted from the membrane and chromatographed on the column in a single step. While this is a rather unorthodox approach, its success might have been predicted since the enzyme was shown to be extractable from membranes by salt. Their preparation was some 800-fold purified but had a specific activity that was approximately 5% that of the other reports, 0.032 unit/mg protein when assayed using liposomes of phosphatidylcholine. The most striking feature of the procedure was that the yield was 1140%, indicating that a rather potent inhibitor was present in the cellular homogenate. This inhibitor did not bind Ca^{2+} since increasing Ca^{2+} above the optimal level of 0.1 mM did not overcome inhibition. This inhibitor was also partially purified and was found to be extractable in organic solvents. While its nature is not fully established, it appears to be a fatty acid(s) or its metabolite. On the other hand, an activator has been described in rat platelet that presumably is a protein (Etienne *et al.*, 1982).

The hydrolysis of phosphatidylcholine liposome by the phospholipase A_2 from rabbit platelets was markedly stimulated both by a neutral detergent— Triton X-100—and anionic detergents—cholate and deoxycholate (Kannagi and Koizumi, 1979). The effect of the anionic detergents was noted below the CMC, whereas Triton X-100 was not effective until the CMC was reached and high concentrations (5 mM) were totally inhibitory. The stimulatory effect of each detergent was at a phosphatidylcholine : detergent ratio of about 1:2, a ratio predicted by Deems *et al.* (1975) to produce the optimal mixed micelle for attack by phospholipases. The surface charge of the liposome was apparently important in enzyme interaction. Acidic phospholipids as well as anionic amphipaths, stearate, and dicetylphosphate stimulated the hydrolysis of phosphatidylcholine up to a 1 : 1 mixture. On the other hand, cationic amphipaths, stearoylamine, and cetyltrimethylammonium bromide inhibited. Likewise, cholesterol inhibited the hydrolysis of dimyristoyl phosphatidylcholine. This substrate was optimally hydrolyzed by the phospholipase A_2 at its phase transition, 23°C, which suggested to the authors that the enzyme functions optimally in domains with structural irregularities (phase separations) (Table 7-3).

This conclusion was further supported by the finding that the hydrolysis

Table 7-3. Hydrolysis of Dimyristoyl Phosphatidylcholine[a]

Percentage cholesterol	Percentage hydrolysis		
	10°C	23°C (TM)	37°C
0	4.8	24.6	7.7
10	3.6	9.4	6.3
20	1.8	3.2	3.2

[a] From Kannagi and Koizumi (1979).

of egg phosphatidylcholine, when mixed in an equal proportion of disaturated phosphatidylcholine, was optimal at the phase transition of the mixture, as determined by fluorescence depolarization.

The results of this work and the work of Ballou and Cheung (1983) suggest that both the nature of the membrane as well as endogenous factors including calmodulin might regulate this enzyme. Further evidence for calmodulin stimulation of phospholipases will be needed to establish its role in phospholipase regulation. Likewise, the stimulation of the enzyme by a low concentration of Ca^{2+}, 0.1 mM (relative to that required by many other phospholipases), suggests that Ca^{2+} flux might regulate hydrolysis as well. The Ca^{2+} did not appear to be mediated through calmodulin directly since calmodulin failed to stimulate the partially purified enzyme.

The question as to the role these purified phospholipases A_2 might play in the arachidonate cascade remains open. While they have some of the required properties, it is still not certain that they directly or indirectly respond to stimuli. This enzyme, like its counterpart in neutrophils, is inhibited by low levels of nonsteroid anti-inflammatory compounds such as meclofenamate and indomethacin, which suggests the enzyme might be that initiating the arachidonate cascade (Jesse and Franson, 1979). In that regard Apitz-Castro *et al.* (1979) showed that the partially purified enzyme from human platelets could block the ADP-induced aggregation of platelets at roughly 0.1% the amount required for comparable inhibitory activity by venom enzymes. While the reason for this is not clear, it is possible that platelet-derived phospholipase A_2 can attack membranes at a higher surface pressure than venom enzymes and therefore can more readily degrade platelet phospholipid.

7.5. Phospholipase A_2 of Macrophages

Phospholipases have been recognized to be present in macrophages for quite some period of time. Some of the initial studies centered on the granule phospholipases since they were thought to be involved in bacterial degradation (Franson and Waite, 1973). The granular phospholipases A_1 and A_2 had a number of properties in common with the liver lysosomal enzymes, suggesting that the liver enzymes could be in the Kupffer cells (Franson *et al.*, 1971). More recently, a phospholipase A_1/A_2 in the alveolar macrophage has been suggested to be involved in surfactant turnover (Rao *et al.*, 1981) and in the arachidonate cascade. Indeed, Hsueh *et al.* (1981) and Wightman *et al.* (1981*a*) have shown the activity of two phospholipases A_2 in macrophages. In this case, the activities of the two enzymes were functionally separated and identified with different stages of phagocytosis (Hsueh *et al.*, 1979). Consistent with these conclusions were the findings of Humes *et al.* (1980), who found two sources of arachidonate that appeared to be released in response to different stimuli in mouse peritoneal macrophages. Tetradecanoyl phorbol acetate and lipopolysaccharide stimulated the cyclooxygenase pathway, whereas zymosan stimulated both lipoxygenase and cyclooxygenase activities. Various types of challenge were shown to stimulate the deacylation of phosphatidylinositol (Emilsson and Sundler, 1984). Although the position of attack was not determined, it does demonstrate that acyl hydrolases in addition to phospholipase C, the phosphodiesterase, are capable of degrading phosphatidylinositol. Evidence was presented that phosphatidylinositol phosphate might also be a substrate for deacylation.

A membrane-associated phospholipase A_2 that is optimally active at pH 7.0 and requires Ca^{2+} for activity was solubilized by H_2SO_4 extraction (Lanni and Franson, 1981). Like the phospholipase A_2 isolated from the neutrophil, this macrophage enzyme is associated with the granule fraction; none was found in the plasma membrane. The enzyme, purified about 1300-fold over the homogenates, was optimally stimulated by less than 1 mM $CaCl_2$. While the purity of this preparation is unknown, its activity on *E. coli* membranes is over 20 times that of the phospholipase A_2 isolated from platelets by the same workers, about 0.75 units/mg protein. Little else is known about the mechanism of action or the molecular properties of the enzyme at present.

As with many other systems, the function and regulation of the macrophage phospholipases are poorly understood but are under rigorous study. The phospholipases A_2 in mouse peritoneal macrophages degrade phosphatidylethanolamine more rapidly than phosphatidylcholine or phosphatidylinositol (Wightman *et al.*, 1981*a*). Furthermore, at acid pH mixtures of phosphatidylethanolamine and phosphatidylcholine were degraded more rapidly than when the two were assayed separately. This was not the case at pH 8.5. The two phospholipases A_2 described in macrophages released more than enough arachidonate for prostaglandin synthesis (Bonney *et al.*, 1978; Wightman *et al.*, 1981*a*, 1981*b*). Hsueh *et al.* (1981) found that secretion of prostaglandins and arachidonate required internalization of phagocytized particles and was most closely associated with the release of the soluble granular phospholipases A_2. Arachidonate released by this enzyme for eicosanoid synthesis would require that it become attached to a membrane containing the appropriate substrate. The arachidonate thus released would then have to associate with the cyclooxygenase or lipoxygenase for further metabolism. While these events are possible, they will be difficult to demonstrate.

Activation of alveolar macrophages by injection of rabbits with heat-killed Bacillus Calmette-Guerin (BCG) led to a decrease in eicosanoid production (Hsueh *et al.*, 1982; Cochran *et al.*, 1985, 1986). Likewise, Humes *et al.* (1980) found that mouse peritoneal macrophages elicited by thioglycolate or BCG had reduced capacity to produce eicosanoids and to deacylate membrane lipid.

While it was tentatively concluded by Hsueh *et al.* (1982) that deacylation in alveolar macrophages was decreased, Cochran *et al.* (1985, 1986) found that the total deacylation was near normal but that the source of arachidonate was different, when compared with normal cells. Cochran and her coworkers found that lyso(bis)phosphatidate and phosphatidylinositol deacylation was correlated with eicosanoid synthesis while deacylation of phosphatidylcholine and phosphatidylethanolamine was not. BCG-activated cells were found to have only 25% of the lyso(bis)phosphatidate of control cells which might account for the lack of eicosanoid synthesis by the activated macrophages.

The deacylation process was enhanced by incubation of cell homogenates at 37°C in the presence of Ca^{2+} and ATP. This suggested the action of a protein kinase in activating the phospholipase A_2 (Wightman *et al.*, 1982). This point is supported by blockage of phospholipase A_2 activation by an ATPase stimulator or β-γ-methylene ATP. Also, the activation was increased by the addition of a cAMP-dependent protein kinase (Fig. 7-2). The properties

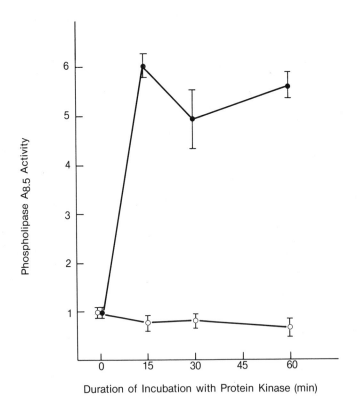

Duration of Incubation with Protein Kinase (min)

Figure 7-2. Rapid protein kinase activation of phospholipase $A_{8.5}$ in preincubated macrophage sonicates. Macrophage sonicate was incubated for 16 hr at 37°C in 10 mM Tris-HCl, pH 7.5, containing 15 mM NaCl and 4 mM $CaCl_2$. Eight 0.5 ml aliquots were removed and 5 pM units of catalytic subunit were added to four of the aliquots. The aliquots with catalytic subunit (●) and without catalytic subunit (○) were then incubated at 37°C and phospholipase $A_{8.5}$ activity was measured. Results are reported as described in the legend to Figure 7-3. (From Wightman *et al.*, 1982.)

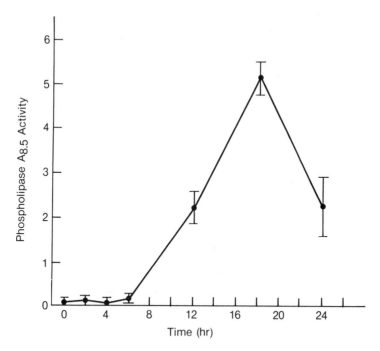

Figure 7-3. Time dependence of phospholipase $A_{8.5}$ activation. Macrophage sonicates were incubated and assayed for phospholipase $A_{8.5}$ activity. Phospholipase $A_{8.5}$ activity is defined as dpm [^{14}C]arachidonate hydrolyzed/(hr × μg protein) × 10^2. Results reported are the mean ± standard deviation; $n = 3$. (From Wightman *et al.*, 1982.)

of the activated enzyme and basal activity are the same, which suggests that an inactive form of enzyme is being activated rather than the properties of an already active enzyme being altered.

While these results provide important insights into possible regulatory processes, some questions must first be answered before a definite conclusion can be drawn about the meaning of these observations. At present, the major question relates to the kinetics of activation in the absence of added protein kinase. As shown in Fig. 7-3, no activation was noted for 6 hr and 18 hr were required for maximal activation. The magnitude of activation was quite impressive, however.

7.6. Phospholipase A₂ of Erythrocytes

The capacity of erythrocyte ghosts to reacylate lysophospholipids was recognized prior to the discovery of phospholipase activity. It was a puzzle initially why this cell should have a reacylase system without having a source of substrate within the membrane, although it was thought that serum lysophospholipid might serve as a substrate. Polonovski's group (Paysant *et al.*, 1970) demonstrated that hemolysates of human erythrocytes had phospholipase activity but that activation by a trypsin preparation was required for activation. It was later found that a platelet-derived factor, released during incubation of the cells, could promote a serum phospholipase activation (Etienne *et al.,* 1980). The most active preparations of phospholipase A_2 were obtained in the membranes of sheep and other ruminants. Importantly, the erythrocytes from this class of animals have a very low content of phosphatidylcholine (Zwaal *et al.,* 1974).

The phospholipase A_2 from sheep erythrocytes has now been purified to homogeneity following extraction of the enzyme with SDS from extensively washed membranes (Kramer *et al.,* 1978). This detergent reversibly inhibited this enzyme but could be replaced by cholate using a column exchange procedure. Cholate, unlike SDS, greatly increased enzyme activity yet was ineffective in solubilizing the enzyme from the membrane. In addition, the detergent exchange procedure gave about a 40-fold purification. The final purification was achieved using a dialkyl phosphatidylcholine affinity column giving another 40- to 50-fold increase in specific enzyme activity (2000-fold over homogenate) and an overall yield of nearly 70%. These workers took advantage of the approach developed by Rock and Snyder (1975) by binding the enzyme to the affinity column in the presence of Ca^{2+}; removal of the Ca^{2+} with EDTA released the enzyme. The final preparation contained a rather high level of phospholipid, roughly 40 molecules per protein molecule. They found a molecular weight by gel filtration lower than by SDS–PAGE, which suggested that some retardation occurred in the gel filtration step and that the molecular weight obtained by SDS–PAGE, 18,500 daltons, is correct.

Detailed studies on the substrate and activation requirements of this enzyme have been helpful in determining the physiologic function of the phospholipase A_2 in the erythrocyte. Under the appropriate condition of detergent and Ca^{2+}, both phosphatidylcholine and phosphatidylethanolamine in mixed micelles were degraded by membrane preparations (Frei and Zahler, 1979). The rate of degradation and Ca^{2+} requirement (50 μM) for the degradation of the two substrates were nearly equivalent. On the other hand, the purified enzyme showed a marked preference for phosphatidylcholine in the absence of detergent and had a preference for very-long-chain polyunsaturated fatty acids when assayed on phospholipids with mixed acyl composition (Jimeno-Abendano and Zahler, 1979).

The orientation of the enzyme in the membrane was examined by studying the susceptibility of the enzyme to proteolytic inactivation. It was shown that roughly 80% of the activity was lost by chymotrypsin treatment of either sealed or "leaky" ghosts, which indicated an exterior localization. The observation that 20% of the activity remained could have been the result of a different orientation of this portion of the enzyme such that it was shielded from proteolysis.

It was possible to block phospholipase A_2 activity in the intact sheep erythrocyte with EGTA (Borochov *et al.,* 1977). Under these conditions, plasma lipid could exchange from serum into the erythrocyte that elevated the phosphatidylcholine to sphingomyelin ratio some fivefold. Concomitant with the shift in phospholipid composition was a marked decrease in the microviscosity of the membrane, as measured by fluorescence depolarization (Fig. 7-4).

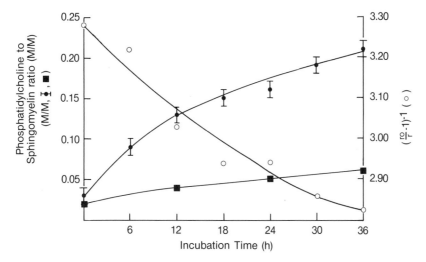

Figure 7-4. The rate of increase in phosphatidycholine : sphingomyelin mole ratio of isolated sheep erythrocyte membranes incubated at 37°C in human plasma containing 2 mM EGTA (♦) and in human plasma containing 1.5 mM free Ca^{2+} (■). The concomitant changes in membrane microviscosity at 20°C, presented by the parameter $(r_o/r - 1)^{-1}$, following the plasma–EGTA treatment, are also presented (○———○). (From Borochov *et al.*, 1977.)

Since both the shift in ratio of phosphatidylcholine to sphingomyelin and the fluidity change was blocked if Ca^{2+} were present, it was concluded that phosphatidylcholine exchanged into the membrane was degraded by the phospholipase in the presence of Ca^{2+}, thereby maintaining the low level of phosphatidylcholine and more rigid membrane character. Once the membrane-associated phosphatidylcholine was split, the products were returned to the blood. This provides an efficient means of supplying essential fatty acids to peripheral cells.

7.7. Phospholipase A₂ of Liver Mitochondria

The phospholipase A_2 in rat liver mitochondria was one of the first intercellular phospholipases recognized. The enzyme in mitochondrial membranes attacked phosphatidylethanolamine preferentially and required high pH values (9.0) and Ca^{2+} for activity (Waite and van Deenen, 1967). The enzyme was capable of attacking both endogenous as well as exogenous lipids (Waite *et al.*, 1969). This or a similar enzyme has been described in the mitochondria of a number of tissues and indeed could be an obligatory enzyme common to mitochondria of all species. It is generally agreed that the enzyme is localized in both the inner and outer membranes of the mitochondria, although the specific enzymatic hydrolysis is higher in the outer membrane fraction (Waite, 1969; Nachbaur *et al.*, 1972).

The initial studies on the isolation of this phospholipase A_2 employed delipidation of the enzyme with ammoniacal acetone (Waite and Sisson, 1971;

deWinter *et al.*, 1982). Treatments that were successfully used for other phospholipases such as H_2SO_4 or butanol extraction failed in this case. The enzyme from rat liver as well as pig liver mitochondria rapidly aggregated in salt buffers that thwarted the complete purification initially. deWinter *et al.* (1982), however, found that glycerol and high salt concentrations maintained the enzyme in a stable and nonaggregated condition which allowed the complete purification of the enzyme, some 3300-fold over the mitochondrial preparation (Table 7-4). Other procedures were successfully used by Natori *et al.* (1983) to obtain a 1400-fold purification. The final specific activity of this preparation was about 3.7 units/mg protein, in good agreement with specific activity of the 3300-fold purified enzyme, 8.1 units/mg protein (deWinter *et al.*, 1982). An improved method for purification appeared when Aarsman *et al.* (1984) reported that a modification of the affinity chromatographic procedure (Rock and Snyder, 1975) circumvented nonspecific hydrophobic interactions. Aarsman *et al.* (1984) coupled 10-*O-p*-toluene-sulfonyl-1-*O*-phosphocholine to Sepharose that yielded an affinity column with a single alkyl chain rather than the dialkyl affinity matrix previously used. The phospholipase A_2 purified by these procedures had an apparent molecular weight of 15,000 daltons but its behavior on gel filtration suggested that some retardation of the enzyme by the matrix had occurred. It could not be determined if phospholipases A_2 in the inner and outer membranes were the same because of the lack of complete recovery of the enzyme at various stages. Now that the problems of stability have been overcome, it should be possible to do sufficient purification to compare the phospholipase(s) from the two sources.

Substrate specificity studies with the purified enzyme confirm the earlier studies that showed phosphatidylethanolamine aggregates are degraded more rapidly than other substrates tested. These studies were not rigorously pursued and a number of questions concerning the interaction of this enzyme with its aggregated substrate remain. This is borne out by the finding that phosphatidylserine is preferentially attacked at a lower pH, relative to the attack on phosphatidylethanolamine (Waite and Sisson, 1971). The enzyme showed some specificity for the type of acyl chain in position 2 of the glycerol

Table 7-4. Purification of Phospholipase A_2 from Rat Liver Mitochondria[a]

Purification step	Protein (mg)	Activity (units)	Specific activity (units/mg)	Recovery (%)	Purification (-fold)
Mitochondria	4,338	10,411	2.4	100	—
Extract	1,518	7,288	4.8	70	2.0
First Ultrogel AcA54	212	5,830	27.5	56	11.5
Second Ultrogel AcA54	5.3	4,685	884	45	368
Hydroxyapatite	1.38	4,060	2,942	39	1,225
Matrex gel Blue A	0.34	2,740	8,058	26	3,357

[a] From de Winter *et al.* (1982). One munit represents release of 1 nmole fatty acid/min.

(Waite and Sisson, 1971; Natori *et al.*, 1983) but again, more detailed studies using defined mixtures of phosphatidylethanolamine species are required. While the enzyme is absolutely specific for position 2 of the glycerol, it is not known at this time if the enzyme can attack the 2-acyl lysophospholipid.

Apparently, some conformational change occurs during delipidation that leads to instability. Detergents, phospholipids, and probably glycerol can overcome this stability problem (Nachbaur *et al.*, 1972; Natori *et al.*, 1983). The purified enzyme probably has a histidine in the active site since p-bromphenacyl bromide inhibits activity (de Winter *et al.*, 1982). Other active-site inhibitors such as diisopropylphosphate, N-ethylmaleimide, iodoacetimide, and 5-5'-dithiobis(2-nitrobenzoic acid) were without effect. These results suggest some analogies between this enzyme and the pancreatic and venom enzymes (see Chapters 9 and 10). It is generally thought that those groups of enzyme do not have functional thiols or serines in the active site. While it will be difficult to obtain sufficient quantities of the mitochondrial enzyme to compare structural features with the digestive phospholipases, chemical modification studies should allow some interesting comparisons.

The function and regulation of this enzyme remain obscure, especially as to the differences one might expect in the inner and outer membranes. Evidence is available that the phospholipase A_2 associated with the inner membrane is located on both sides of the membrane (Zurini *et al.*, 1981). It is possible that it functions in the deacylation–reacylation cycle and is responsible for regulating the acyl composition of the mitochondrial membrane. However, its close tie to the energy state of the mitochondria suggests that it is somehow involved in the regulation of membrane-associated functions. If so, it as well as the system that reacylates lysophospholipids in the outer mitochondria could be regulated by the energy state of the mitochondria (Waite *et al.*, 1970). It appears that Ca^{2+}, either directly or indirectly, plays a role in the activation of the phospholipase A_2. Cationic amphipaths such as local anesthetics inhibit the phospholipase A_2 under the appropriate conditions, presumably by displacing Ca^{2+} from the enzyme–substrate complex. Perhaps an analogous regulation occurs naturally (Waite and Sisson, 1972). Two activity plateaus were found when increasing concentrations of Ca^{2+} were used to stimulate the purified enzyme: one at 6–20 μM Ca^{2+} and one at 1.4–12.0 mM Ca^{2+} (deWinter *et al.*, 1984). Calmodulin was shown not to regulate this phospholipase A_2. While these authors raised the possibility that two binding sites for Ca^{2+} could be present on the enzyme, or that two phospholipases A_2 were present, it is possible that the Ca^{2+} was influencing the physical state of the substrate. Table 7-5 shows that Sr^{2+} can stimulate both the purified and mitochondrial-associated phospholipase A_2, unlike the pancreatic and venom phospholipases A_2.

The finding that metal ions such as Mn^{2+}, Mg^{2+}, and Ba^{2+} gave some stimulation of the mitochondrial-associated enzyme is consistent with the concept that metal ions might alter membrane structure, which favors phospholipase activation and hydrolysis. In that case, endogenous Ca^{2+} (16 μM) would be sufficient for phospholipase A_2 activity.

Table 7-5. *Effect of Divalent Cations on Mitochondrial Phospholipase A_2 Activity*[a]

Metal ion	Assay conditions		
	Purified enzyme	Mitochondria exogenous substrate	Mitochondria endogenous substrate
Ca^{2+}	100	100	100
Ba^{2+}	1	15	17
Cd^{2+}	0	0	0
Co^{2+}	0	0	0
Mg^{2+}	0	26	24
Mn^{2+}	0	4	24
Pb^{2+}	1	7	0
Sr^{2+}	23	33	23
Zn^{2+}	0	1	0

[a] Enzyme assays contained either 0.06 μg purified enzyme, 120 μg mitochondrial protein (with 0.2 mM exogenous 1-acyl-2-[1- ^{14}C]linoleoylphosphatidylethanolamine), or 1.0 mg mitochondrial protein (when hydrolysis of endogenous phosphatidyl[^{14}C]ethanolamine was measured). Metal ions were added to give a final concentration of 8 mM. Mean values are expressed as the percentage of the activity found in the presence of Ca^{2+}.

7.8. Phospholipase A_2 of Lung Exudate

A unique phospholipase A_2 has been purified using a two-step procedure from pulmonary secretions in patients with alveolar proteinosis (Sahu and Lynn, 1977). The enzyme, after extraction of the secretions with acetone/butanol/ether, was purified 120-fold in a single step that employed DEAE–cellulose chromatography. The enzyme had optimal activity in the alkaline pH range and required Ca^{2+} and deoxycholate for activity. The molecular weight of the exudate enzyme, established by SDS–PAGE and gel filtration, was 75,000 daltons. A number of the enzyme's properties determined thus far, including the amino acid composition, indicated that it is not of bacterial origin and apparently is not found in lung cells studied thus far. It was found to be active optimally on the phospholipid in egg-yolk emulsions and to have maximal activity higher than most other mammalian phospholipases, about 1500 units/mg protein. A number of interesting and important clinical as well as enzymologic questions arise from this study including whether or not this phospholipase A_2 with such high specific activity attacks membranous lipid and, if so, what the pathologic consequences are.

7.9. Phospholipases A_2 in Cultured Tumor Cells

Phospholipase activity in tumor cells is a particularly important area of research since a number of the properties of tumors seem to be influenced by membrane structure and associated activities. Indeed, the membranes of tumors have unusual composition that might be related to altered functioning of phospholipases. Little, however, has been done in this area of research.

One of the few phospholipases in tumor cells to be studied in detail was isolated from ascites hepatoma cells (Natori *et al.*, 1980). This plasma membrane-bound phospholipase A_2 was solubilized and maintained in soluble form with detergents. Despite the very-high-fold purification, 13,000 over homogenates, it still was not homogeneous. Under optimal conditions at pH 7.0 with 10 mM $CaCl_2$, the enzyme degraded phosphatidylethanolamine and, to a lesser extent, phosphatidylglycerol. Other phospholipids were not degraded, although it would be of interest to test the effect of detergents on the hydrolysis of phosphatidylcholine. Likewise, it would be of interest to compare this enzyme with the phospholipase A_2 found in rat liver plasma membranes (Victoria *et al.*, 1971). Despite some differences noted in the relative hydrolytic rates on phosphatidylglycerol and phosphatidylethanolamine, such a comparison should shed light on the function (or dysfunction) of a phospholipase on the surface of a tumor cell. Significantly, the phospholipase A_2 activity of a dimethylbenzanthracene-induced mammary tumor was found to be higher than that found in control tissue, suggesting a role for the phospholipase in cell proliferation (Rillema *et al.*, 1980).

Krebs II ascites cells were shown to have both phospholipases A_1 and A_2 (Record *et al.*, 1977). When assayed at low pH values, both are present in the cytosolic fraction even though they have some characteristics comparable to liver lysosomal phospholipases. The phospholipase A_2 active in the neutral pH range was mainly in fractions corresponding to endoplasmic reticulum and plasma membranes.

In an interesting study, Suzuki and Matsumoto (1974) found that cultured human amniotic cells underwent a change in the distribution of lysosomal phospholipases A_1 and A_2 when infected with measles virus. The kinetics of appearance of these enzymes in the cytosol paralleled the cytopathic effect, the formation of multinucleated giant cells. This relation was not specific for lysosomal phospholipases and acid phospholipase exhibited a similar shift. Even though this shift probably reflected a disruption of the lysosome and the release of lysosomal enzymes, the released phospholipases might play a unique role in cytopathology.

7.10. Intestinal Phospholipase A₂

The major emphasis on the degradation of dietary phospholipid traditionally has focused on the pancreatic phospholipase A_2, even though there have been numerous reports of intestinal phospholipase activity. The intestine itself was recently shown to have a very active phospholipase A_2 that degraded phosphatidylglycerol primarily. Since this enzyme has been well characterized and compared with the pancreatic enzyme, it will be emphasized here (Mansbach *et al.*, 1982; Verger *et al.*, 1982). In this study, great care was taken to eliminate contamination by the pancreatic as well as bacterial phospholipases A_2. The enzyme of the mucosa, which could be acid extracted, was most active in the distal section and in the cryptal cells, compared with villus-tip cells. The phospholipase A_2 was first detected 11 days postpartum, increased to a maximum at 24 days, and then decreased by 50%. Both starvation and feeding a low-fat diet produced some decrease in activity (Mansbach, 1984).

A phospholipase A_2 found in the lumenal fluid was active on both phosphatidylcholine and phosphatidylglycerol, unlike the mucosal enzyme that was active only on phosphatidylglycerol. Although not proven, these workers argued that the enzymes from the mucosa and the lumen are the same and that some modification such as proteolysis had occurred that altered substrate specificity. Table 7-6 shows the phospholipase A_2 activity on phosphatidylglycerol in various species.

The enzyme purified from delipidated, washed pig intestine was 200 times more active on phosphatidylglycerol than phosphatidylcholine. Since the substrate–Ca^{2+} complex precipitated substrate liposomes, a monolayer system was used for assay. Figure 7-5 shows the pressure–activity relation for four different substrates.

Considerable information has been published on the intestinal enzyme with emphasis on its relation to the pig pancreatic phospholipase A_2. It had a molecular weight of 15,800 daltons, based on its amino acid composition, and a pI greater than 9.6 (Table 7-7).

Table 7-7 gives the amino acid compositions of the two types of phospholipases A_2, including the 14 cysteines. Considerable homology exists between the two; however, some notable differences were seen. The ratio of basic-to-acidic amino acids was 1.33, compared with 0.53 for the pancreatic enzyme, which accounts for its strongly cationic character. Also, considerable differences exist between the serine and asparate content. Not presented here, some striking features were noted in the sequence of the two. While regions that are invariant in the pancreatic enzyme were common to the intestinal enzyme, the glutamate found in the pancreatic enzymes was replaced by asparagine. This is particularly significant in understanding the differences in substrate specificity between the pancreatic and ileal phospholipases. As will be discussed in more detail in Chapter 9, a hydrogen bond between the

Table 7-6. Intestinal Mucosal Phospholipase: Activity in Various Animal Species[a]

Animal species	Duodenum (nmoles/min/mg protein)	Ileum (nmoles/min/mg protein)
Rat ($n = 5$)	62 ± 11	185 ± 33
Pig ($n = 2$)	67	153
Dog ($n = 1$)[b]	14	0
Cat ($n = 2$)	N.A.[c]	0
Man ($n = 2$)[d]	670	N.A.[c]
Sheep ($n = 4$)	80 ± 13	110 ± 77
Ox ($n = 1$)	140	150

[a] From Mansbach *et al.* (1982). Didodecanolphosphatidylglycerol was used as substrate at a surface pressure of 20 dynes/cm.

[b] No activity was found in the Thiery–Vella loop (made from ileum) or in secretions from the loop. The activity reported in the table is from ileum distal to the loop.

[c] Not available.

[d] Duodenal biopsies obtained for diagnostic purposes when histologically normal.

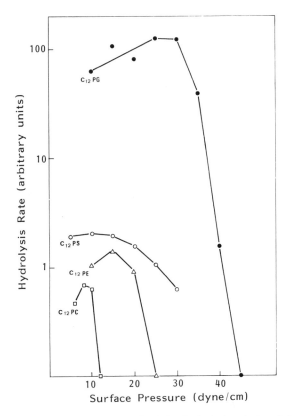

Figure 7-5. Variation with surface pressure of relative hydrolysis velocity of didodecanoylphosphatidylcholine (PC), -ethanolamine (PE),-serine (PS), and -glycerol (PG) monolayers by pure intestinal phospholipase A_2 (3.75, 2, 1, and 0.025 μg for PC, PE, PS, and PG, respectively) injected in the subphase of the special zero-order trough. (From Barochev *et al.*, 1982.)

α-amino group and glycine 4 has been proposed to be essential in the pancreatic enzyme for catalysis. Substitution of this glycine by a norleucine residue eliminated activity of the pancreatic enzyme. It was therefore suggested that some other modification in composition will permit phosphatidylglycerol hydrolysis by the intestinal enzyme. It will be of interest in the future to compare the relation between the molecular properties and the catalytic activity of the enzyme from the mucosa and the lumenal fluid. Such a study should clearly delineate the compositional characteristics that define substrate specificity.

The function of this enzyme will be covered in greater depth in Chapter 11 in the discussion of phospholipid digestion. One of the intriguing facets of this enzyme and its substrate specificity is its complementation with the pancreatic phospholipase. Since plants are rich in phosphatidylglycerol, the intestinal enzyme could be targeted toward the digestion of plant phospholipids. Likewise, the intestinal enzyme could be involved in the clearance of bacterial phospholipid rich in phosphatidylglycerol that occurs in the normal turnover of the intestinal flora of bacteria. Together in the lumen, the two

Table 7-7. Amino Acid Composition of Phospholipases Isolated from Horse, Pig, and Ox Pancreas and Pig Ileum[a]

	Pancreas			Pig intestine (ileum)	
Amino acid	Horse[b]	Pig[c]	Ox[d]	Experimental values	Next integers
Aspartic acid	19	22	25	10.08	10
Threonine	6	6	4	6.18	6
Serine	11	10	10	4.05	4
Glutamic acid	11	7	8	10.98	11
Proline	6	5	5	6.70	7
Glycine	6	6	6	9.58	9–10
Alanine	9	8	6	11.26	11
Cysteine	14	14	14	13.68	14
Valine	5	2	4	5.86	6
Methionine	1	2	1	1.89	2
Isoleucine	4	5	5	5.38	5
Leucine	6	7	8	13.04	13
Tyrosine	7	8	7	6.88	7
Phenylalanine	6	5	4	5.29	5
Lysine	8	9	11	15.85	16
Histidine	1	3	2	1.98	2
Arginine	4	4	2	9.77	10
Tryptophan	1	1	1	1.04	1
Total number of amino acids	125	124	123		139–140
Molecular weight	13,928	13,981	13,783		15,814 ± 39

[a] From Verger *et al.* (1982).
[b] Evenberg *et al.* (1977*a*).
[c] Puijk *et al.* (1977).
[d] Fleer *et al.* (1978).

phospholipases A_2 should be capable of digesting all dietary phospholipid present.

7.11. Heart Phospholipase A_2

Most of the heart phospholipases A_2 are described in the papers on heart phospholipases A_1. The major phospholipase A_2 in rat heart was reported to be in mitochondria and lysosomes (Weglicki *et al.*, 1971; Franson *et al.*, 1972; Palmer *et al.*, 1981; Nalbone and Hostetler, 1985). It is known to have optimal activity at high pH values and an absolute requirement for Ca^{2+}. It therefore is either similar or identical to the liver mitochondrial phospholipase A_2. It appears to be present in both the subsarcolemmal and interfibrillar mitochondria (Nalbone and Hostetler, 1985), but thus far no information is available on its distribution within the mitochondria. It will be of considerable interest to determine whether or not it is similar to the liver phospholipase A_2.

7.12. Spleen Phospholipase A₂

Rat spleen contains acid-active phospholipases A_1 and A_2 that are similar to those found in the lysosomes of many other tissues (Lloveras and Douste-Blazy, 1973). In this case, the apparent activity of the phospholipase A_1 was slightly higher than that of the phospholipase A_2. A phospholipase A_2 active at neutral pH values was purified to near homogeneity from rat spleen (Rahman *et al.*, 1973). This enzyme had a molecular weight of 15,000 daltons, absolutely required Ca^{2+}, and had a pI of 7.4. While it was thought to originate from a membrane fraction, no evidence for this was provided. Since the purification procedure did not employ organic solvents or detergents, the enzyme probably is of cytosolic origin.

A novel approach for the purification of a spleen phospholipase A_2 was recently reported by Tojo *et al.* (1984). They employed a three-step procedure to obtain a 23,000-fold purification over the starting cytosolic fraction with an overall yield of 70%. Key to this procedure was reverse-phase high-pressure liquid chromatography using an acetonitrile–H_2O gradient. This step gave about a 500-fold purification alone. The enzyme had a molecular weight of about 14,800 daltons and an absolute requirement for Ca^{2+} (Teramoto *et al.*, 1983). Only its pH optimum, pH 8.0–10.5, thus far distinguishes it from that reported by Rahman *et al.* (1973). This procedure should prove to be of considerable interest to phospholipase enzymologists, especially in the study of phospholipases that are stable in organic solvents.

Chapter 8

Phospholipases C and Phospholipases D of Mammalian Cells

As with Chapters 6 and 7, which covered mammalian acyl hydrolases, emphasis is on the characterization and purification of the mammalian phosphodiesterases; studies on their physiologic roles are covered in Chapter 11.

8.1. General Considerations

A brief overview of the wealth of information on the functions of phospholipase C introduces this chapter and gives the reader some feeling for the importance of this enzyme prior to reading about its purification and characterization. This class of phospholipases is now known to be central in regulatory processes and work on phospholipases C is providing invaluable information on agonist–receptor function and transmembrane signaling. Some of the recent breakthrough discoveries have done much to clarify riddles in biochemistry and cell biology that have been under investigation for decades.

Conversely, very little evidence has been obtained thus far for significant phospholipase D action in mammalian cells. Numerous studies have been done that would detect phospholipase D, which leads to the general conclusion that mammalian cells have adopted lipid catabolic pathways that do not significantly involve phospholipases D. This is distinctive to plant tissues that have a significant role for phospholipases D in their phospholipid metabolic pathways. There are some exceptions to this generality that will be described and perhaps future work will relegate the generalities made here about phospholipases D to the "dust bin."

8.2. Mammalian Phospholipases C: Historical Background

Most information on the molecular characteristics of the phospholipases C is derived from studies on bacteria as described in Chapter 3. On the other hand, much work has been done to determine the physiologic function of these enzymes since they have been implicated as mediators in a wide range of stimulus–response activities. Most, but not all, phospholipases C in mammalian tissues described thus far are specific for phosphatidylinositol or its phosphorylated derivatives. In most cells, the substrate for phospholipase C is phosphatidylinositol rich in stearate at position 1 and arachidonate in position 2 of the glycerol. Phospholipase C active on phosphatidylinositol initiates cellular response via one or more of the reaction products and involves a metabolic cycle that reutilizes the reaction product. Indeed, its role in the *phosphatidylinositol (PI) cycle* has been recognized for nearly three decades (Scheme 8-1) [for review of earlier work on the phosphatidylinositol cycle, see Hokin and Hokin (1960) and, more recently, Hokin (1985), Abdel-Latif (1983), Fain (1982), and Bleasdale *et al.* (1985)].

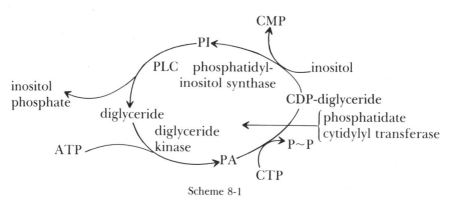

Scheme 8-1

Alternatively, the phospholipase C can be coupled with a lipase to release arachidonate, leading to the *arachidonate cascade* (Scheme 8-2) (see Section 7.1). It is now known that there is a close relation between these two functions of the phospholipase C as discussed in Chapter 11.

Phosphatidylinositol $\xrightarrow{\text{phospholipase C}}$ diacylglycerol + inositol phosphate

Diacylglycerol $\xrightarrow{\text{lipase}}$ free fatty acids (arachidonate) + glycerol

Arachidonate $\xrightarrow{\text{lipoxygenase/cyclooxygenase}}$ bioactive eicosanoids

Importantly, myoinositol is the naturally occurring form and all studies thus far have used phosphatidylmyoinositol. Two key points in these pathways should be emphasized: First, the substrate for degradation by phospholipase C is phosphatidylinositol (or a phosphorylated derivative, phosphatidylinositol-4-phosphate or phosphatidylinositol-4,5-bisphosphate); and second, in most cases, both pathways can be stimulated (Schemes 8-1 and 8-2) by an external agent plus Ca^{2+}. The role of Ca^{2+} in hydrolysis, as well as its possible relation to the phosphatidylinositol cycle, has been the focus of intensive study since investigators have implicated the cycle in Ca^{2+} metabolism.

A major interest in this cycle concerns the substrate specificity of the phospholipases C. As mentioned above, purified phospholipase C from the cytosolic fraction degrades phosphatidylinositol as well as its phosphorylated derivatives. Recent evidence, however, suggests that the plasma membrane of neutrophils and other cells has a phospholipase C that is relatively specific for phosphatidylinositol bisphosphate (C. D. Smith *et al.*, 1985; Verghese *et al.*, 1985; Jackowski *et al.*, 1986; Snyderman *et al.*, 1986). This enzyme(s) has been studied as a part of signal-coupling mechanisms and will be covered in more detail in Chapter 11.

One of the early reports of a phospholipase C activity in mammalian tissue came from Sloane-Stanley (1953) who demonstrated that guinea pig brain contained an enzyme(s) that released inositol and phosphate from phosphatidylinositol phosphate. Subsequent work showed the reaction to be stimulated by Ca^{2+} and that two enzymes were involved, phospholipase C and phosphomonoesterase (Rodnight, 1956). A wide variety of tissues as well as different sections of the brain have phospholipase C activity, indicating the widespread importance of this enzyme. The activities measured were quite high; for example, initial rates were as high as 60–70 μmoles of substrate hydrolyzed per gram of tissue per hour (measured manometrically as CO_2 released from $NHCO_3$ buffer as acid was released). A second enzyme, a phosphomonoesterase, was found to release the second phosphate from the inositol bisphosphate following the action of the phospholipase C (Dawson and Thompson, 1964). In brain, the phospholipase C was recovered in the cytosol free from phosphomonoesterase activity (Thompson and Dawson, 1964). Overall, the sequence of events in the degradation of phosphatidylinositol bisphosphate involves removal of the phosphates of the inositol ring by phosphomonoesterases followed by the action of the phospholipase C.

The following sequences of hydrolyses for the various phosphatidylinositols have now been shown (Cooper and Hawthorne, 1975; Akhtar and Abdel-Latif, 1980; Graff *et al.*, 1984; Moore and Appel, 1984; Siess and Binder, 1985):

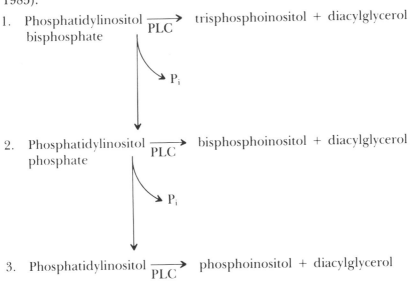

1. Phosphatidylinositol bisphosphate $\xrightarrow{\text{PLC}}$ trisphosphoinositol + diacylglycerol
   $\searrow P_i$

2. Phosphatidylinositol phosphate $\xrightarrow{\text{PLC}}$ bisphosphoinositol + diacylglycerol
   $\searrow P_i$

3. Phosphatidylinositol $\xrightarrow{\text{PLC}}$ phosphoinositol + diacylglycerol

Scheme 8-2

As can be seen, there are multiple points of entry of the diacylglycerol into the phosphatidylinositol cycle. The actual pathway appears to be dependent on the tissue and stimulant employed since, as will be described, in many cells phosphatidylinositol bisphosphate is initially hydrolyzed by phospholipase C with subsequent degradation of inositol trisphosphate by phosphomonoesterases.

8.2.1. Heart Phospholipase C

It now appears that multiple forms of enzyme can hydrolyze all three phospholipase C reactions in bovine heart (Low and Weglicki, 1983; Low *et al.*, 1984). Four distinct enzyme forms were separated by ion-exchange chromatography and had molecular weights ranging from 40,000 to 120,000 daltons. Most were optimally active in the acid pH range; however, addition of deoxycholate markedly enhanced the activity above pH 7.0. Similar activities were found in a variety of tissues and some indications were presented that proteolysis of a high molecular weight enzyme from platelets might regulate its activity (Low *et al.*, 1984).

A novel phospholipase C has been partially purified from dog myocardium (Wolf and Gross, 1985). This enzyme, originating from the cytosolic fraction, preferentially hydrolyzes phosphatidylcholine. Table 8-1 shows the substrate specificity of the enzyme when assayed as sonicated aggregates in the presence of Ca^{2+}.

Unlike many phospholipases C, this enzyme does not appear to have an absolute requirement for Ca^{2+}, and Mg^{2+} or Co^{2+} gave activity comparable to that found with Ca^{2+} whereas Zn^{2+} or Cd^{2+} inhibited. The authors suggested that the main reason this enzyme had not been identified previously was that a potent inhibitor is present in the cytosolic fraction that almost completely blocked activity. For example, a 30-fold increase in total phospholipase C activity was achieved by ion-exchange chromatography raising the specific enzymatic activity from 0.04 to 6.7 munits/mg. While the nature of this inhibitor remains obscure, its potential role in regulating the synthesis of diacylglycerol is significant. A similar type of phospholipase C has been

Table 8-1. Substrate Specificity of Dog Heart Phospholipase C[a]

Substrate	Specific activity (nmoles/mg/hr)
Plasmenylcholine	87
Phosphatidylcholine	109
Phosphatidylethanolamine	33
Phosphatidylinositol	<1
Sphingomyelin	<1

[a] The Sephadex G-75 eluate (68 μg) was incubated with radiolabeled plasmenylcholine, phosphatidylcholine, phosphatidylethanolamine, phosphatidylinositol, or sphingomyelin at a concentration of 10 μM for all substrates. Sphingomyelin was assayed in the absence or in the presence of six concentrations of Triton X-100 from 0 to 1%.

described in cultured Madin–Darby canine kidney cells (Daniel *et al.*, 1986*b*) to respond to tetradecanoyl-phorbol-acetate stimulation. This raises the possibility that this phospholipase C active on the choline- and ethanolamine-containing phospholipids might stimulate protein kinase C by the production of the required diacylglycerol. This would differ from the response of the phosphatidylinositol-specific phospholipase C that also can give rise to inositol trisphosphate that is thought to be a second messenger in Ca^{2+} mobilization (Berridge, 1983; Berridge *et al.*, 1983).

8.2.2. Brain Phospholipase C

Over a period of years, the group of Dawson investigated the properties of phospholipases C in brain tissue. They recognized that both the lysosomal and cytosolic fractions contained active but distinct phospholipases C (Irvine *et al.*, 1978; Hirasawa *et al.*, 1982*b*). The two enzymes were differentiated by their sensitivity to EDTA and their pH optima. The lysosomal phospholipase C, like other phospholipases, was inhibited by divalent cations and acted preferentially but not exclusively on phosphatidylinositol. Similar activity was found in the lysosomes of rat liver (Richards *et al.*, 1979; Hostetler and Hall, 1980).

Similar to the bacterial phospholipase C from a variety of sources, the cytosolic phospholipase C produced high levels of inositol 1 : 2 cyclic phosphate, leading Dawson *et al.* (1971) to suggest an additional name that more accurately reflects the nature of the reaction catalyzed, phosphatidylinositol-2-inositol phosphotransferase (cyclizing). This name has not been generally adopted. The ratio of the two products was dependent on pH. At pH 4.4, 85–90% of the product was cyclized, while at pH 6.8 there was an equal mixture of the two. While the cause of this change in the ratio was not fully understood, it was concluded that increased hydroxyls competed with the cyclization reaction.

Crude brain cytosolic preparations had four types of phospholipase C activity active from pH 5.0 to pH 8.5 (Hirasawa *et al.*, 1982*b*). The enzyme active at pH 8.5 was lost on ammonium sulfate fractionation, a technique commonly used in the isolation of these enzymes. The isoelectric points for the brain enzyme(s) active at pH 5.5 ranged from 3.8 to 7.4, while the enzyme active at 7.0 focused sharply with a pI of 4.6–4.8. This contrasted with a pI of 5.5 for phospholipase C_I from ram seminal vesicles (Hofmann and Majerus, 1982*a*). As with many phospholipases, surface characteristics of the substrate aggregate affected the enzyme activity markedly; the ratios of Ca^{2+}, substrate, detergent, and $[H^+]$ influenced the rate of reaction. At the optimal pH, 5.5–6.0, amphipathic cations stimulated activity of the crude cytosolic phospholipase C. Activity on monomolecular films of substrate was found at pressures up to 33 dynes/cm; above this pressure, activity ceased unless anionic phospholipids were added (Hirasawa *et al.*, 1981*b*). Phosphatidic acid showed two stimulatory effects in this system: (1) It increased the maximal surface pressure at which hydrolysis could occur from 33 to about 40 dynes/cm; and (2) it decreased the lag period prior to hydrolysis by about 50% at 30–33 dynes/cm. Phosphatidylcholine decreased the maximal surface pressure permitting hydrolysis as well as increasing the lag period (Figs. 8-1, 8-2, and

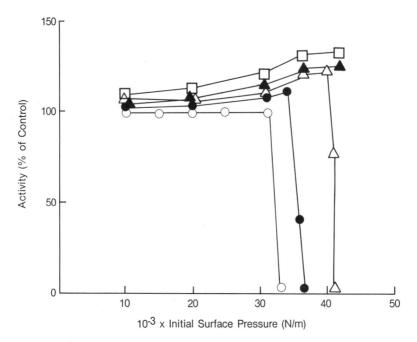

Figure 8-1. Effect of the surface pressure of a [^{32}P]phosphatidylinositol and phosphatidic acid mixed monolayer on phosphatidylinositol phosphodiesterase activity. Phosphatidylinositol monolayers contained various amounts of phosphatidic acid: ○, 0% (pure phosphatidylinositol); ●, 5%; △, 10%; ▲, 20%; and □, 40%. Results are expressed as percentage of routine controls using pure phosphatidylinositol monolayers at 25 × 10^{-3} N/m and represent this activity plotted against the initial surface pressure of the film. (From Hirasawa *et al.*, 1981*b*.)

8-3). The regulation of phospholipase C, including the brain enzyme, is described further in Section 8.2.3. These observations plus the observation that proteolysis of the phospholipase C increased the sensitivity to Ca^{2+} (Hirasawa *et al.*, 1982*a*) led to the postulate that the phospholipid environment and enzyme modification provide a very sensitive regulation of this enzyme.

8.2.3. Ram Seminal Vesicle Phospholipase C

Until recently, the detailed characterization of the cytosolic phospholipase C has been hampered by the lack of purified preparations. The recent emphasis on phospholipase C as a key enzyme in agonist response mechanisms has prompted the purification and characterization of the enzyme from seminal glands that are quite active in prostaglandin synthesis.

Recent evidence obtained during purification of the ram seminal vesicle glands (Hofmann and Majerus, 1982*a*) showed that phospholipase C activity is actually catalyzed by a family of proteins of unknown relations, similar to that described in the previous sections for the cytosolic fraction of the brain. Two immunologically distinct phospholipases C were purified from crude extract of seminal vesicles, one (phospholipase C$_I$) to homogeneity. The phos-

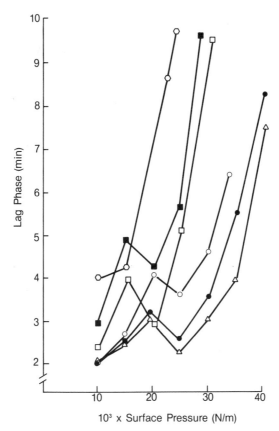

Figure 8-2. The relation between lag phase of enzymic hydrolysis and initial surface pressure of monolayer. Phosphatidylinositol monolayers contained various amounts of phosphatidic acid [○, 0% (pure phosphatidylcholine); ●, 5%; △, 10%] or of phosphatidylcholine (□, 5%; ■, 10%, ◐, 20%) (From Hirasawa *et al.*, 1981*b*.)

pholipase C_I could be further fractionated by AH–Sepharose affinity chromatography. Both subfractions, upon further purification, had a final specific activity of about 25 units/mg protein and a molecular weight of 65,000 daltons, similar to the enzyme from rat liver that had a molecular weight of 68,000 daltons (Takenawa and Nagai, 1981). It appears that the phospholipase C_I corresponds to a minor form separated by isoelectric focusing from rat brain. Indeed, the antibody against phospholipase C_I precipitated only about 15% of the total phospholipase C activity in sheep brain cytosol, suggesting that the other 85% of the activity in sheep brain could be the other three forms identified in rat brain cytosol. The relation of phospholipase C_I to the phospholipase(s) reported by Dawson's group in the cytosol of brain is unclear at present. Phospholipase C_{II}, only partially purified, was somewhat larger, 85,000 daltons as determined by gel filtration, and had a specific activity only one-third that of phospholipase C_I. The purified phospholipase C_I had a rather high percentage of hydrophilic amino acids, as would be expected of a cytosolic

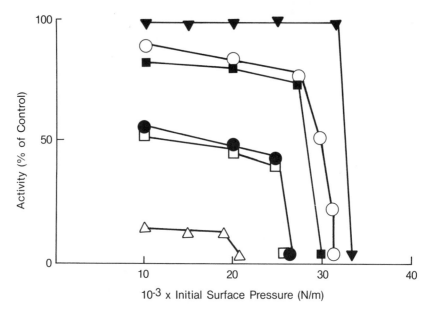

Figure 8-3. Effect of initial surface pressure of [^{32}P]phosphatidylinositol and phosphatidylcholine or palmitoylcholine mixed monolayers on phosphatidylinositol phosphodiesterase activity. Phosphatidylinositol monolayers contained various amounts of phosphatidylcholine [▼, 0% (pure phosphatidylinositol); ○, 5%; ●, 20%; △, 40%] or palmitoylcholine (■, 10%; □, 25%). Data are expressed as percentages of controls using pure phosphatidylinositol at 25 × 10^{-3} N/m. (From Hirasawa *et al.*, 1981*b*.)

protein. Likewise, it appeared to be devoid of carbohydrate. Both phospholipase C$_I$ and phospholipase C$_{II}$ were specific for phosphatidylinositol and absolutely require Ca^{2+} for activity (K_a or Ca^{2+} = 2.2 μM). The requirement for Ca^{2+} was identical to that found for the enzyme purified from rat liver (Takenawa and Nagai, 1981). The pH optimum was dependent on the conditions employed. When pure phosphatidylinositol was used, the optimal pH for hydrolysis was 5.3–5.4. However, the addition of deoxycholate stimulated activity 10-fold at pH 7.0 at a ratio of 3 : 1 (detergent : substrate). Similar results have now been obtained with a crude phospholipase C preparation from brain (Hirasawa *et al.*, 1982*b*) and with more highly purified preparations from platelets (Hakata *et al.*, 1982) and neutrophils (Smith and Waite, 1986).

8.2.4. Rat Liver Phospholipase C

A phospholipase C has been purified 4300-fold from the supernatant fraction of rat liver. This enzyme with a molecular weight of 70,000 daltons had a number of properties similar to other cytosolic phospholipases; it required Ca^{2+}, is active in the neutral pH range, and is activated by unsaturated fatty acids (Takenawa and Nagai, 1981). They examined the effect of phosphatidate on the rate of hydrolysis since it had been proposed by Irvine *et al.* (1979) that phosphatidate, potentially generated from the released diacylglycerol, could also function as an autostimulator. They reported that the purified liver phospholipase could be stimulated only by phosphatidate that contained two unsaturated acyl chains; all others were inhibitory.

8.2.5. Platelet Phospholipase C

A rather high molecular weight (143,000 daltons) was reported for the phospholipase C purified from bovine platelets that requires Ca^{2+} and is stimulated by high concentrations of arachidonate. The purified enzyme was not inhibited by phenylmethylsulfonyl fluoride, unlike the effect of this inhibitor on a crude preparation of the enzyme (Walenga *et al.*, 1980). These results further suggest that proteolysis of the platelet phospholipase C might be involved in its activation. The effect of arachidonate is unclear at this point and could be a nonspecific charge effect and possibly related to the Ca^{2+} concentrations used. It is known that the mixture of lipid present, phosphatidylinositol (or polyphosphate derivative) plus other phospholipid, will dictate the amount of Ca^{2+} required. Furthermore, the presence of deoxycholate and presumably other anionic amphipaths alters the Ca^{2+} requirement. Many of its properties are comparable to other phospholipases C, with the exception of its apparent molecular weight. Interestingly, this preparation was not stimulated by calmodulin or inhibited by indomethacin.

8.2.6. Lysosomal Phospholipases C

A phospholipase C from lysosomes has been described recently that differs from the cytosolic enzyme in that it is optimally active at pH 4.5 and has a broad substrate specificity (Richards *et al.*, 1979; Matsuzawa and Hostetler, 1980). This enzyme is widely distributed although the relative activity in different tissues could not be assessed owing to the activity of competing phospholipases A_1 and A_2 and lipases (Hostetler and Hall, 1980). It would appear, however, that intestine is an unusually rich source of the enzyme. Indeed, the state of the substrate markedly affected the relative rates of degradation by the enzymes of crude lysosomal preparations so that it has not yet been possible to quantitate accurately the various lipolytic activities in the lysosome (Irvine *et al.*, 1978; Richards *et al.*, 1979; Matsuzawa and Hostetler, 1980). The hydrolysis of phosphatidate apparently is catalyzed by an enzyme distinct from the phospholipase C (Irvine *et al.*, 1978). The phospholipase C has some of the properties described for the phospholipases A of lysosomes. For example, it is inhibited by divalent cations and cationic amphipaths such as chlorpromazine and mepacrine. A difference is noted, however; the lysosomal phospholipase C degradation of phosphatidylinositol is inhibited by Triton WR1339 (Irvine *et al.*, 1978) whereas stimulation is noted when the purified phospholipase A_1 is used (Robinson and Waite, 1983).

8.2.7. Some Aspects of Phospholipase C Regulation

The regulation of phospholipase C in both the crude and purified states is intriguing but complex. Information exists that suggests its activity can be regulated by the membrane milieu. The product of the reaction, diacylglycerol, plus compounds that are known to be metabolically derived from diac-ylglycerol, unsaturated fatty acid and phosphatidate, potentially stimulate hy-drolysis of phosphatidylinositol (Dawson *et al.,* 1980; Hofmann and Majerus, 1982*b*; Takenawa and Nagai, 1982) which led to the postulate that the system could be self-stimulatory. This is an appealing concept since the triggering of activity in response to a stimulant should be rapid but under tight regulation. Some of the features of regulation by the lipid composition stem from the observation that phosphatidylinositol in microsomal membranes is barely hy-drolyzed (Hofmann and Majerus, 1982*b*) unless dioleoylglycerol was added. Similar results were obtained with lipids isolated from those membranes which were used as substrates. This effect on phosphatidylinositol hydrolysis was dependent on the composition of the diacylglycerol when pure phosphati-dylinositol was the substrate (Fig. 8-4).

Under a variety of conditions, a 7 : 1 ratio of phosphatidylinositol : 1,2-dioleoylglycerol caused marked stimulation of hydrolysis and correlates with the ability of the diacylglycerol to disrupt the bilayer in some yet undefined manner (Dawson *et al.,* 1984). Under similar conditions, a 1 : 3 ratio of phos-phatidylinositol : deoxycholate gave comparable rates of hydrolysis. Phos-phatidylcholine was found to inhibit the phospholipase C seminal vesicles although this inhibition could be overcome by the addition of phosphatidyl-serine. It appears that the inhibition by phosphatidylcholine is at the stage of enzyme–substrate binding. This is based on the observation that the enzyme

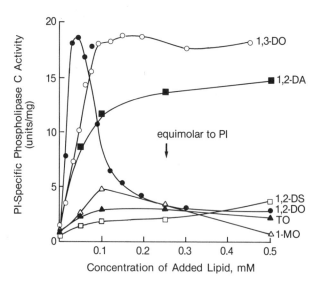

Figure 8-4. Effect of neutral lipids on phosphatidylinositol-specific phospholipase C activity in sonicated dispersions with phosphatidylinositol. The phosphatidylinositol concentration was 0.25 mM and the enzyme concentration was 0.68 µg/ml. The buffer contained 100 mM Tris-maleate, pH 6.7, 100 mM NaCl, 1 mM CaCl₂, and 0.5 mg/ml of bovine serum albumin. Assays were conducted for 10 min at 37°C. Values represent data from the average of two or more experiments giving similar results. ●, 1,2-Diolein (1,2-DO); ○, 1,3-diolein (1,3-DO); ■, 1,2-diarachidonin (1,2-DA); □, 1,2-distearin (1,2-DS); △, 1-monoolein (1-MO); ▲, triolein (TO). (From Hofmann and Majerus, 1982*b*.)

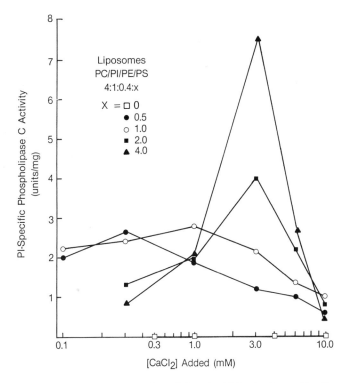

Figure 8-5. Effect of increasing the proportion of phosphatidylserine in small unilamellar vesicles containing phosphatidylcholine, phosphatidylinositol, and phosphatidylethanolamine (4 : 1 : 0.4) on the calcium concentration dependence of phosphatidylinositol-specific phospholipase C activity. The phosphatidylinositol concentration in the assays was fixed at 60 μM. The phosphatidylserine : phosphatidylinositol ratio was varied: □, 0; ●, 0.5; ○, 1; ■, 2; and ▲, 4. Values represent data from two experiments giving similar results. Enzyme concentration was 0.46 μg/ml. Assays were performed for 2 min at 37°C. (From Hofmann and Majerus, 1982*b*.)

does not bind to vesicles of a mixture of phosphatidylcholine–phosphatidylethanolamine–phosphatidylinositol. The conclusion that phosphatidylcholine blocks the physical interaction with the aggregate rather than at the interaction of the substrate at the enzyme's active site is demonstrated by the stimulation of hydrolysis by dihexanoyl phosphatidylcholine that disrupts bilayers (Dawson *et al.,* 1980, 1984). All these effects are influenced to a considerable extent by the concentration of Ca^{2+} present (Hofmann and Majerus, 1982*b*) (Fig. 8-5).

Table 8-2 summarizes some of the factors that stimulate or inhibit phospholipase C action.

From the physiologic point, perhaps one of the most interesting aspects of Table 8-2 is the phosphatidate stimulation of phosphatidylinositol hydrolysis when mixed with phospholipid present in the inner but not outer leaflet of the membrane. Experiments of this nature that include other components such as cholesterol and glycolipid would be useful in more accurately assessing the physiologic impact of this work.

Table 8-2. Factors That Regulate Phospholipase C[a]

Addition	Effect	Substrate mixture[b]	Effect of high Ca^{2+} [b]
Fatty acids			
Unsaturated	+	PI	+
Saturated	−	PI	N.D.
Phospholipid			
Long-chain phosphatidylcholine	−	PI	N.D.
Short-chain phosphatidylcholine	+	PI	N.D.
Long-chain lysophosphatidylcholine	−	PI	N.D.
Short-chain lysophosphatidylcholine	+	PI	N.D.
Phosphatidate	0	PI:PE:PC:PS:Sph[c] (outer leaflet)	N.D.
	+	PI:PE:PC:PS:Sph[c] (inner leaflet)	N.D.
Lysophosphatidate	+	PI	N.D.
Phosphatidylserine	+	PC:PI:PE	+
Dioleoylglycerol	+	PI	N.D.
	+	PC:PI:PE:PS	+
	0	PC:PI:PE	

[a] From Hofmann and Majerus (1982*b*).
[b] Abbreviations: PI, phosphatidylinositol; PE, phosphatidylethanolamine; PC, phosphatidylcholine; PS, phosphatidylserine; Sph, sphingomyelin; N.D., not determined.
[c] Percentage composition constituted to represent outler leaflet (11:19:41:10:19) and inner leaflet (24:24:13:23:16) (Hofmann and Majerus, 1982*b*).

Most studies done thus far have been at a fixed pH value, 6.7–6.8. However, one of the effects of acidic phospholipids is to shift the pH of the aqueous phase required for optimal activity (Dawson *et al.*, 1980) (Fig. 8-6).

The complexity found in these studies is indeed difficult to interpret, mainly because of the lack of precise information on the physical nature of the substrate mixtures. While information gleaned from the literature suggests that structural changes are occurring, precise data on the substrates used in these studies are lacking. The various workers on this problem suggest, however, that phase separation of the lipids is important in permitting penetration of the enzyme into the lipid aggregate (bilayer?). This would help explain the stimulatory effect of high Ca^{2+} concentrations when acidic phospholipids were used.

8.3. Sphingomyelinases

Although sphingomyelinases catalyze the same phosphodiesterase activity as phospholipase C, the difference in substrate specificity (glycero- vs. sphingolipid) puts the two types of enzyme in separate classes, as described in the Introduction. As will be seen in this section, their functions are clearly different, as far as can be determined at this time. The sphingomyelinases appear

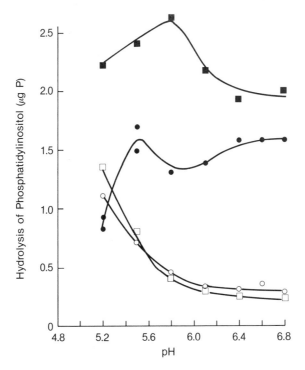

Figure 8-6. Stimulation of phosphatidylinositol phosphodiesterase by lysophosphatidic acid at various pH values between 5.2 and 6.8. The open symbols (○,□) represent the control values and contrast with the activities measured in the presence of 0.115 μmoles of either oleoylglycerophosphate (●) or ovophosphatidic acid (■). The enzyme was assayed in a range of KOH maleic acid buffers. (From Dawson *et al.*, 1980.)

to have a digestive as opposed to a regulatory function, especially those localized in lysosomes.

The capacity of mammalian tissue to degrade sphingomyelin by removal of the phosphoryl and choline moieties was recognized as early as 1940. This activity found in calf liver (Thannhauser and Reichel, 1940) and pig spleen (Fujino, 1952) had very low activity and was not studied in great detail until the early-to-mid-1960s when it was recognized that the lack of this lysosomal enzyme in human spleen was involved in both the classical (Type A) and visceral (Type B) Niemann–Pick disease, a sphingomyelin storage disorder (Brady *et al.*, 1966; Schneider and Kennedy, 1967). Interestingly, a second sphingomyelinase active at pH 7.4 that required Mg^{2+} was found in high levels in the microsomal fraction of spleens from Types A and B patients. This indicated that the lack of the lysosomal rather than the microsomal sphingomyelinase accounts for the accumulation of sphingomyelin in this pathologic state. Biologic distribution and developmental changes noted indicate that the lysosomal and microsomal enzymes are distinct. For example, their distribution between sections of the rat brain differs markedly (Spence and Burgess, 1978) (Table 8-3). Likewise, the specific activity of the micro-

Table 8-3. Distribution of Sphingomyelinases in Brain[a]

Section	Sphingomyelinase (nmoles/mg protein/hr)		
	pH 5.0 (lysosomal)	pH 7.4[b] (microsomal)	Ratio pH 7.4 : pH 5.0
Cerebral gray matter	61.7	163.0	2.6
Cerebral white matter	46.5	88.8	1.9
Medulla	42.6	34.8	0.8
Cerebellum	65.0	41.7	0.6
Sciatic nerve	8.2	10.6	1.3

[a] From Spence and Burgess (1978).
[b] Assayed in the presence of 6 mM $MgCl_2$.

somal sphingomyelinase increased severalfold from day 0 to day 40 following birth of the rat, while the lysosomal enzyme did not change significantly. In this section, the microsomal and lysosomal sphingomyelinases will be covered separately rather than describing the enzymes according to the tissue of origin, followed by studies on erythrocyte sphingomyelinase.

8.3.1. Lysosomal Sphingomyelinase

The lysosomal sphingomyelinase is widely distributed in mammalian tissues (Rao and Spence, 1976) and probably functions in the turnover of ingested sphingomyelin by the lysosome. Some of the early studies on sphingomyelinase involved partial purification of the enzyme from rat liver (Heller and Shapiro, 1966) and brain (Barenholz et al., 1966). These preparations still had very low activity under optimal conditions at pH 4.5–5.5, usually less than 30 munits/mg protein. No metal ion was required. The rat liver enzyme degraded the *erythro* isomers to phosphorylcholine and dihydroceramide several times more rapidly than the *threo* isomers of DL-dihydrosphingomyelin (Heller and Shapiro, 1966). Phosphatidylcholine was not degraded. Since those early reports, purification of the lysosomal enzyme has been pursued with the formidable goal of enzyme replacement therapy for patients with Niemann–Pick disease. For example, more recently the enzyme has been purified 10,000-fold over the homogenates of human placentas but remained heterogeneous (Pentchev et al., 1977). This preparation had a final specific activity of about 4 units/mg protein and was recovered in an aggregated form that contained 8–10 monomers with molecular weights of 36,800 and 28,300 daltons. Substrate specificity studies showed the placental enzyme is specific for both the phosphorylcholine and ceramide moieties of sphingomyelin although 2-hexadecanoylamino-4-nitrophenyl-phosphocholine was also degraded. The latter finding verifies the suitability of using this chromogenic substrate for enzyme quantitation in cells from carriers of Niemann–Pick disease.

A part of the difficulty in purifying the lysosomal enzyme from brain was the removal of contaminating enzymes in lysosomes from human brain (Yamanaka *et al.*, 1981). While this has been achieved to a certain degree, the complete removal of other lysosomal hydrolases was not a simple matter as they copurified to varying degrees with the sphingomyelinase. Two different procedures were compared: one that employed Triton X-100 and one that did not. The final preparations reported for the human brain were heterogeneous even though they were purified 1700-fold. Some differences were noted in the isoelectric point of the sphingomyelinase preparations, depending on the presence of Triton X-100 during one or more of the purification steps. The enzyme not exposed to Triton had a pI = 4.75, whereas that exposed to Triton had a pI ranging from 5.2 to 6.6 (Yamanaka *et al.*, 1981). The reason for this discrepancy was not apparent. Likewise, the molecular weight of the brain enzyme cannot be assigned with accuracy. The lowest molecular weight of about 20,000 daltons has been reported (Gatt and Gottesdiner, 1976; Yamanaka *et al.*, 1981).

As with many lipolytic enzymes, little is known about the regulation of the lysosomal sphingomyelinase. Generally, the enzyme has been assayed using detergent-emulsified sphingomyelin. Small amounts of a wide range of lipids could stimulate the activity on ultrasonically dispersed bilayers of sphingomyelin as shown in Table 8-4 (Gatt *et al.*, 1973).

Bile salts are also good stimulators probably under conditions that lead to mixed micelle formation (Yadgar and Gatt, 1980). While the details of the interaction of the enzyme and the substrate are not yet established, it is clear that the stimulation of activity is not dependent on surface-charge effects.

8.3.2. Microsomal Sphingomyelinase

The microsomal sphingomyelinase that is active at pH 7.4 and requires Mg^{2+} appears to be restricted to the brain (Gatt, 1976). Although it is termed microsomal here, its subcellular distribution has not been thoroughly determined. At present, little work has been done on the purification and char-

Table 8-4. Effect of Various Amphipaths on Sphingomyelinase Activity[a]

Addition	Additive : sphingomyelin[b]	Percent reaction rate
None	—	100
Phosphatidylcholine	0.15	102
Lysophosphatidylcholine	0.10	50
Phosphatidate	0.15	230
Dicetylphosphate	0.15	223
Octadecylamine	0.10	137
Methyloleate	0.25	242
Cholesterol	0.15	151
Triton X-100	0.20	152
Triton X-100	2.70	415

[a] From Gatt *et al.* (1973).
[b] Molar ratio of the two lipids present in the assay of the lysosomal sphingomyelinase from rat brain.

acterization of the microsomal sphingomyelinase nor has any role been as-cribed to it. It is interesting to note the finding of Gatt *et al.* (1973) that patients with Type A Niemann–Pick disease have normal levels of brain sphingo-myelin. It is therefore possible that the microsomal sphingomyelinase which is normal in these patients could serve in the same capacity as the lysosomal enzyme in other tissues, namely, to degrade excess sphingomyelin. If the excess sphingomyelin is interlysosomal, it is somewhat difficult to visualize the mode of attack of the microsomal enzyme.

8.3.3. Erythrocyte Sphingomyelinase

Avian erythrocytes have a sphingomyelinase that is latent and becomes active upon hypotonic lysis of the cell. This enzyme degrades sphingomyelin within the avian erythrocyte as well as that in lysed erythrocytes of other species that did not have the sphingomyelinase (Record *et al.*, 1980). The Ca^{2+} ionophore A23187 can likewise stimulate the chicken erythrocyte to degrade 20–30% of its sphingomyelin despite the fact that the enzyme does not require Ca^{2+} (Allan *et al.*, 1982). In this case, it is thought that the Ca^{2+} caused some microvesiculation that brings the membrane-associated sphingomyelinase to an internal pool of sphingomyelin, thereby promoting hydrolysis. This model is based on the observations that (1) the product phosphorylcholine remains in the cell, 2) externally added sphingomyelinase degrades only 65–70% of the total cellular sphingomyelin under nonlytic conditions, and (3) osmotically lysed cells have 100% of the sphingomyelin degraded by the endogenous enzyme, even in the presence of EGTA. The latter observation indicates that the endogenous enzyme can degrade the entire pool of cellular sphingomye-lin, if given the opportunity, and that Ca^{2+} stimulates the breakdown of the smaller internal pool by a mechanism other than a direct stimulation of the enzyme.

8.4. Phospholipase D

The first suggestion that mammalian tissues might contain a phospholi-pase D came from the laboratory of Hubscher in the early 1960s (Dils and Hubscher, 1961; Hubscher,1962). The possibility of such activity was inferred from base exchange rather than hydrolytic measurements of phosphatidate formation. The microsomal fraction of rat liver was found to incorporate serine, inositol, and choline into their respective phospholipids via an energy-independent, Ca^{2+}-stimulated exchange process. Since it was recognized that plant tissue contained phospholipase C (now termed D) that catalyzed a similar base-exchange reaction, it was suggested that a phospholipase D might be responsible for this base exchange in mammalian tissue. Furthermore, since the incorporation of each base group had some unique requirement, it was thought that distinct enzymes could be involved. However, the relation of the exchange to hydrolytic activity was not established since only radiolabeled base groups were employed that could measure hydrolysis. Also, attempts to demonstrate hydrolytic activity were thwarted by the inability of membrane-associated phospholipase D to attack exogenous phosphatidylcholine. Kanfer and his coworkers have described the partial purification and some characteristics of a phospholipase D and its differentiation from base-exchange enzymes. The latter enzymes are thought not to be the same as transphosphatidylases such as found in plants (Chapter 5) since no phosphatidate–enzyme intermediate has been proposed.

8.4.1. Brain Phospholipase D

The recent attempts to unravel the relation of base-exchange and hydrolytic phospholipase D activities have come from the laboratory of Kanfer who measured phospholipase D activity in a solubilized preparation from rat brain microsomes. Initially, they reported that the base-exchange and hydrolytic activities might be one and the same (Saito and Kanfer, 1973). Subsequent studies of the two activities pointed out some significant differences, however, notably the pH optimum. Whereas the exchange incorporation of choline into phosphatidylcholine was optimally active at pH 7.2, hydrolysis was maximal at pH 6.0. Similar to the plant phospholipase D, the hydrolysis was sensitive to sulfhydryl agents (Saito and Kanfer, 1975).

Phospholipase D, partially purified from rat brain extracts, was found to be free of exchange activity although the preparation did have transphosphatidylation activity which may be the result of the action of a single enzyme (Taki and Kanfer, 1979; Chalifour and Kanfer, 1980, 1982). In that case, glycerol was the acceptor forming a racemic mixture of phosphatidylglycerols from phosphatidylcholine. Of the various phosphatidyl donors, only phosphatidylsulfocholine equaled phosphatidylcholine as substrate, a rather surprising finding since the head group does influence activity. These authors made a distinction between base exchange, as originally described by Hubscher, and the transphosphatidylation catalyzed by phospholipase D, including the plant enzymes. This differentiation was based on pH optimum, K_m of the soluble substrate, specificity of the phospholipid under attack, and metal ion requirement (Table 8-5).

Its function in brain was proposed to be a mechanism by which choline is made available for acetylcholine synthesis in the synapse (Hattori and Kanfer, 1984). The activity of the phospholipase D in synaptosomes was stimulated dramatically by oleate. A review and discussion of these comparative activities have been prepared by Chalifour *et al.* (1980) and Kanfer (1980). It will be

Table 8-5. Comparison of the Base-Exchange and Phospholipase D Activities in Brain[a]

	Base exchange	Phospholipase D
pH optimum	8.5	6.0
Effect of Ca^{2+}	Required	Slight inhibition
K_m of acceptor	1.9×10^{-4} to 4.2×10^{-5} M	0.2 M
Phospholipid preferred	Phosphatidylethanolamine	Phosphatidylcholine

[a] See Kanfer (1980) for a review on this subject.

of considerable interest to have purified preparations of this enzyme for a more complete understanding of its functioning and mechanism of action.

8.4.2. Eosinophil Phospholipase D

An enzyme similar to the rat brain phospholipase D was partially purified from human eosinophils (Kater *et al.*, 1976). The major emphasis of this work was the activity that the preparation had on platelet-activating factor (PAF) activity. The authors were testing the possibility that eosinophils could be involved in the immediate hypersensitivity response and showed copurification of the phospholipase D activity as measured using phosphatidylcholine as substrate and the inactivation of PAF in a bioassay system. The chemistry of the degradation of PAF could not be studied since no radiolabeled substrate was available at that time. As a consequence, it is possible that deacylation of the acyl group at position 2 of the glycerol took place prior to the removal of the choline. However, since the two base-removal activities copurified on isoelectric focusing, ion-exchange chromatography, and gel filtration, it appears likely that base removal was the means of PAF inactivation. It is of interest to note that the eosinophil had many characteristics in common with the plant phospholipase D; its pH optimum was 4.5–6.0 and its pI was 4.8–6.2. The molecular weight, however, was roughly one-half that of the plant system, about 50,000 daltons. Owing to the physiologic role that this enzyme might play and the apparent ease with which it can be solubilized, further study of this system should be forthcoming.

8.4.3. Lysophospholipase D

In what might be related studies on the metabolism of phospholipids with an ether linkage in position 1 of the glycerol moiety, Wykle and coworkers found evidence for a phospholipase D that acted preferentially on compounds with a free hydroxyl at position 2 (Wykle and Schremmer, 1974; Wykle *et al.*, 1977, 1980). The activity was highest in the microsomal fraction of rat brain, liver, and testes and also was detected in kidney, intestine, and lung. This enzyme could be differentiated from a base-exchange enzyme (transphosphatidylate) in the cytosolic fraction by the Mg^{2+} requirement of the lysophospholipase D. If the hydroxyl at position 2 of the glycerol were acetylated or if the alkyl group were replaced by an acyl group, the activity was reduced by more than 90%. No real difference was detected between the activity on the choline- and ethanolamine-containing substrates.

The function and regulation of the lysophospholipase D are unknown. In most tissues, the degradation of lyso-alkyl phosphatidylcholines occurs with the necessity of base group removal. The possibility remains, however, that the enzyme could be involved in the metabolism of PAF, perhaps via a type of transphosphatidylation reaction. Such a pathway, while distinct from other transphosphatidylations, would allow the formation of lyso-PAF from ethanolamine phospholipids, known to be rich in ether linkages in position 1.

This is not unreasonable since it is a sulfhydryl enzyme and can act on 1-hexadecyl-*sn*-glycero-3-phosphoethanolamine.

8.5. Summary

The last three chapters (6–8) have covered the cellular phospholipases of vertebrates. Although a large number have been identified, few are well studied and by and large we know little of their functions. There has been considerable interest lately on those involved in the receptor-coupling mechanism, however, and progress in this research is being made. When we compare our understanding of this class of enzymes with what is known about venom and pancreatic phospholipases, we understand how much more is to be learned. Indeed, the work described in Chapters 9 and 10 provides an excellent example of what to look for in our work on cellular phospholipases.

Chapter 9

Pancreatic and Snake Venom Phospholipases A₂

Chapters 6–8 covered the mammalian phospholipases that are cellular for the most part. This chapter and the following chapters also include some mammalian phospholipases A_2; however, these are extracellular and originate from the pancreas of different species. Since the pancreatic phospholipases are of a general class of phospholipases A_2 that are closely related to the enzymes from venoms, they are covered together. There is reason to believe that some cellular phospholipases A_2 might have properties in common with the extracellular digestive phospholipases A_2. As pointed out in Chapter 7, however, insufficient information is available at this time to speculate further on this point.

Chapters 9 and 10 are organized to cover sequentially purification and structure of the enzymes and Ca^{2+} and substrate interaction with the enzymes. Chapter 9 emphasizes the purification and structure of the enzymes, whereas Chapter 10 covers enzyme interaction with Ca^{2+} and substrate as well as the mechanism of action of the enzymes. These are important considerations from a number of points of view. First, we have the most detailed knowledge of this group of phospholipases and, as a consequence, the studies described here serve as models for the study of other phospholipases. Second, structure–function relations are evolving from studies in which systematic comparison of the three-dimensional structure is related to ligand binding (also a model for protein–lipid interaction). Third, these enzymes are used for the development of drugs aimed at specifically blocking the action of cellular phospholipases, in particular in the development of anti-inflammatory agents. Fourth, comparison of enzyme structure has been a powerful tool in the evolutionary grouping of species, especially venomous snakes. Other such examples will emerge throughout Chapters 10 and 11.

9.1. General Considerations

The members of this group of phospholipases A_2 are structurally and functionally quite similar and most often are considered together. As will be described, the amino acid sequence and crystallographic structures appear to be highly conserved throughout evolution and, based on their structures, two basic classes of phospholipases A_2 have been proposed (Heinrikson *et al.*, 1977). One class is composed of the phospholipases A_2 in *Elapidae, Hydrophidae,* and pancreas, while the second class is made up of the enzymes in *Viperidae* and *Crotalidae.* The division of the two classes is based on the positions of the disulfide bridges and the length of carboxyl terminus. The classification has been expanded further on the basis of amino acid composition surrounding the active site (Dufton and Hider, 1983). Despite these divisions, great commonality exists in the active sites and hydrophobic regions, even including the recently described enzymes that have lysine at position 49 and that from bee venom (Maraganore *et al.*, 1986*a*). Such commonality is not surprising when one considers that the basic function of the enzymes is digestion of dietary phospholipids and that modifications in structure have evolved for toxic and hemolytic activities. A wide variety of other eukaryotes have venoms that contain phospholipases A_2; these will be covered separately.

Detailed studies now exist on the structural organization of the protein involved in binding of Ca^{2+} and both monomeric and aggregated substrates to the enzyme. These studies include chemical modification directed toward reactive residues in the protein, semisynthetic substitution of specific amino acids, and physical analysis using a variety of spectroscopic and crystallographic techniques. As will be detailed in this and the following chapter, some generalities can be made even though complete information is not always available. Basically, it appears that these phospholipases A_2 have two lipid binding regions: (1) the active site where hydrolysis occurs and (2) an activator site that interacts with bulk lipid, a point of some controversy. It is well established that Ca^{2+} binds in the active site and participates in catalysis. It also is clear that minor modifications in the primary and secondary structures alter the tertiary structure profoundly with concomitant changes in ligand binding and catalytic properties. A number of excellent and lively reviews recently have been published with cogent arguments for particular models of enzyme action (Verger, 1980; Verheij *et al.*, 1981*b*; Slotboom *et al.*, 1982; Verger and Pattus, 1982; Dennis, 1983; Dennis and Pluckthun, 1985).

9.2. Purification, Activation, and Some Characteristics of the Pancreatic Phospholipase A_2

Pancreatic phospholipases A_2 are found in most but not all mammals studied thus far. Their function, once secreted into the intestine and activated by proteolytic cleavage, is to digest the emulsified phospholipid in the intestine that originates either from the diet or from bile juice. Since this enzyme is synthesized as a soluble zymogen, it is somewhat different from most other phospholipases A_2. These enzymes were probably the first phospholipases to be observed, when Bokay found over a century ago that phosphatidylcholine was degraded by pancreatic juice (Bokay, 1877–1878). This observation undoubtedly was important since it was published in the first volume of the prestigious *Zeitschrift der Physiologica Chemie*. One of the earliest reports of phospholipase in pancreatic extracts came in 1932 when Belfanti and Arnaudi (1932) and Nukuni (1932) demonstrated the conversion of phosphatidylcholine to lysophosphatidylcholine. Partial purification of the enzyme was achieved as early as 1936: advantage was taken of the enzyme's heat stability and solubility in 50% ethanol (Gronchi, 1936). The review by Zeller (1951) is quite useful for a historical perspective on this early work in the field.

9.2.1. Purification of Pancreatic Phospholipase A_2

Rimon and Shapiro (1959) found that the aging of beef pancreas led to the activation of the phospholipase A_2 by, as we now know, the conversion of the proenzyme to the enzyme. These preparations, purified 40-fold over pancreatic homogenates, had a rather broad substrate specificity and interestingly lost activity on phosphatidylcholine but not other phospholipids upon heating at 90°C. For reasons not totally understood, the inactivation of phosphatidylcholine hydrolysis was reversed in about an hour. Ca^{2+} was required for the hydrolysis of all phospholipids tested except phosphatidic acid.

The enzyme from human pancreas was purified later using an alkaline glycerol solution for solubilization (Magee *et al.*, 1962). Likewise, it was found that deoxycholate is better than diethyl ether as an activator of hydrolysis although diethyl ether was effective. The 40- to 150-fold purified preparation was found to be free of lipase activity, resolving existing questions about the relation of lipolytic activities in pancreas. Optimal activity was found between pH 8.0 and 9.0.

The specificity of the pancreatic phospholipase A_2 was established by deHaas *et al.* (1963), who used a series of chemically synthesized mixed acid phospholipids. This enzyme, partially purified from human pancreas, was absolutely specific for position 2, regardless of the acyl composition at positions 1 and 2, similar to their earlier findings using snake venom phospholipase. They established that the enzyme was stereospecific for the acyl ester adjacent to the phosphate ester. The *sn*-3 but not *sn*-1 phospholipids were attached. Also, the acyl group in position 1 of the glycerol is not essential since 1-acyl-*sn*-2-ethyleneglycolphosphocholine was degraded. They also established at that time that a net negative charge on the lipid aggregate was required for hydrolysis.

The question existed as to whether it was better to purify the proenzyme or the activated enzyme. It was recognized in the initial procedure that the aging process led to greater yields of the active enzyme through the tryptic cleavage of a heptapeptide from the proenzyme, similar to the activation of other pancreatic enzymes, chymotrypsin and trypsin. Purification of the proenzyme was preferred since the activation by trypsin could be regulated more easily. Two closely related forms of the proenzyme were purified that differed by four amino acids in activation peptide (form I: Ser-Ser-Arg vs. form II: Glu-Glu-Gly-Ile-Ser-Ser-Arg) (Nieuwenhuizen *et al.*, 1973; Dutilh *et al.*, 1975). Elastase could convert form I to II by specifically cleaving the Ile₄-Ser₅ peptide bond. The proenzyme was purified to homogeneity in three steps with a 70% yield. The purified phospholipase had a pI of 7.3, whereas proenzyme I was more basic, pI = 7.5, and proenzyme II was more acidic, pI = 6.5. No carbohydrate was detected. The procedure used initially involved heating at low pH, salt precipitation, and ion-exchange chromatography. Newer procedures include affinity and hydrophobic chromatography, selective precipitations, and antibody chromatograph (Verheij *et al.*, 1981*b*). The human enzyme had the same number of amino acids as the equine (125) which was one more than the porcine and two more than the bovine enzymes. There is a high degree of homology in the pig, cow, and horse phospholipases A_2, differing only at 26, 28, and 32 residues, respectively (Verheij *et al.*, 1983). The purified enzymes with a molecular weight of about 14,000 daltons were monomeric, unlike many of the venom enzymes. In general, the same approaches have been used for the purification of the pancreatic phospholipases A_2 from all species examined. The desirability of this system for detailed mechanistic and structural studies is obvious. For example, 75 kg of pork pancreas could yield roughly 25 g of pure enzyme with a specific activity of 1300 units/mg using an egg-yolk Ca^{2+}–deoxycholate–substrate mixture (van Wezel and deHaas, 1975). Two isomers of the porcine prephospholipase A_2 were separated in this large-scale preparation: a major form termed α (95% of total) and a minor form termed β or iso. These yielded distinct active enzymes upon tryptic cleavage. In this case, both proenzymes had identical heptapeptide activation sequences but differed primarily by two amino acids in the activated enzyme; the β form lacked His-24 and Met-27. These deletions resulted in a lower pI, 5.9, and a specific activity equal to 60–70% of that expressed by the α form. Other workers have reported the existence of isophospholipases although the full significance of these observations is not fully understood at this point. For example, the multiple forms found in pork pancreatin (Tsao *et al.*, 1973) could result from proteolytic cleavage of the active enzyme (van Wezel and deHaas, 1975). On the other hand, such isoenzymes could represent genetic variants, since different mammalian species have phospholipases with distinct catalytic properties. Likewise, five different activation peptides have been identified (Evenberg *et al.*, 1977*b*).

9.2.2. Activation of the Prophospholipase A_2

The activation of the proenzyme is efficiently catalyzed by trypsin (Abita *et al.*, 1972). Invariant in this activation is the Ala at the N-terminus of the activated enzyme that is essential for activity of the enzyme on aggregated substrate. Figure 9-1 shows that whereas both the pro- and activated phospholipases A_2 hydrolyze phospholipids in the monomeric short-chain phosphatidylcholine below the CMC, the zymogen form has no activity on aggregated substrate (Pieterson *et al.*, 1974). When the activated enzyme was assayed using a monomolecular film of phosphatidylcholine, a distinct lag was observed that was not seen using the phospholipase A_2 from bee venom (Verger *et al.*, 1973) (Fig. 9-2). The lag period is thought to reflect the time required for physical interaction of the enzyme with the substrate. The now classic model developed to explain these kinetics is given in Fig. 9-3.

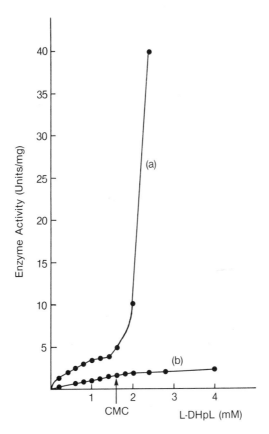

Figure 9-1. Michaelis curves showing the activity of phospholipase A (curve a) and its zymogen (curve b) as a function of diheptanoyllecithin concentration. Assay conditions: 0.5 mM NaAc, 0.1 M NaCl, pH 6.0, 40°C. CMC stands for critical micelle concentration range. (From Pieterson *et al.*, 1974.)

The initial rate of the reversible absorption or penetration is given by

$$E \underset{k_d}{\overset{k_p}{\rightleftharpoons}} E^*$$

where k_p is the rate of penetration and k_d is the rate of desorption; E^* is the catalytically active enzyme. Evidence exists to promote the idea that a conformational change occurs when E^* is formed that favors active-site alignment with substrate. The proenzyme, however, was thought not to bind organized substrate aggregates and therefore did not undergo this change to the activated E^*. Recent evidence, however, has led to a better understanding of the binding properties of the proenzyme (Volwerk *et al.*, 1984). Little is published on the rates of the individual steps in catalysis other than that the sequence is not ordered, unlike the enzyme from snake venom. This and the proposed mechanism of catalysis are covered in Chapter 10.

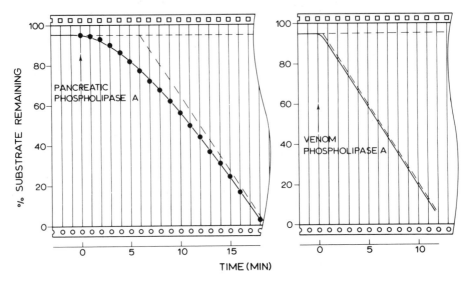

Figure 9-2. Kinetics of the hydrolysis of a 3-*sn*-dinonanoyllecithin film upon injection of phospholipase A_2 from different sources. Surface pressure, 5 dynes/cm; Tris buffer, 10 mM, pH 8.0; NaCl, 0.1 M; $CaCl_2$, 20 mM. *Left:* Injection of 5.9 μg of pancreatic phospholipase A_2. *Right:* Injection of 14 μg of bee venom phospholipase A_2. The continuous curves are the tracings from the barostat recorder. The points on the continuous curve as well as the dashed lines are computed values. (From Verger *et al.*, 1973.)

Upon activation, the Trp at position 3 undergoes a change in its fluorescence that indicates a shift from a polar to apolar environment. Likewise, the α amino group, thought to stabilize the activated form of the enzyme, becomes shielded. It is proposed that this conformational change exposes a hydrophobic binding sequence in the N-terminal region that is required for activity on aggregated substrates. [See Slotboom *et al.* (1982) for detailed description.]

The peptide sequence of 37 phospholipases or phospholipaselike proteins has been determined, at least partially, including five pancreatic enzymes. From these data, plus x-ray diffraction patterns, the two-dimensional picture of the bovine proenzyme shown in Fig. 9-4 has been proposed (Slotboom *et*

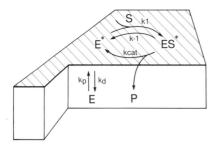

Figure 9-3. Model for the action of a soluble enzyme at an interface. (From Pieterson *et al.*, 1974.)

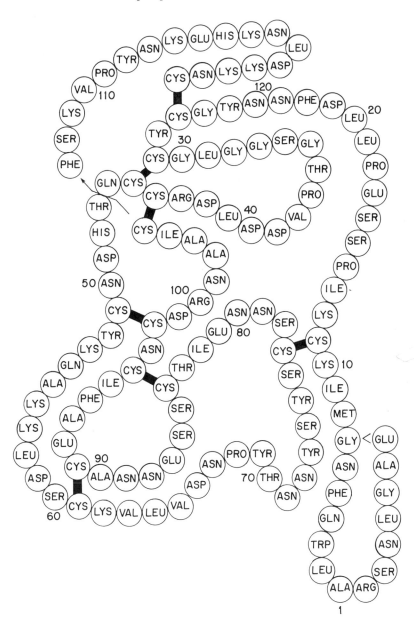

Figure 9-4. Amino acid sequence of bovine prophospholipase A₂ and the connection of the disulfide bridges. (From Slotboom *et al.*, 1982.)

al., 1982). Like most, the pancreatic phospholipase A₂ has seven disulfide bridges that provide the known stability of these enzymes.

The amino acid sequence of dog and rat pancreatic phospholipases A₂ have been deduced from the cloned cDNA that encodes for these enzymes (Ohara *et al.*, 1986). Figure 9-5 shows the comparison of their deduced amino

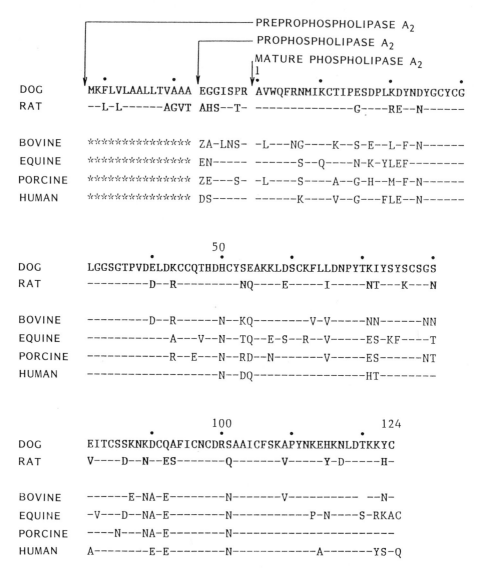

Figure 9-5. Comparison of the amino acid sequences of pancreatic phospholipases A_2. The amino acid sequences of pancreatic phospholipases A_2 from cow, horse, pig, and human are aligned to demonstrate the homology among these enzymes. Amino acid residues are indicated as one-letter abbreviations. (From Ohara *et al.*, 1986.)

acid sequences with those determined by direct analysis. It was possible from this analysis to suggest the sequences of the signal peptides for these two phospholipases A_2, as indicated in Fig. 9-5.

9.2.3. General Characteristics of Pancreatic Phospholipases A_2

First, it should be noted that sulfhydryl agents do not inactivate the enzyme since all 14 cysteines are cross-linked. Likewise, serine does not appear to be involved since diisopropylfluorophosphate used to inactivate proteases during the purification did not inactivate the phospholipase A_2 in either the zymogen or activated form (deHaas *et al.*, 1968). Under the appropriate conditions, the pig pancreatic phospholipase A_2 can be completely reduced with 2-mercaptoethanol. Nearly full activity and all native characteristics were restored upon reoxidation in the presence of O_2 and cysteine (van Scharrenburg *et al.*, 1980). Since the zymogen form can be treated in the same manner, it would appear that the additional peptide of the proenzyme did not place structural restraints on the protein that could alter the favorable matching of all seven disulfide bonds.

Some comparative data on the catalytic properties of the phospholipases A_2 from the pancreas of three species have been published. Despite the remarkable similarities in the structures of the pancreatic phospholipases, their catalytic activities were found to be different (Evenberg *et al.*, 1977*b*) (Table 9-1).

Another striking difference between the ruminant (including sheep) and the pig is the pH for optimal hydrolysis, 8.0 versus 6.0, respectively. Also, the ruminant enzyme has low capacity to penetrate organized interfaces such as monolayers and long-chain lipids such as egg-yolk phosphatidylcholine (Evenberg *et al.*, 1977*b*). Since there is a high degree of homology in the N-terminal region of these phospholipases, it was suggested that the three-dimensional structure rather than the primary structure of the N-terminal region is responsible for the differences in their catalytic properties.

Interestingly, two forms of enzyme were isolated from the horse pancreas but only one form, the more acidic, was recovered from horse pancreatic juice. This presumably was the result of proteolytic cleavage. It would appear therefore that processing of the enzyme altered its activity somewhat even though the enzyme is well designed to live in the hostile environment of proteases. Similar modification by proteolysis is thought to occur with snake venom phospholipases.

9.3. Snake Venom Phospholipases A₂

It was recognized as early as 1902 that cobra venom had hemolytic activity directed toward the lipid of the cell membrane (Kyes, 1902, 1907). The lytic

Table 9-1. Comparative Data on the Catalytic Properties of the Phospholipases A_2 from the Pancreas of Three Species[a]

| Species | V (μequiv/min/mg) | | K_m (mM) |
	DOPC[b]	EPC[b]	DOPC
Horse	12,000	2,200	5
Pig	4,000	1,400	1
Cow	2,500	60	1

[a] From Evenberg *et al.* (1977*b*).
[b] Abbreviations: DOPC, dioctanoyl phosphatidylcholine; EPC, egg-yolk emulsion.

component was shown to be a partial hydrolysis product of the phosphatidyl-choline, differing from the parent compound in fatty acid composition (Wills-tatter and Ludecke, 1904). In 1914, this compound produced by the action of the venom was isolated and termed *lysocithin* (Delezenne and Fourneau, 1914). To determine which *lecithins* in egg yolk were degraded, Levene and coworkers chemically isolated the lysophosphatidylcholine and lysophospha-tidylethanolamine from 48 dozen eggs that had been degraded by 250 mg of cobra venom (Levene and Rolf, 1923; Levene *et al.*, 1923–1924). The products were isolated by precipitation techniques and the methyl esters of the fatty acids characterized by molecular weight and melting and boiling points. It was thus established that cobra venom could remove a fatty acid from more than a single phospholipid. The lack of cobras in Canada led King and Nolan (1933) to extend the number of species by studying venom of another snake, the fer-de-lance (*Bothrops atrox*). The activity in three venoms on other sources of phospholipids demonstrated that the relative capacity to degrade lipids was dependent on the source of venom, which suggested that while all were phos-pholipases some differences in the enzyme existed (Chargaff and Cohen, 1939). Studies to understand better the interaction of the enzyme with a defined lipid interface were carried out by Hughes (1935) who introduced the use of monomolecular films to lipid enzymology. In this case, changes in the surface potential were measured to monitor hydrolysis of the monolayer.

The innovative use of moist ether (Hanahan, 1952) and short-chain syn-thetic phosphatidylcholines (Roholt and Schlamowitz, 1961) permitted better enzyme assays that led to purification and characterization of these phospho-lipases. The latter group using the crude venom of *Crotalus durissus terrificus* found that the optimal pH for hydrolysis was 8.0 and that Ca^{2+} but not other metal ions was required for activity. Also important was the observation that a pH stat could be used to follow the release of protons during hydrolysis. Tween 20 stimulated the hydrolysis of long-chain but not short-chain phos-phatidylcholines. This observation was capitalized on by Dennis (1973*a*) who has developed a kinetic analysis of the *Naja naja naja* enzyme using mixed micelles of Triton X-100 and long-chain phosphatidylcholines.

The phospholipase in venoms of many species of snake has now been purified and very thoroughly studied. As described earlier, there are a number of similarities between the pancreatic and venom enzymes. Similarities within this group of phospholipases would be expected from the embryological and evolutionary points of view. As pointed out by Heinrikson *et al.* (1977), pan-creatic, salivary, and venom glands contain acinar cells and have a common embryological derivation. A difference in developmental stage between the elapids and the vipers and crotalids exists, the former being more advanced and therefore closer to the pancreatic system. In addition to their digestive function, these enzymes have varying degrees of toxicity that will be covered in Chapter 11. In general, those that are acidic or neutral have limited toxicity when compared with the basic enzymes that are highly toxic. The enzyme from a single source is often heterogeneous, originally assumed to be the result of proteolysis that could occur in the gland. Durkin *et al.* (1981) dem-onstrated the electrophoretic heterogeneity of some 162 variants of enzyme from 53 species of snake. While they concluded that the patterns were too complex to allow classification at the family or subfamily level, electrophoretic analysis might be useful at the subspecies level. They recognized that multiple causes of enzyme heterogeneity exist (genetic, polymerization, proteolysis, etc.), and therefore a more detailed understanding of the enzymes is required before their further use in taxonomy. For more detailed descriptions of venoms, the paper of Heinrikson *et al.* (1977) and reviews on venoms are recommended (Tu, 1977; Habermann and Breithaupt, 1978; Glenn and Straight, 1982).

9.3.1. Purification of Snake Venom Phospholipases A$_2$

The purification of venom phospholipases was initially attempted by Slotta in the late 1930s (Slotta and Fraenkel-Conrat, 1938). It was not until nearly three decades later, however, that the venom phospholipases were purified on a large-scale basis, which permitted detailed kinetic and structural analysis. It is beyond the scope of this volume to describe the purification of phospholipases from the numerous venoms studied thus far. The references for the amino acid sequence data cover the purification of most studied. A few of the more commonly used systems will be described briefly. It is of interest to note that most need only 15- to 20-fold purification of the enzyme to achieve homogeneity. A two-step procedure involving gel filtration and ion-exchange chromatography has been devised that yields a family of phospholipases that differ in the charge characteristics (Maraganore *et al.*, 1984). In general, it is found that anionic phospholipases tend to be dimeric, whereas more positively charged enzymes are monomeric.

9.3.1.1. Crotalus adamanteus

Two forms of enzyme have been isolated from this species. Initially, these were purified to roughly 80% purity and found to have similar molecular weights, 30,000 daltons (now known to be the dimer), but to differ in their isoelectric points (Saito and Hanahan, 1962). Homogeneity was achieved in a 21-fold purification and with an overall yield of 35% (Wells and Hanahan, 1969). Using the moist ether assay system, a final specific activity of 3200 units/mg was observed. Two recent procedures have been developed that further simplify the isolation of these enzymes. Wells (1975) precipitated the enzymes from 50% isopropanol with neodynium chloride to increase the yield twofold over the previous procedure. Rock and Snyder (1975) used a phospholipid analog affinity column based on the Ca^{2+} requirement of the enzyme to bind lipid. The enzyme, when bound in the presence of Ca^{2+}, could be released from the affinity columns by the addition of EDTA. This procedure has been used successfully for a number of phospholipid and Ca^{2+}-requiring enzymes.

The spectral properties of this enzyme have been studied in detail by Wells in a series of papers with special emphasis on Ca^{2+} binding and chemical modification to be covered in Section 9.5.

9.3.1.2. Crotalus atrox

The purification of this enzyme is based on the procedure worked out for the *C. adamanteus* enzyme (Hachimori *et al.*, 1977). Although it had somewhat lower specific activity than the *C. adamanteus* enzyme, 1500 units/mg, it has the same monomeric molecular weight, 14,700 daltons, and exists as a dimer even at concentrations as low as 2×10^{-9} M.

9.3.1.3. Crotalus scutulatus scutulatus

Although this enzyme is from a member of the rattlesnake family, it has characteristics that are quite different in that it is composed of two unequal subunits with a combined molecular weight of 24,310 daltons (Cate and Bieber, 1978). A toxin assay system was used during the purification, and although the usual degree of purification data was not given, the procedure was relatively simple and gave a homogeneous preparation with a pI of 5.5. The two subunits were separated by DEAE chromatography in 6 M urea. The acidic subunit has a molecular weight of 9593 daltons and a pI of 3.6, while the cationic subunit is 14,673 daltons and has a pI of 9.6. The amino acid composition of the cationic subunit was that predicted for this class of enzymes. Perhaps the most interesting feature of this system was the report that only the basic protein had the phospholipase activity and that its combination with the acidic component inhibited phospholipase activity but potentiated its toxicity. This enzyme is similar to other known toxins, β-bungarotoxin, notexin, taipoxin, and crotoxin. The regulation of toxins and associated phospholipases by regulatory proteins has been studied in detail over the past few years (Chapter 11).

9.3.1.4. Naja naja naja

The enzyme from *N. naja naja* purified 14-fold to homogeneity using a procedure that included $HClO_4$ precipitation (Deems and Dennis, 1975). This procedure, based on the techniques of Braganca *et al.* (1969), gave a 30% yield. More recently, Hazlett and Dennis (1985) developed a new procedure using Affi-Gel Blue and two ion-exchange columns to achieve purification without the harsh $HClO_4$ treatment. This procedure also was more rapid and yielded a more highly purified preparation. Interestingly, they showed that the enzyme heterogeneity found by isoelectric focusing existed for the venom from a single snake. This result ruled out the origin of heterogeneity arising from the pooling of venoms from several snakes.

The purified enzyme had a specific activity apparently somewhat lower than that of the *C. adamanteus*, 537 units/mg, using mixed micelles as substrate. Because of the differences in the assay systems employed, comparison of the activity of the *N. naja naja* enzyme relative to that of the enzyme from *C. adamanteus* is not possible. The *N. naja naja* is somewhat smaller than most phospholipases A_2 of this class, 11,000 daltons. There is some degree of hetereogeneity of the preparation as shown by isoelectric focusing; the major form had a pI of 4.95 while the minor form had a pI of 4.6–4.7. Like the *C. atrox* enzyme, it formed dimers at a concentration of 4.5×10^{-6} M (dimer formation of *C. atrox* phospholipase A₂ was at a concentration of 2×10^{-9} M).

9.3.1.5. Agkistrodon halys blomhoffii

The venom of this snake contains three phospholipases that differ in their isoelectric points, pI 4.9, pI 6.9, and pI 8.7 (Kawauchi *et al.,* 1971; Hanahan *et al.,* 1980). These were purified 8-, 21-, and 3-fold and had a final specific activity of 43, 105, and 15 units/mg, respectively, using the moist ether assay system. The combined recovery of the three amounts to about 50% of the original activity. The molecular weight is about 13,500 daltons for each. Somewhat surprisingly, the acidic enzyme retained some activity in the presence of EDTA, unlike the other two from this source and from other species. Perhaps related to this observation was the finding that the acidic enzyme was stable at 100°C without added Ca^{2+} while the others were not. The three differ markedly in their ability to hemolyze erythrocytes. The basic enzyme was most active while the acidic was only weakly active.

9.3.1.6. Bungarus multicinctus and Bungarus caeruleus

Five distinct β-bungarotoxins have now been purified and their amino acid sequence determined (Abe *et al.,* 1977; Kondo *et al.,* 1978, 1982*a*, 1982*b*). All five were shown to have phospholipase A₂ activity and to be neurotoxic in mice. The molecular weight of these was about 21,000 daltons and had a pI of 9. All these were composed of two subunits; A had a molecular weight of 14,000 daltons and B was about 9000 daltons. The separation of these two required reduction of intrapeptide disulfide bonds. In that regard, the bungarotoxin is similar to notoxin and taipoxin but different from crotoxin since in that case no disulfide bridge exits. The A chain was shown to have phospholipase activity. Since the A chain is a phospholipase A₂ and the toxic factor, it would appear that there is a direct correlation between toxicity and phospholipase activity. The function of the B chain is not known and it is argued that it is not a proteinase inhibitor by comparison with known inhibitors of these enzymes. The possibility that it is a phospholipase inhibitor also appeared unlikely.

9.3.1.7. Vipera berus orientale

A basic phospholipase A₂, pI = 9.2, was purified from this venom that had potent anticoagulant activity (Boffa *et al.,* 1976). It had a molecular weight of 13,400 daltons and was reported to have six disulfide bonds. Unlike many other phospholipases, it formed trimers. The anticoagulation activity will be covered in Chapter 11.

9.3.1.8. Bitis gabonica

The purification of this enzyme was achieved in two steps. The unusual feature of this enzyme is its reported Ca^{2+}-dependent dimerization (Botes and Viljoen, 1974). The molecular weight of the monomeric form is 13,347 daltons, based on amino acid analysis. It was subsequently reported that a site other than histidine is responsible for binding Ca^{2+} with a pK = 6.0 (Viljoen and Botes, 1979).

Considerable effort has been expended by the group in Pretoria to purify and sequence a number of venom enzymes (Joubert and Haylett, 1981). This reference cites a number of their studies as well as the other purifications not cited in Slotboom *et al.* (1982).

9.3.1.9. Venom Phospholipases with Lys-49

Recently, a new class of phospholipases A_2 has been isolated and characterized from the venoms of *A. piscivorus piscivorus* (App) and *B. atrox* (Ba) that have Lys-49 rather than Asp-49 (Maraganore *et al.*, 1984). Until this report, the Asp-49 was thought to be an invariant essential residue at the active site. These enzymes are strongly cationic and have similar kinetic properties (Table 9-2). As shown in Table 9-2, the rate constant k_{cat}/K_m and the kinetic dissociation constant for Ca^{2+}, $K_m^{Ca^{2+}}$, were found to be about equal regardless of the residue present in position 49. The major difference to be noted is the apparent reverse order of binding. In the absence of substrate, the Lys-49 enzyme did not bind Ca^{2+} (K_d^{Ca} value) while the Asp-49 could not bind substrate in the absence of Ca^{2+} (K_s value). Other alterations in the secondary structure were shown to occur to compensate for conformational changes introduced by the substitution of Lys at position 49.

9.3.1.10. Bee Venom Phospholipase A_2

The phospholipase A_2 of bee (*Apis mellifera*) venom has been purified in a five-step procedure (Shipolini *et al.*, 1971). This phospholipase A_2 comprises about 12% of the dry weight of the venom and is the predominant enzyme present (Shipolini *et al.*, 1974b). Unlike the snake venom phospholipases discussed thus far, it is a glycoprotein having 14 carbohydrate moieties: fructose, galactose, mannose, and glucosamine (1 : 1 : 8 : 4). This enzyme shares with many other phospholipases a high pI, 10.5, and requires another venom

Table 9-2. Kinetic Properties of Phospholipases with Aspartate or Lysine at Position 49[a]

Enzyme[b]	pI	$k_{cat}/K_m \times 10^{-6}$	$K_m^{Ca} \times 10^3$	$K_d^{Ca} \times 10^3$	$K_s \times 10^{-5}$
App-Asp-49	9.6	9.0	0.19	1.6	$>10^3$
App-Lys-49	>10.0	3.5	0.15	>100	1
Ba-Lys-49	11.0	2.7	0.50	>100	—

[a] From Maraganore *et al.* (1984).

[b] Abbreviations: App, *A. piscivorus piscivorus;* Asp, asparagine; Lys, lysine; Ba, *Bothrops atrox.*

protein, melittin, for optimal activity. It has been shown that the action of melittin is on the lipid bilayer, making substrate more accessible to the enzyme (Mollay and Kreil, 1974).

9.3.1.11. Phospholipases from Other Species

Reports have appeared that describe phospholipase A activity in the venoms of other organisms. However, little if any work has been done to purify or characterize these enzymes. Indeed, in many instances, insufficient information is available to determine the positional specificity of the phospholipase. Table 9-3 lists some sources and gives the appropriate references. At least in one case, the wasp, activator peptides were shown to be present in the venom. These peptides, termed mastoparans, have some but not all properties comparable to bee venom melittin (Argiolas and Pisano, 1983).

9.4. Structure of the Phospholipases A₂

Over the past decade, much information has appeared on the comparative structures of pancreatic and venom phospholipases arising from sequence analysis, crystallographic studies, and spectroscopic measurements.

9.4.1. Sequence Analysis of Phospholipases A₂

The complete amino acid sequence of nearly 40 phospholipases from snake venom and pancreas has now been completed. Table 9-4 shows the comparative sequences. The venom proteins taipoxin γ chain and B chain of β-bungarotoxin were included by the authors because of their homology to the phospholipases. Notably, all contain seven disulfide bonds with the exception of the phospholipase A₂ from *B. gabonica* and the two nonphospholipases from taipoxin and β-bungarotoxin. However, these disulfide bonds that are essential for the stability are not identical in all cases. As pointed out by Heinrikson, those enzymes in group 1 (pancreas, *Elapidae*, and *Hydrophidae*) have a unique disulfide bond between positions 11 and 69, while the enzymes

Table 9-3. Phospholipases D in Venoms of Other Species

Organism	Reference
Brown recluse spider	Kurpiewski *et al.* (1981)
(*Loxosceles reclusa;* phospholipase D)	
Bulldog ant (*Myrmecia pyriformis*)	Lewis and de la Lande (1967)
Scorpion (*Heterometrus scaber*)	Kurup (1966)
Portuguese man-of-war (*Physalia physalis*)	Stillway and Lane (1971)
Wasp (*Polistes humilis; Ropalidia revolutionaris*)	Owen (1979)
Oriental hornet (*Vespa orientalis*)	Rosenberg *et al.* (1977)
Yellow hornet (*Vespula arenaria*)	Reisman *et al.* (1984)
Yellow jacket (*Vespula maculifrons*)	King *et al.* (1978)

Table 9-4. Comparison of Amino Acid Sequences of Phospholipases from Various Sources[a]

	1		10		20		30		40

```
          1         10        20        30        40
 1  A L W Q F R S M I K C A I P G S H P L M D F N N Y G C Y C G L G G S G T P V D E L D R C C E
 2  * V * * * * * * Q * T * N * K * Y L E * * D * * * * * * * * * * * * * * * * * A * * Q
 3  * * * * * N G * * * K * * S * E * * L * * * * * * * * * * * * * * * * * D * * * * Q
 4  * * * * * * * * * T * * * D * * L * * * * * * * * * * * * * * * * * * * * * * * *
 5  * V * * * * K * * * V * * * * D * F L E Y * * * * * * * * * * * * * * * * * K * * Q
 6  N * V * * S N L * Q * N V K * * R A S Y H Y A D * * * * * A * * * * * * * * * * * * K
 7  N * V * * T Y L * Q * * N S * K R A S Y H Y A D * * * * * A * * * * * * * * * * * * K
 8  N * V * * S Y L * Q * * N T * K R A S Y H Y A D * * * * * A * * * * * * * * * * * * K
 9  N * V * * S Y V * T * N H N R R S S L * Y A D * * * * * A * * * * * * * * * V K K
10  N * V * * S Y L * Q * * N H * K R * T W H Y M D * * * * * A * * * * * * * * * * * * K
11  N * V * * S Y L * Q * * N H * R R * T R H Y M D * * * * * W * * * * * * * * * * * * K
12  N * V * * S N * * Q * * N H * * R * S L A Y A D * * * * S A * * * * * * * * * * * * K
13  N * Y * * K N * * * * T V * S - R S W W H * A * * * * * * R * * * * * * * D * * * * Q
14  N * Y * * K N * * H * T V * N - R * W W H * A * * * * * * R * * K * * * * D * * * * Q
15  N * Y * * K N * * Q * T V * N - R S W W H * A * * * * * * R * * * * * * * D * * * * Q
16  N * Y * * K N * * H * T V * N - R S W W H * A * * * * * * R * * * * * * * D * * * * Q
17  N * Y * * K N * * H * T V * S - R * W W H * A D * * * * * R * * K * * A * D * * * * Q
18  N * Y * * K N * * H * T V * S - R * W W H * A D * * * * * R * * K * * A * D * * * * Q
19  N * Y * * K N * * H * T V * S - R * W W H * A D * * * * * R * * K * * * * D * * * * Q
20  N * Y * * K N * * H * T V * S - R * W W H * A D * * * * * R * * K * * * * D * * * * Q
21  N * Y * * K N * * * * T V * S - R S W L * * A * * * * * * R * * * * * * * D * * * * Q
22  N * Y * * K N * * Q * T V * N - R S W W * * A D * * * * * R * * * * * * * D * * * * Q
23  N * Y * * K N * * Q * T V * N - R S W W N * A D * * * * * R * * * * * * * D * * * * Q
24  N * Y * * K N * * Q * T V * S - R S W W * * A D * * * * * R * * * * * * * D * * * * Q
25  N * L * * G F * R * N R R * R * V W H Y M D * * * * * * K * * * * * * * D * * * * Q
26  N * V * * G K * E * * R N R R * A L * M * * * * * * * * K * * * * * * * D * * * * K
27  D F E * * S N * Q * T * * C(G/S)E C L A Y M D * * * * * * P * * * * * I * D * * * * K
28  N * I N * M E * * R Y T * * C E K T W G E Y A D * * * * * A * * * R * I * A * * * * Y
29  N * I N * M E * * R Y T * * C E K T W G E Y A D * * * * * A * * * R * I * A * * * * Y
30  N * I N * M E * * R Y T * * C E K T W G E Y A T * * * * * A * * * R * I * A * * * * Y
31  N * Y * * K N * V * * G T - - R * W I G Y V * * * * * * A * * * * * * * * * * * * Y
32  N * I * G N * S - * M T * - K S S L A Y A S * * * * * W * K * Q * K * D T * * * * F
33  D * T * * G N * N - K M - * - Q S V F * Y I Y * * * * * W * K * K * I * A T * * * * F
34  S * V * * E T L * M - K V A K R S G * L W Y S A * * * * * W * H * R * Q * A T * * * * F
35  S * V * * E T L * M - K * A * R S G * L W Y S A * * * * * W * H * L * Q * A T * * * * F
36  H * L * * N K * * * - * F Z T R K B A V P F Y A F * * * * * W * Z * R * K B A T B * * * F
37  S                                    I              L
38  H * M * * E T L * M - K * A * R S G V W W Y G S * * * * * A * * Q * R * Q * P S * * * * F
```

```
          50        60        70        80
 1  T H D N C Y R D A - K N L D S C K F L V D N P Y T E S Y S Y S - C S N T E I T - C N S K N N
 2  V * * * * * T Q * - * E * S * * R * * * * * * * * * K F * - * G * * V * - * S D * * *
 3  * * * * * * K Q * - * K * * * * * V * * * * * * * N N * - * * * * * * * * S * E * *
 4  * * * * * * * * * - * * * * * * * * * * * * * * * N * * - * * * * * * * * * * * *
 5  * * * * * D Q * - * K * * * * * * L * * * * * H T * * * * - * G(B A I T - C S S K B K)
 6  I * * * * * G E * E * - M - G * - Y - - - - * K W T L * T * E S * T D * S P - - * D E * T -
 7  I * * * * * G E * E * - M - G * - Y - - - - * K L T M * N * Y - * G T Q S P - - * D D * T -
 8  I * * * * * G E * E * - M - G * - Y - - - - * K L T M * N * Y - * G T Q S P - - * D * T -
 9  I * * D * * G E * E * - Q - G * - Y - - - - * K M L M * D * Y - * G S N G P Y - * R N V K K
10  I * * D * * D E * G * - K - G * - * - - - - * K M S A * D * Y - * G E N G P Y - * R N I K K
11  I * * D * S * * E * - K - G * - S - - - - * K M S A * D * Y - * G E N G P Y - * R N I K K
12  * * * D * * A R * T * - S Y S * - T - - - - * W T L * W Q - * I E K T P - - * D * * T -
13  * * * * * * S * E * - I S G * - R - - - - * F K T * * D - * T K G K L * - * K E G * *
14  I * * K * * D E * E * - I S G * - W - - - - * I K T * T * E S * - Q G T L * - * K D G G -
15  I * * * * * G E * E * - I S G * - W - - - - * I K T * T * E S * - Q G T L * S * G A N * -
16  I * * * * * G E * E * - I S G * - W - - - - * I K T * T * D S * - Q G T L * S * G A A * -
17  V * * * * * G E * E * - * - G * - W - - - - * L T L * K * E - * * Q G K L * - * S G G * *
18  V * * * * * G E * E * - * - G * - W - - - - * L T L * K * E - * * Q G K L * - * S G G * *
19  V * * * * * E K * G * - M - G * - W - - - - * F T L * K * K - * * Q G K L * - * S G G * S
20  V * * * * * E K * G * - M - G * - W - - - - * L T L * K * K - * * Q G K L * - * S G G * S
21  I * * * * * N E * G * - I S G * - W - - - - * F K T * * E - * * Q G T L * - * K G D * *
22  V * * * * * D E * E * - I S R * - W - - - - * F K T * * E - * * Q G T L * - * K N G * *
23  V * * * * * D E * E * - I S G * - W - - - - * F K T * * E - * * Q G T L * - * K G G * *
24  V * * * * * N E * E * - I S G * - W - - - - * F K T * * E C S * Q G T L * - * K G G * *
25  V * * E * * G E * V R - R F G * - A - - - - * W T L * * W K - * Y G K A P * - * * T * T -
26  V * * E * * A E * E * - H - G * - Y - - - - * S L T T * T W E - * R Q V G P Y - * * * T -
27  * * * E * * A E * G * - * S A * * S V L S E * N N D T * * E - * N E G Q L * - - * D D N D
28  V * * * * * G * * E * - K H K * - - - - - * * K * S Q * * K - L T K R T * I - * Y G A A G
29  V * * * * * G * * E * - K H K * - - - - - * * K * S Q * * K - L T K R T * I - * Y G A A G
30  V * * * * G * D A A - I R D * - - - - - * * K * S Q * * K - L T K R T * I - * Y G A A G
31  V * * * * * G E * E * - I P G * - - - - - - * K * K T * * T - * T K P N L * - * T D A A G
32  V * * C * * G K * D * - - - - * - - - - - S * K M I L * * * K - F H * G N * V - * G D * * -
33  V * * C * * G K M G T - - - - Y - - - - - - D T K W T * * N * E - I Q * G G * D - * D E D P -
34  V * * C * * G K * T N - - - - * - - - - - - * K * V * * T * S - E E * G * * V - * G G D D -
35  V * * C * * G K * T D - - - - * - - - - - - * K * V * * T * S - E E * G * * I - * G G D D -
36  V * B C * * G K L A * - - - - * - - - - D T K * W B I * R * S - L K S G Y * * - * G K G T -
37
38  V * * C * * G K V T G - - - - * - - - - - * T K D * F * T * T - E E E G A * S - * G G N(D -
```

(continued)

Table 9-4. (Continued)

```
      90              100             110             120             130

 1  A C E A F I C N C D R N A A I C F S K A - - P Y N K E H K N L D T K - K Y C
 2  * * * * * * * * * * * * * * * * * * * * - - * * * P * N * * * * S * - R K * A
 3  * * * * * * * * * * * * * * * * * V - - * * * * * * * * * * - * - * N *
 4  * * * * * * * * * * * * * * * * * * * - - * * * * * * * * * * * * - * * *
 5  E * * * * * * * * * * * * * * * * * * - - * * * * A * * * * * * - Y S * Q
 6  G * Q G * V * A * * L E * * K * * A R S - - * * * N K N Y * I * * S - * R * K
 7  G * Q R Y V * A * * L E * * K * * A R S - - * * * N K N Y * I * * S - * R * K
 8  G * Q R Y V * A * * L E * * K * * A R S - - * * * N K N Y * I * * S - * R * K
 9  K * N R K V * D * * V A * * E * * A R N - - A * * N A N Y * I * * - * R * K
10  K * L R * V * D * * V E * * F * * A * - - * * * N A N W * I * * * * R * Q
11  K * L R * V * D * * V E * * F * * A * - - * * * N A N W * I * * * * R * Q
12  G * Q R * V * D * * A T * * K * * A * - - * * * * N Y * I * P * - * R * Q
13  E * A * * V * K * * L * * * * A G * - - H * * D N N N Y I * L A - R H * Q
14  K * A * S V * D * * * V * N * A R * - - T * * D K N Y * I * F N - A R * Q
15  K * A * S V * D * * * V * N * A R * - - T * * D K N Y * I * F N - A R * Q
16  N * A * S V * D * * * V * N * A R * - - * * I D K N Y * I * F N - A R * Q
17  K * * * A V * * * * L V * N * A G * - - * * I D A N Y * V N L * - E R * Q
18  K * A * A V * * * * L V * N * A G * - - R * I D A N Y * I N L * - E R * Q
19  K * G * A V * * * * L V * N * A G * - - R * I D A N Y * I N F * - * R * Q
20  K * G * A V * * * * L V * N * A G * - - R * I D A N Y * I N F * - * R * Q
21  S * A * S V * D * * * L * * * A G * - - * * * N D N Y * I N L * - A R * Q
22  * * A * A V * D * * * L * * * A G * - - * * * N N N Y * I * L * - A R * Q
23  * * A * A V * D * * * L * * * A G * - - * * * N N N Y * I * L * - A R * Q
24  - * A * A V * D * * * L * * * G G * - - * * * D N N N Y I * L * - A R * Q
25  R * Q R * V * R * * A K * * E * * A R S - - * * Q N S N W * I N * - A R * R
26  Q * * V * V * A * * F A * * K * * A Q E - - D * P A * S * I N * G - E R * K
27  E * K * * * * * * * * T * V T * A G * - - * * D D L Y * I G M I - E - * H K
28  T * G R I V * D * * * T * L * G Q S - - D * I E G * * * I * A - R F * Q
29  T * A R I V * D * * * T * L * G Q S - - D * I E R * * * I * * K - R H * R
30  T * A R V V * D * * * T * L * G Q S - - D * I E G * * * I * A - R F * Q
31  T * A R I V * D * * * T * * * A A A - - * * * I N N F M I S S - T H * Q
32  * K K K V * E * * * V * * * * A A S K H S * * - K N L - W R Y P S S K * T G T A E K C
33  Q - K K E L * E * * * V * * * * A N N R N T * * - S N Y - F G H S S S K * T G T - E Q C
34  P * G T Q * * E * * K A * * * * R D N I P S * D - N K Y - W L F P P * D * R Q E P E P C
35  P * G T Q * * E * * K A * * * * R D N I P S * D - N K Y - W L F P P * D * R E E P E P C
36  W * Z Z Z * * Z * B * V * * Z * L R R B L S T * K - B Z Y - M F Y P D S R * R G P S E T C
37                                                          G
38  P) * L K E V * E * * L A * * * * * R D N L N T * D S K K Y - W M F P A * N * L E S E E P C
```

a From Slotboom *et al.* (1982). Sequences completed are: (1) pig; (2) horse; (3) ox; (4) iso-pig; (5) human; (6–8) *Laticauda semifasciata*, fractions I, III, and IV; (9) *Enhydrina schistosa;* (10) *Notechis scutatus*, notexin; (11) *N. scutatus*, fraction II-5; (12) *N. scutatus*, fraction II-1; (13) *Hemachatus haemachatus;* (14–16) *Naja melanoleuca*, fractions DE-I, DE-II, and DE-III; (17–19) *N. mosambica*, fractions CM-I, CM-II, and CM-III; (20) *N. nigricollis*, basic; (21) *N. n. oxiana;* (22,23) *N. n. kaouthia;* (24) *N. n. atra;* (25–27) *Oxyguranus scutallatus*, α and β chain and the γ chain starting at residue 9; (28–30) *Bungarus multicinctus*, β-bungarotoxin, A1, A2, and A3 chains; (31) *B. multicinctus*, phospholipase; (32) *Bitis caudalis;* (33) *B. gabonica;* (34) *Crotalus adamanteus*, fraction α; (35) *C. atrox;* (36) *C. durissus terrificus;* (37) *C. durissus terrificus*, microheterogeneity; (38) *Trimeresurus okinavensis*. Gaps (—) have been introduced to obtain alignments of half-cysteines and maximal homology. Residues identical to the corresponding residue in porcine pancreatic phospholipase A are indicated with an asterisk. The numbering has been based on horse pancreas phospholipase A; note that gaps introduced in the pancreatic model do not affect the numbering. Note also that the numbers used here do not necessarily correspond to the numbers used in the original publications. The IUPAC one-letter notation for amino acids has been used.

in group 2 (*Viperidae* and *Crotalidae*) have a disulfide bond between the half-cystine at the C-terminus of the polypeptide that is extended by six residues with the half-cystine at position 50 [Maraganore *et al.* (1984); see this reference for explanation of amino acid residue numbering].

Until the discovery of the venom phospholipases that contain Lys-49, it had been shown that minimally 32 amino acids were conserved and that 29 amino acids were substituted by amino acids comparable in size, charge, and hydrophobicity. It has generally been accepted that the major reasons for this high degree of conservation in secondary structure is twofold (Slotboom *et al.*, 1982): (1) to provide the essential arrangements of the sites for substrate and Ca²⁺ binding and (2) to ensure essential structural functions such as that provided by the disulfide bridges. The sequence of Lys-49 phospholipase A₂ has now been completed (Maraganore *et al.*, 1984) (Fig. 9-6). Based on the positioning of its disulfide bridges and substitutions at sites thought to be

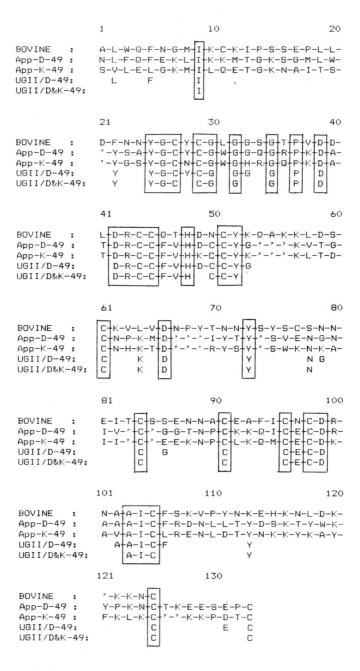

Figure 9-6. Comparison of the amino acid sequences of bovine pancreatic **phos**pholipase A_2, App-D-49, App-K-49, Unanimous Group II/Asp-49, and Unanimous Group II/**Asp**- and Lys-49. The sequences are aligned to maximize homology. Boxes are drawn about amino acid residues that are conserved in the five sequences. (From Maraganore *et al.*, 1984.)

invariant, it does fit into group II. For example, concomitant with the introduction of Lys-49 are substitution of the highly conserved Glu-Phe at positions 4 and 5 by Glu-Leu and changes in the *invariant* region between residues 25 and 33 (Maraganore *et al.*, 1984). These alterations in the sequence produce changes in the Ca^{2+}-*binding loop* [Verheij *et al.* (1981*a*); covered in Chapter 10] and apparently cause a complete shift in the sequence of Ca^{2+} and substrate binding (see Section 9.3.1.9). While there is considerable homology in the sequence between the phospholipases, some key differences are to be

			1	2	3	4	5	6	7	8	9	10	11	12	13	14	15	16	17	18	19	20	21	22	23	24	25	26	27	28	29
1	Pig		0	28	19	4	68	68	67	71	65	65	58	60	69	68	68	64	66	66	66	59	59	60	63	65	70	83	91	85	83
2	Horse		28	0	32	27	71	71	71	74	69	70	63	65	68	67	68	61	63	63	63	62	57	58	62	66	74	86	96	89	88
3	Ox		19	32	0	17	72	73	73	73	71	71	62	59	68	67	67	61	63	63	63	56	58	58	61	65	72	84	92	84	83
4	Iso-pig		4	27	17	0	68	69	68	71	66	66	58	59	68	67	67	63	65	65	65	57	58	59	62	63	69	83	91	84	82
5	L. semif.	I	68	71	72	68	0	18	18	41	42	43	39	57	51	48	50	53	53	54	54	55	53	52	57	69	86	73	79	81	81
6	ibid.	III	68	71	73	69	18	0	3	34	36	38	40	60	54	52	53	54	54	55	54	56	54	53	59	70	65	72	81	82	81
7	ibid.	IV	67	71	73	68	18	3	0	33	35	37	40	60	54	52	53	54	54	55	54	56	54	53	59	58	65	71	82	82	81
8	E. schist.		71	74	73	71	41	34	33	0	26	26	46	62	56	54	56	58	57	58	58	54	52	54	59	72	66	74	83	80	80
9	Notexin		65	69	71	66	42	36	35	26	0	7	41	60	55	56	58	56	56	54	54	55	52	52	58	67	66	79	86	85	84
10	N. scut.	II-5	65	70	71	66	43	38	37	26	7	0	42	58	57	57	59	57	57	56	56	57	54	54	59	68	65	78	86	83	82
11	ibid.	II-1	58	63	62	58	39	40	40	46	41	42	0	58	56	56	57	57	57	55	55	53	51	51	56	62	66	74	85	86	85
12	H. haem.		60	65	59	59	57	60	60	62	60	58	58	0	33	32	32	37	35	37	38	25	24	22	22	60	62	74	84	87	88
13	N. mel.	I	69	68	68	68	51	54	54	56	55	57	56	33	0	10	13	32	30	32	32	27	25	25	29	63	68	74	82	86	87
14	ibid.	II	68	67	67	67	48	52	52	54	56	57	56	32	10	0	6	34	32	35	35	25	24	24	27	62	66	73	82	85	86
15	ibid.	III	68	68	67	67	50	53	53	56	58	59	57	31	13	6	0	33	32	34	34	26	26	26	29	64	64	74	85	85	86
16	N.m mos.	I	64	61	61	63	53	54	54	58	56	57	57	37	32	34	33	0	3	13	12	31	29	26	30	62	67	74	82	85	86
17	ibid.	II	66	63	63	65	53	54	54	57	56	57	57	35	30	32	32	3	0	11	10	30	28	25	29	62	66	74	82	85	86
18	ibid.	III	66	63	63	65	54	55	55	58	54	56	55	37	32	35	34	13	11	0	1	32	32	29	33	64	66	72	81	81	82
19	N. nigri.		66	63	63	65	54	54	54	58	54	56	55	38	32	35	34	12	10	1	0	33	33	30	34	64	66	72	81	81	82
20	N.n. oxian.		59	62	56	57	55	56	56	54	55	57	53	25	27	25	26	31	20	32	33	0	14	12	16	55	65	73	80	84	85
21	N.n.kaouth.	I	59	57	58	58	53	54	54	52	52	54	51	24	25	24	26	29	28	32	33	14	0	4	11	57	62	72	81	85	86
22	ibid.	III	60	58	58	59	52	53	53	54	52	54	51	22	25	24	26	26	25	29	30	12	4	0	9	57	61	72	82	84	85
23	N.n. atra		63	62	61	62	57	59	59	59	58	59	56	22	29	27	29	30	29	33	34	16	11	9	0	61	62	75	84	86	87
24	Taip.		65	66	65	63	69	70	68	72	67	68	62	60	63	62	64	62	62	64	64	55	57	57	61	0	75	82	86	98	89
25	β-bung.		70	74	72	69	66	65	65	66	66	65	66	62	68	66	64	67	66	66	66	65	62	61	62	75	0	76	87	82	81
26	B. Caud.		83	86	84	83	73	72	71	74	79	78	74	74	74	73	74	74	74	72	72	73	72	72	75	82	76	0	50	61	61
27	B. Gabon.		91	96	92	91	79	81	82	83	86	86	85	84	82	82	85	82	82	81	81	80	81	82	84	86	87	50	0	65	64
28	C. Adam.		85	89	84	84	81	82	82	80	85	83	86	87	86	87	87	87	87	81	81	84	85	84	86	90	82	61	65	0	6
29	C. Atrox.		83	88	83	82	81	81	81	80	84	82	85	88	87	86	86	86	86	82	82	85	86	85	87	89	81	61	64	6	0

Figure 9-7. Sequence difference matrix for phospholipases from various sources. Sequences were aligned as shown in Table 9-4 and the comparison is based on a total number of residues (including deletions) of 138. The values shown are the number of positions (including deletions) where a change has occurred; a value of 69 in the figure therefore indicates a 50% homology. For the full names of the phospholipase sources see Table 9-4. (From Verheij *et al.*, 1981*b*.)

noted. In fact, one would expect that there would be some structural differences since a number of venom enzymes serve more than a digestive function, in particular, neurotoxicity and hemolysis. Therefore, structural features that regulate the catalytic function of these enzymes are under intense investigation in a number of laboratories (see Chapter 11).

A noticeable variation has been reported for the number of disulfide bridges present in phospholipases A_2. While most venom enzymes having seven disulfides have definitive assignment of the pairings, some are known to have fewer than seven. For example, the β-bungarotoxin A chain, taipoxin γ chain, and *B. gabonica* have six and the acidic enzyme from *A. halys blomhoffii* has approximately three although this has been challenged (Heinrikson *et al.*, 1977). Likewise, there is considerable variation in the length of the protein molecules ranging from the smaller *N. naja naja* enzyme that has approximately 102 residues and 11 half-cysteines to the *Crotalus* enzymes with 130. Figure 9-7, a sequence difference matrix taken from Verheij *et al.* (1981*b*), summarizes the relation between the enzymes of various groups of species. As would be expected, those most closely related have the highest degree of homology.

The organization of the intramolecular disulfide bond for β-bungarotoxin has been proposed but is yet unverified (Kondo *et al.*, 1982*b*) (Fig. 9-8).

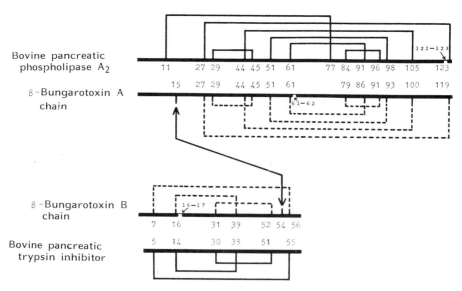

Figure 9-8. Tentative intrachain and interchain disulfide bonds in β-toxin. A deletion of five residues is located between residues 61 and 62 in the amino acid sequence of the A chain of β-toxin from comparison with that of bovine pancreatic phospholipase A_2. A one-residue deletion is located between residues 16 and 17 in the amino acid sequence of the B chain of β-toxin from comparison with that of bovine pancreatic trypsin inhibitor in which two deletions are located before the N-terminus. Broken lines show the tentative intrachain disulfide bonds in the respective chains and an arrowed line shows the tentative interchain disulfide bond between the A and B chains of β-toxin. (From Kondo *et al.*, 1982*b*.)

Table 9-5. Comparison of the Amino Acid Sequence of Bee Venom with Proposed Ancestral Enzyme[a]

```
                          10          20          30          40        50
ANCESTOR   N L W Q F R K M I Q C K M T G S N P I L E Y N D Y G C Y C G P G G S A - P P K D - - - - - A P D R C C F V H D M C
           D V L                   G N             S A                       R           L E                 D L - - - S
                 T I P         N S A F               I Y                       L S         L   Q               T - - - M
                               K R                   A Y - - -                 R                               S
                                                     L D S - - -               W                              M

                                          10          20          30
BEE VENOM  K V Y Q W F D L - - - - - - - - - - I Y P G T L - - W C G H G N K S S G P N E L G R F K H T D A C C R T H D M C
           II8


                       50          60              70          80              90
ANCESTOR   - - - - - - Y G E A D K L T G - - C K F L V D N P N T D T Y S Y E S C N D K D I T C N E D N N T C - - Q T
                       Q V E           S - - - - - - - - - - - R             N R       W K N T E         S   K E
                           K                                     K         Y K I -       I   S R A         E K A   E A
                                                                   Y               E M       G

                       40          50          60              70          80              90
BEE VENOM  P N V M S A G E S K H G L T D T A S R L S C N N - N D F Y K N S A D T I S S Y F V G K M Y F N L I N T K C Y K L E


                       100             110          120          130
ANCESTOR   H - I C D G - D R T A A I C F R K T P N N N N N Y N S N S K T N P C K E K S E T C
           Q L D         E         E L D T Y K K K K W D         - I K E A         G M L P
           E   V Y       A - - - - - E N Y   K A D   K W D - - - M A Q     L   - - L - P
             F           - - - - - - - - - - - - - - - - - - - - - - - - - G     F   A
                                                                                 A

                       100
BEE VENOM  H P V T G C G E R T E G R C
                                   107
```

[a]From Maraganarc *et al.* (1986*a*).

In this case, the proposed assignment of the disulfide in the A chain is similar to the bovine pancreatic enzyme and the half-cysteine at the N-terminus is bridged to the half-cysteine at position 54 of the B chain.

Until recently, the structure of the bee venom phospholipase A₂ has been considered to have little if any homology with the pancreatic and snake venom phospholipases even though the molecular weight, 14,555 daltons (based on amino acid composition), is similar to that of the snake venom and pancreatic enzyme. Table 9-5 gives the sequence of the bee venom phospholipase A₂ as

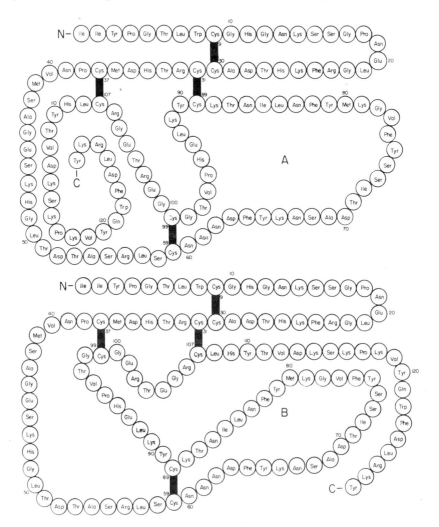

Figure 9-9. The amino acid sequence and proposed disulfide patterns of the bee-venom phospholipase A₂. (A) The proposed disulfide pattern for the bee-venom enzyme linking half-cysteines at residues 9, 31, 37, and 59 to those at positions 30, 89, 107, and 99, respectively (Shipolini *et al.*, 1974*b*). Our efforts to align the amino acid sequence of the bee-venom enzyme to those of other venom and pancreatic phospholipases A₂ suggest that the disulfide pattern of Shipolini and coworkers may be incorrect. (B) The proposed half-cysteines at positions 9, 31, 37, and 59 are cross-linked to those at positions 30, 107, 99, and 89, respectively (Maraganore *et al.*, 1986*b*). (From Maraganore *et al.*, 1986*a*.)

aligned with the proposed ancestral progenitor enzyme (Maraganore *et al.*, 1986*a*). The sequence of the ancestral phospholipase A_2 was derived from alignment of amino acid sequences of venom and pancreatic enzymes, considerations of required base substitutions, and the construction of a phylogenic tree. When arranged in this manner, a possible alignment of disulfide bridges and amphipathic α-helical regions could be identified. In the latter case, the α helix was thought to replace a segment missing in the N-terminal region of the bee venom enzyme.

The precise assignment of the disulfide bridges remains open. Based on the suggested alignment of the amino acids, bridges exist between half-cysteines at residues 9, 31, 37, and 59 and 30, 107, 99, and 89, respectively (Maraganore *et al.*, 1986*a*) (Fig. 9-9B). On the other hand, Shipolini et al. (1974*a*, 1974*b*) proposed bridges between residues 9, 31, 37, and 59 and 30, 89, 107, and 99, respectively (Fig. 9-9A). The former proposal (Fig. 9-9B) allows for a three-dimensional structure closer to that of the other venom phospholipases A_2.

As described earlier, two broad categories of phospholipases A_2 could be defined based on the positioning of the disulfide bridges. Further comparison of the polypeptide chains and the similarity of amino acid residues surrounding the active site led to the construction of two dendrograms (evolutionary trees). Scheme 9-1 shows the dendrogram constructed from a difference ma-

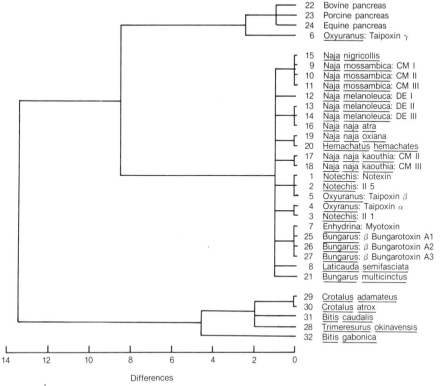

Scheme 9-1. Dendrogram of 32 phospholipases A_2 based on minimal mutation distances. (From Dufton and Hider, 1983.)

trix for 32 phospholipases based on the minimum mutation distance between the positions of 25 residues. This differs from that constructed from chain length between disulfide bridges because different aspects in the evolution of phospholipase were employed. As pointed out, the scheme presented here compares the *cores* of the enzyme and therefore comparative studies on substrate specificity should be of considerable value. Likewise, the inclusion of the Lys-49 enzymes in such a dendrogram would be of interest.

9.4.2. Crystallographic Studies of Phospholipases A₂

While progress has been made in x-ray crystallographic probing of the phospholipases A_2, difficulty has arisen in obtaining enzyme crystals of sufficient quality from a number of the sources of enzyme now available. Also, it would be desirable to obtain crystals with substrate bound to the active site but thus far this has not been possible.

9.4.2.1. Pancreatic Phospholipase A₂

Detailed x-ray diffraction studies have been carried out on Ca^{2+}-containing crystals of the bovine enzyme at a resolution of 1.7 Å (Dijkstra *et al.*, 1981). This detailed analysis has allowed a good understanding of the Ca^{2+} and H_2O binding sites, the interaction of the N-terminal region with the active site, the secondary structure, and the rigidity of the active site and has been most useful in projecting how the substrate monomer and micelle bind to the enzyme. Figure 9-10, taken from Dijkstra *et al.* (1978), shows the stereographic diagram of the enzyme obtained from data obtained with 2.4 Å resolution.

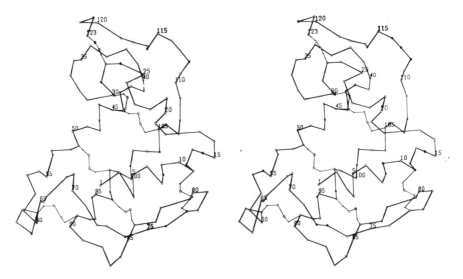

Figure 9-10. Stereo diagram showing the conformation and disulfide bridges of the bovine pancreatic phospholipase molecule. (From Dijkstra *et al.*, 1978.)

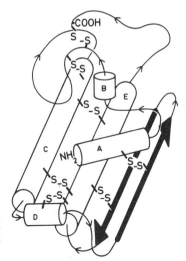

Figure 9-11. Schematic drawing showing the helices and β-structure of the phospholipase molecule. (From Dijkstra *et al.*, 1978.)

Studies with higher resolution on specific regions will be given in Chapter 10 on Ca^{2+} and micelle binding.

The overall dimensions of the monomeric bovine enzyme are roughly 22 Å × 30 Å × 42 Å and it has approximately 50% α helix and 10% β structure. Furthermore, the structure is stabilized by the disulfide bonds plus extensive hydrogen bonding as shown by the schematic drawing in Fig. 9-11. In that regard, it is thought that the active site is rather rigid. Five regions make up the helical structure, three of which are rather long (residues 1–13, 39–58, 89–108). The β structure is localized primarily between two antiparallel strands running from residues 74–78 and 81–85.

9.4.2.2. Venom Phospholipase A₂

The crystal structure of dimers of the phospholipase from *C. atrox* at 2.5-Å resolution closely resembles the structure of the monomeric enzyme from bovine pancreas (Keith *et al.*, 1981). Its crystal size is considerably larger than that of the pancreatic enzyme, 54 Å × 99 Å × 49 Å (Pasek *et al.*, 1975). The scheme in Fig. 9-12 proposes that pairs of intramolecular ionic bridges exist between the monomers and involves Asp-49 (Brunie *et al.*, 1985). Since Ca^{2+} binds at Asp-49, destabilization could occur when Ca^{2+} binds in its site. This orientation also shields the active site in the absence of Ca^{2+} which interferes with substrate entering the active site. It was proposed by Brunie *et al.* (1985) that entry of substrate may be facilitated by the presence of Ca^{2+}.

The overall structure is an oblate ellipsoid that has extensive intramolecular interaction at the Ca^{2+} binding and active site. This provides cooperativity between these sites in the two monomers but lacks an obvious entrance for the phospholipid molecule. Also, the assignment of the lipid interfacial sites on opposite sides of the dimer obscures visualization of an organized lipid phase with these two sites. It seems most likely that some sort of con-

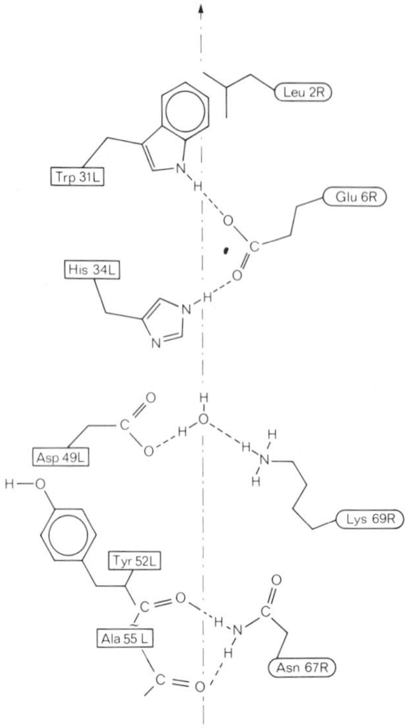

Figure 9-12. Dimer stabilizing interactions. Schematic showing the set of interactions that passes *in front* of the molecular dyad; a symmetrical set passes *behind* the dyad but is not shown. R, right; L, left. (From Brunie *et al.*, 1985.)

formational change occurs with the binding of Ca^{2+} and/or substrate to one monomer favoring the activity of the second. This concept is particularly attractive since the closely related enzyme from *C. adamanteus* is active only in the dimeric form. In fact, it is possible that all phospholipases A_2 are active in some degree of aggregation. Detailed comparison of the three-dimensional structures of the bovine and *C. atrox* phospholipases A_2 was provided by Dennis (1983). As shown in Fig. 9-13, the two show a high degree of superimposability. Some differences are noted in the region of the Ca^{2+}-*binding loop*, Tyr-28, Gly-30, Gly-32, and Asp-49. In part, this may result from the fact that, unlike the pancreatic phospholipase A_2, the *C. atrox* enzyme does not bind Ca^{2+} upon crystallization. It will be of considerable interest to compare this structure with that obtained using the Lys-49 enzymes that do not have the same Ca^{2+}-*binding loop*.

Figure 9-13. Superposition of the x-ray crystal structures of phospholipase A_2 from bovine pancreas and the "right" subunit from the dimer of *Crotalus atrox*. These structures are shown in stereo pairs for (A) one view of the entire molecule, (B) blowup of the active-site region, and (C) the calcium-binding loop. This superposition was carried out by Professor P. Sigler on the Molecular Modeling System at the Research Resource Computer Facility at the Department of Chemistry, University of California, San Diego, by using coordinates kindly supplied by S. Brunie, J. Bolin, P. Sigler, and Professor J. Drenth. The snake venom enzyme is shown in black and the pancreatic enzyme in gray. The α-carbon backbone is shown, and the following side chains are indicated: Phe-5, His-48, and Tyr-52. (From Dennis, 1983.)

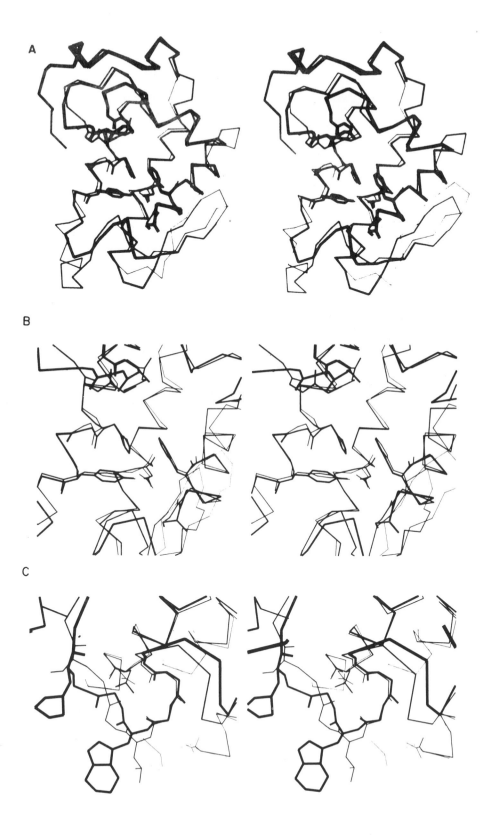

A

B

C

Other phospholipases A_2 from venoms have been crystallized and studied by x-ray diffraction. For example, notoxin was studied with 1.8 Å resolution and shown to have six molecules per unit cell with dimensions 75 Å × 75 Å × 49 Å (unit cell) (Kannan *et al.*, 1977). By comparison, the phospholipase A_2 from *C. adamanteus* has group spacing of 108 Å × 79 Å × 64 Å.

Dufton *et al.* (1983) utilized the known crystallographic analyses, circulation dichroic spectra, and secondary-structure predictions to compare the tertiary structures of several phospholipases. These structures were general although considerable differences were noted in the ratio of hydrophobic to hydrophilic area. Also, the α-helical region designated D (in Fig. 9-11) is absent in all but one of the venom phospholipases, taipoxin γ. The D region α helix therefore probably confers a unique characteristic to the pancreatic enzymes. The comparable region in the venom enzymes should have sufficiently similar physical properties to accommodate a similar mechanism of action for all these phospholipases.

9.5. Functionality of Amino Acids in Venom and Pancreatic Phospholipases A_2: Enzyme Modification

These enzymes have been probed by many techniques in order to understand the structural features that regulate its interaction with organized lipid interfaces and Ca^{2+} as well as its catalytic active site. In many cases, comparisons have been made with monomeric and aggregated lipid as substrate or a protecting agent. Recent evidence that monomers of substrate cause enzyme polymerization and micelle formation suggests that the interpretation of some experiments using this comparison is not as simple as first thought. This is particularly true since polymers of phospholipases A_2 might be the active species. These techniques include, but are not limited to, (1) chemical modifiers or replacement of amino acids, (2) x-ray diffraction, and (3) spectroscopy. Many of the early studies relied on chemical modification. More recently, emphasis has been placed on semisynthetic approaches coupled with structural analysis.

9.5.1. Chemical Modification

The use of chemical modifiers is thoroughly described for both the pancreatic and venom enzymes (Slotboom *et al.*, 1982) and therefore only a few of the newer key highlights are covered here. Studies in which a single amino acid or group of amino acids is chemically modified require that the modified protein be separated from any nonmodified enzyme and that the specific sites of modification be determined. However, this was not possible or done in all cases and a number of uncertainties exist from experiments on enzyme modifications. Table 9-6 is a summary of amino acids that have been implicated in the one or more requirements for enzyme function. Usually, the difference between lipid binding and catalysis was determined by comparing activity of monomers versus emulsions of substrate.

Table 9-6. The Effect of Amino Acid Modification on Phospholipase A₂–Ligand Interaction

Amino acid modified[a]	Modifying agents[a]	Pancreas		Venom
		Position	Function[b]	Position
His	BPB,M-*p*-NBS,BO-2-O	48	b,c	48
Asp (Lys)	DBE,EDC	49	b	49
Tyr	TNM,I₂	19,69	a,b,c	
Ala (α amino)	GA,Cu²⁺	1	a	1
Trp	NBS,HNB		a	31,70
Met	MI	8,20	a (?)	
Glu		71	d	
Lys	PSP,AA,SA		a	4–6 Residues
Lys	EOFA			1 Lys per dimer
Arg	CHD,PG		a,o	16

[a] Abbreviations: AA, acetic anhydride; BO-2-O, 1-bromo-octan-2-one; BPB, bromphenacyl bromide; CHD, 1,2-cyclohexanedione; DBE, *N*-diazoacetyl-*N'*-(2,4-dinitrophenyl)ethylenediamine; EDC, 1-ethyl-3-(*N,N*-di-methyl)amino propyl carbimide; EOFA, ethoxy formic acid anhydride; GA, glyoxylic acid; HNB, 2-hydroxy-5-nitrobenzyl bromide; IB, interfacial binding; MI, methyl iodide; M-*p*-NBS, methyl *p*-nitrobenzene sulfonate; NBS, *N*-bromosuccinimide; PG, phenylglyoxal; PSP, pyridoxal-5-phosphate; SA, succinic anhydride; TNM, tetranitromethane.
[b] Functions: a, interface binding; b, high-affinity Ca²⁺ binding; c, monomer binding; d, low-affinity Ca²⁺ binding; o, none.

9.5.1.1. Pancreatic Phospholipase A₂

9.5.1.1a. Histidine. His-48 of the horse phospholipase reacts specifically with radiolabeled bromphenacyl bromide to give a stoichiometric loss in cat-alytic activity (Volwerk *et al.*, 1974; Verheij *et al.*, 1980*a*). Since both Ca²⁺ and reaction products and monomers of nonhydrolyzable substrate analogs block bromphenacyl bromide binding and inactivation, it was concluded that His-48 is a part of monomeric substrate and Ca²⁺ binding. Likewise, the proen-zyme that degrades only monomeric substrates can be protected from inac-tivation by these agents. Binding of the enzyme modified by the various agents listed in Table 9-6 does not influence its binding to water–lipid interfaces, showing that these two functions are carried out by distinct sections of the enzyme molecule. Methylation of the histidine blocked all catalytic activity but not the binding of substrate monomers and Ca²⁺. It appears therefore that His-48 participates in the catalytic event. The presence of Ca²⁺ lowered the apparent pK of His-48 from 6.5 to 5.7 as shown by proton NMR and ultraviolet spectroscopy (Williams, 1981; Slotboom *et al.*, 1982).

9.5.1.1b. Aspartate. Reaction of the bovine phospholipase A₂ with 1-ethyl-3-(*N,N*-dimethyl)amino propyl carbimide (EDC) and semicarbazide inacti-vated the enzyme activity on both monomers and micellar phosphatidylcholine (Fleer *et al.*, 1981). Ca²⁺ was able to protect against EDC inactivation and to prevent carboxylate modification of Asp-49. It was concluded therefore that Asp-49 is also involved in Ca²⁺ binding. The apparent pK of the aspartate is thought to be 5.25, in agreement with binding studies of Ca²⁺. This as-signment has been supported using other procedures such as ⁴³Ca²⁺–NMR spectroscopy (Andersson *et al.*, 1981). Other metal ions can bind Asp-49 including Ba^{2+}, Sr^{2+}, Eu^{3+}, Tb^{3+}, and Gd^{3+}; only Gd^{3+} could support catalytic activity. The results show that the Ca^{2+} site has some flexibility in metal binding but rather rigid requirements for the spacing of substrate, metal ion, H_2O, and active-site residues in catalysis.

It has been suggested that another carboxylate in the pig pancreatic enzyme could bind Ca^{2+}, perhaps that of Asp-99 (Dinur *et al.*, 1981). While this question remains open (Slotboom *et al.*, 1982), it is of interest to note that Asp-99 is proposed as a part of the proposed proton relay system (see Chapter 10).

9.5.1.1c. Tyrosine. Nitration of Tyr in the prophospholipases A$_2$ from horse, pig, and cow caused a rapid partial inactivation of activity which was enhanced rather than blocked by lysophosphatidylcholine and Ca^{2+} (Meyer *et al.*, 1979*a*, 1979*b*). Common to the enzymes from all three was the nitration of Tyr-69. In addition, Tyr-124 was modified in the pig enzyme and Tyr-19 in the horse enzyme was modified. All three purified, mononitrated enzymes (activated by trypsin) retained 15–50% of activity on a micellar substrate. Reduction of these nitrate groups to the amine followed by dansylation of Tyr-19 and Tyr-69 enhanced binding of the enzyme to the water–lipid interface, indicating that these two groups in the active but not proenzyme are involved in bulk lipid interaction. Also, Ca^{2+} was shown to interact with nitrated Tyr-69 and iodination of the enzyme occurred at Tyr-69 primarily enhancing activity on egg-yolk emulsions and penetration into monolayers of didecanoyl phosphatidylcholine.

9.5.1.1d. Alanine (α-Amino Group). The phospholipases A from pig and horse were rapidly inactivated by transamination with glyoxylic acid and Cu^{2+} (Verheij *et al.*, 1980*a*; van Scharrenburg *et al.*, 1984). The enzyme from cow also could be modified; however, the rate of reaction was much slower. The result of this reaction is the replacement of the N-terminal amino group with a carbonyl moiety. The addition of a denaturing agent such as urea led to the inactivation of the proteolytically activated enzymes. The proenzymes, however, when treated in the same fashion, were fully activated by trypsin, showing protection of Ala-1 by the activator peptide. Micelles of substrate analogs could also protect the activated enzymes from modification, showing that the Ala-1 is involved in interfacial lipid binding. However, Ca^{2+} or monomers of substrate failed to protect, suggesting that the binding of aggregates was at a site distinct from the active site. As would be predicted, the transaminated enzyme had properties similar to the proenzyme. The finding that the bovine enzyme was less readily attacked by glyoxylate suggests that Ala-1 on this enzyme is more internally oriented than that of the phospholipase from the other two species.

More recent studies on the Ala-1 transaminated phospholipase A$_2$ from the pancreas of pig (including the iso form), horse, human, sheep, and cow showed that inactivation was first order, as measured by their ability to bind micelles of substrates (Dijkstra *et al.*, 1984). X-ray structural analysis of the bovine enzyme showed that the N-terminal region as well as the 63–72 sequence region became more flexible upon transamination and resembled those regions in the proenzyme. It was concluded that transamination disrupts the hydrogen bonding of the α amine with oxygen in Gln-4 and Asn-71. It was proposed that this bonding, while not influencing the active site, is responsible for binding neutral micelles. It will be of interest to determine if the transaminated enzyme can bind negatively charged aggregates as well as the proenzyme (Volwerk *et al.*, 1984).

  9.5.1.1e. Tryptophan. Modification of Try-3 of the pig phospholipase A with *N*-bromosuccinimide reduced activity on egg yolk but not micelles of dioctanoyl phosphatidylcholine, which suggested that Try-3 is involved in interfacial binding (Slotboom and deHaas, 1975). It is of interest that the interfacial binding component is less involved in micellar interaction than in the penetration into emulsion of long acyl chain phosphatidylcholines. Comparison of this modification with the transamination of Ala-1, where inhibition on micellar substrate was complete, indicates that the nature of the modification as well as its site can regulate the function under study.

  9.5.1.1f. Methionine. Carboxymethylation of Met-8 of the activated phospholipases from horse, cow, and pig (iso forms) with iodoacetic acid led to the inactivation of the enzyme when assayed with both micellar and monomeric substrates (van Wezel *et al.*, 1976). Binding of Ca^{2+} and a substrate analog still occurred to some extent, showing that the active site was not totally blocked. This was confirmed by the observation that bromphenacyl bromide could react with His-48 following carboxymethylation. Further analysis of the carboxymethylated protein brought to light some difficulties with this approach since this modification introduced charge-induced changes in the tertiary structure in the interior of the protein.

  Methylation and carboxymethylation of Met-20 of the pig phospholipase A₂ did not cause major changes in the activity of the enzyme on monomeric and egg-yolk substrates with the exception that carboxymethylation reduced the activity on egg yolk by 50%. On a monolayer system, the maximal surface pressure allowing activity was reduced, suggestive of an alteration in the interfacial binding site.

  9.5.1.1g. Lysine. By analogy with studies done on venom phospholipases and supported by x-ray crystallographic work, it appears that the Lys are involved in binding to interfaces and, in the case of the enzymes with Lys-49, participate in catalysis. Little, if any, work has been done on Lys modification of the pancreatic enzymes.

  9.5.1.1h. Arginine. Controversy surrounds the functionality of Arg-6. On one hand, Vensel and Kantrowitz (1980) reported that treatment of the pig phospholipase A₂ with phenylglyoxal and other agents blocked activity shown using the egg-yolk assay. Since the inhibition was reduced by dodecyl and hexadecyl phosphorylcholine but not Ca^{2+}, they concluded that the arginine interacts with the phosphate moiety of the substrate or analog. Although they considered the possibility that a transamination of the α amino group could have occurred, their data favored the modification of the Arg. Slotboom *et al.* (1982), on the other hand, reckoned that more transamination had occurred under conditions employed by Vensel and Kantrowitz (1980) and that no conclusion should be drawn. In that regard, Fleer *et al.* (1981) isolated an enzyme modified with 1,2-cyclohexanedione at Arg-6 and demonstrated that this modification did not alter the activity of the enzyme on soluble and aggregated substrates.

9.5.1.2. Venom Phospholipases A_2

Many of the conclusions that can be drawn for essential amino acid residues in the venom phospholipases A_2 are similar to those for the pancreatic enzyme. Only notable differences either in the type of experiment done or the conclusion drawn between the pancreatic and venom enzymes are covered here.

9.5.1.2a. Histidine. Similar to the pancreatic enzyme, bromphenacyl bromide blocks the activity of many venom enzymes by reacting with His-48. Unlike the pancreatic enzyme, several venoms apparently still bound Ca^{2+} after reaction with bromphenacyl bromide (Roberts *et al.*, 1977*a*; Yang and King, 1980). Many of the venom phospholipases A_2 with toxic properties lose both hydrolytic and toxic activities upon reaction with bromphenacyl bromide, supporting the concept that phospholipase A_2 activity is related to the toxicity. Complex toxins such as crotoxin do not react with bromphenacyl bromide unless the subunits are dissociated. It appears that the A chain protects the His-48 in the B chain (Jeng and Fraenkel-Conrat, 1978). It is of interest that Ca^{2+} failed to protect the B chain from inactivation, perhaps due to some structural difference that occurs upon dissociation of the A chain. Although reports exist that substrate or substrate analogs protected several venom phospholipases A_2 from bromphenacyl bromide inhibition, Verheij *et al.* (1981*a*) question the interpretation that this is the result of active-site blockage. They proposed that nonspecific binding of the lipophilic inhibitor by the substrate could have prevented the inhibitor from interacting with the enzyme.

9.5.1.2b. Tryptophan. Oxidation of two Trps by *N*-bromosuccinimde inactivated two venom enzymes when assayed on micellar substrates. The Trp-31 in the *Bitis gabonica* enzyme was implicated as being the crucial residue (Viljoen *et al.*, 1976), although at present it is not clear if Trp-31 is a component of the active site. As pointed out by Slotboom *et al.* (1982), Trp-31 is variable and only Ca^{2+} and micelles protect against inactivation; monomers of substrate do not protect against this type of inactivation. Trp-70 has been implicated to influence the catalytic activity of the enzyme from *Laticauda semifasciata* (Yoshida *et al.*, 1979). Only one of four isozymes from this venom has Trp-70 and its oxidation by *N*-bromosuccinimide caused the kinetic characteristics of the Trp-70 isozyme to shift to those of the Trp-70 deficient isozymes. Modification of two Trp residues per dimer of *C. adamanteus* by 2-hydroxy-5-nitrobenzyl bromide changed neither the activity of the enzyme nor Ca^{2+}-induced structural changes. These studies plus others leave the role of Trp somewhat up in the air as to a definite assignment of function.

9.5.1.2c. Lysine. Modification of one Lys per dimer in *C. adamanteus* with ethoxyformic acid anhydride (EOFA) led to inactivation and dissociation of the dimer. Since the unmodified monomer could reassociate into an active dimer, this was taken as suggestive evidence for a role of the dimer in catalysis (Wells, 1973*b*). Questions about this interpretation have been raised by Slotboom *et al.* (1982) based on analogy with the pancreatic enzyme and on the lack of specificity of the agent used, EOFA. The enzyme from *B. gabonica*, when modified by pyridoxal-5-phosphate and sodium borohydride, appeared to lose its capacity to bind aggregated substrate. In this case, four Lys were modified to a limited degree. It appears that the introduction of a charged group to the ε-amino function is key in inactivation since the extensive modification by various charged but not neutral agents inactivates a number of enzymes. In the case of the enzyme from *N. naja oxiana*, blockage of all ε-amino groups with agents such as acetic anhydride did not reduce activity appreciably (Slotboom *et al.*, 1982). However, agents such as succinic anhydride that produced negatively charged residues caused complete inhibition of hydrolysis. The introduction of one mole of pyridoxal group per mole of enzyme also caused complete inhibition of the *N. naja oxiana*. It was not clear from this study, however, which of the Lys had been modified.

More recently, Forst *et al.* (1982) showed that modification of one mole equivalent of lysine ε-amino groups with either cyanate or formaldehyde did not interfere with catalysis but blocked binding of the basic phospholipase from *A. halys blomhoffii* to *E. coli* membranes. They were able to show this using the dependence of this enzyme on an added leukocyte protein, bactericidal/permeability increasing protein (BPI), to degrade the phospholipid in live *E. coli*. Modification of the ε-amino groups caused the loss of BPI-dependent hydrolysis even though the enzyme was fully active on heat-killed *E. coli* membranes. Subsequently, data from Forst *et al.* (1986) indicated that four of the lysines in amino terminal α-helical *lipid binding domain* binding site have 50% of the total modification, leading to the conclusion that the loss of binding of the enzyme to the intact membranes of live *E. coli* required unmodified lysines in residues 7, 10, 11, and 15. In this case, the change in binding properties was not simply the result of a charge change since reductive methylation was as effective as carbamylation.

Additional evidence for the importance of lysines in enzyme action was provided by Lombardo and Dennis (1985a). They found that manoalide, a nonsteroidal sesterterpenoid isolated from sponges, caused a 50% inhibition of the *N. naja naja* phospholipase A₂ at a concentration of 2×10^{-6} M. In their studies, they found that at a minimum four of the six lysines in this enzyme were modified, probably through the attack of a free aldehyde generated in manoalide at alkaline pH values. Interestingly, manoalide is a potent anti-inflammatory agent that perhaps functions by blocking cellular phospholipases by a mechanism similar to its inhibition of the *N. naja naja* enzyme. It was of interest in this study to find that the binding of manoalide to the enzyme blocked activity on phosphatidylcholine but actually enhanced hydrolysis of phosphatidylethanolamine.

While a total picture of the role of Lys in catalysis is not available, it is clear that the charge provided by Lys is important. Studies on the locations of modification as well as further studies on the essential nature of dimerization in catalysis will help our understanding of this point.

The substitution of Asp-49 in the *Ca²⁺-binding loop* by Lys creates a special interest in the Lys of this class of phospholipases (Maraganore *et al.*, 1984).

Attempts to modify this residue in the Lys-49 phospholipase A_2 with trinitrobenzene sulfonate (TNBS) yielded some rather unexpected results. Unlike the Asp-49 phospholipase A_2 from the same species that is not inhibited by TNBS, derivatization of lysines with one mole of TNBS produced an 80% inhibition but no modification of Lys-49. Sequence analysis showed that Lys-53 rather than Lys-49 had been derivatized and implicated Lys-53 in the formation of the catalytically active complex (Maraganore and Heinrikson, 1985). These workers speculated that Lys-53 "may be involved in the binding of a second, non-substrate phospholipid molecule involved in the recognition of surfaces by all phospholipases A_2." It will be of considerable interest to see how this proposal fits with the Dennis "dual phospholipid" model (Dennis, 1983) and the hydrophobic binding region proposed by the Utrecht group (Slotboom *et al.*, 1982) for the pancreatic phospholipases A_2.

9.5.1.2d. Methionine.   Modification of *C. adamanteus* with 2-bromoacetamido-4-nitrophenol at Met-10 did not cause changes in activity (Wells, 1973a). Similar treatment of the pancreatic enzyme did lead to some inactivation; although some minor changes were noted with Met-20, no major functional role for this amino acid could be assigned.

9.5.1.2e. Aspartate.   The modification of a single carboxylate of an aspartate per dimer of the enzyme from *N. naja oxiana* with *N*-diazoacetyl-*N'*-(2,4-dinitrophenyl)ethylenediamine caused the loss of activity on monomeric substrates (Zhelkovskii *et al.*, 1978). By analogy with the pancreatic enzyme, Asp-49 would be suspected as the site of modification if only a single modification occurred. The finding that only one aspartate per dimer is modified would suggest two points: First, the active site of one monomer in the dimer is more readily attacked than the other (otherwise some unaltered dimers would exist when on average one modification per dimer was reached); and second, the dimer is the active species of the enzyme.

9.5.1.2f. α-Amino Terminus.   As shown in Table 9-4, a number of different residues are located at the N-terminus of venom phospholipases A_2. Likewise, replacement of this group by other residues in the pancreatic enzyme shows that some flexibility in the structure at this position exists. However, modification of the α-amino residue of the phospholipases A_2 from three families, *Naja*, *Vipera*, and *Crotalus*, with the glyoxylic acid led to a reduction in degradation of micellar but not monomeric substrates. Similar to the pancreatic enzymes, the α-amino group is essential to hydrolysis of bulk lipid. Some differences between the venom and pancreatic enzymes exist, however, since direct binding studies showed that the modified venom enzyme still bound bulk lipid. Further studies with the phospholipase A_2 from *N. melanoleuca* showed that modification of the α-amino group caused a decreased affinity for monomeric substrates (van Eijk *et al.*, 1984a). These results led to the conclusion that changes in the bridging between the α-amino group and interior of the enzyme causes an alteration in the active site.

In summary, the studies done thus far using amino acid modifying agents demonstrate that there are many similarities between the snake venom and pancreatic phospholipases A_2. Some differences were found, as would be expected. This is particularly true when differences between enzymes that are isolated and possibly function as monomers are compared with dimers, and when the interaction between two nonidentical subunits is considered. When used vigorously, chemical modification of enzymes has been quite useful in probing these enzymes and future work will certainly further our understanding of their structure–function relation.

9.5.2. Amino Acid Substitution

There are advantages to substituting amino acids by semisynthetic procedures rather than attempting to modify specifically functional groups on amino acids. This approach, however, is technically a completely different and more complex problem. An elegant systematic approach was used to show that the α-helical region of the N-terminal region is involved in lipid binding (Slotboom and deHaas, 1975; Slotboom *et al.*, 1978). Key to this approach was the ε-amidation of Lys residues to protect those groups during further chemical procedures. Amidation of lysines did not affect normal activity. Table 9-7 shows the modifications that have now been made using limited Edman degradations followed by the addition of various amino

Table 9-7. Effect of Amino Acid Substitution on Pancreatic Phospholipase A₂ Activity

Modification	Hydrolysis	
	Monomer	Micelle
None	+ + +	+ + +
ε-Amidated (AMPA)[a]	+ +	+ +
des-Ala-1AMPA	+	0
1,Leu-des-Ala-2AMPA	0	0
1,Leu-2,Trp-des-Ala-3AMPA	+ +	0
1 β Ala-des-Ala-1AMPA[b]	+ +	+ +
1 Gly-des-Ala-1AMPA[b]	+ +	+ +
1 Asn-des-Ala-1AMPA[b]	+ +	+ +
1 Asp-des-Ala-1AMPA[b]	+ +	+ +
1 Nle-des-Ala-1AMPA[b]	+ +	+ +
1 α Amino isobutyric acid-des-Ala-1AMPA[b]		0
1-*N*-methyl-Ala-des-Ala-1AMPA[b]		0
1-Leu-1-des-Ala-1AMPA[b]		0
1-Phe-des-Ala-1AMPA[b]		0
1-*O*-Ala-1-des-Ala-1AMPA		0

	Other specific modification		
Site	Amino acid	Replaced by	Micelle
3	Trp	Gly	0
3	Trp	Phe	+
4	Gln	Glu	+ +
4	Gln	Nle	0
5	Phe	Tyr	0
6	Asn	Arg	+ +

[a] AMPA, ε-amidated phospholipase A₂.
[b] Designates the amino acid residue added to the 1-des-Ala-1AMPA.

acid residues to the partially degraded pig phospholipase A_2 (Pattus *et al.*, 1979*a,b,c*).

These data are interpreted to show that the α-helical hydrophobic nature of the lipid binding domain must be conserved for activity on micellar lipid, although several chemical modifications are allowed. Most modified forms retained their activity on monomers. Only the substitution of Phe-5 by Tyr caused the loss of activity on monomeric substrate despite the fact that the enzyme could still bind to interfaces. It therefore appeared that this replacement produced conformational change, affecting the active site. This is not surprising since the binding to bulk phospholipid must be organized such that the substrate molecule(s) can enter the active site.

Further substitutions have been done with the pancreatic phospholipases A_2. Substitution of Gln-4, Phe-5, and Met-8 with Nle-4, Try-5, and Nle-8 showed the Gln-4 and Phe-5 but not Met-8 are essential for activity on micelles; the former two are invariant amino acids (van Scharrenburg *et al.*, 1982). The analog Nle-4 ε-amidated phospholipase A_2 (AMPA) retained 25% of its activity toward monomeric substrate while Try-5 AMPA was shown by ultraviolet spectroscopy to bind micelles in the presence of Ca^{2+} despite the fact it was catalytically inactive. Further substitutions in the amino-terminal region indicated that substitution of Asn-6 by Arg in the bovine phospholipase A_2 enhances binding of lipid, making it similar to the porcine enzyme that naturally contains Arg-6 (van Scharrenburg *et al.*, 1981). Furthermore, it was found that with the exception of a substitution of Gly-7 by Ser, any substitution achieved in residues 6 and 7 altered catalytic activity (van Scharrenburg *et al.*, 1983). Amino acids enriched in ^{13}C were introduced into the N-terminal position to determine the pK of the α-amino group. Values of 8.4, 8.8, and 8.9 were found for the pig, horse, and cow enzymes, respectively (Janssen *et al.*, 1972; Jansen *et al.*, 1978*a*; van Scharrenburg *et al.*, 1984). It must be concluded from these studies that the interplay of the α-helical N-terminal region of pancreatic phospholipases with the active-site region centered at residues 48 and 49 is very complex and that assignment of absolute functions to the residues of the N-terminus is difficult.

Much less work has been done on amino acid substitution in the venom phospholipases A_2. It has been shown that CNBr cleavage of the N-terminal octapeptide of the enzyme from *T. flavorides* retained some activity and remained as a dimer (Kihara *et al.*, 1981).

If sufficient protein becomes available, the approach of semisynthesis should be a very powerful tool to study the Lys-49 enzymes. Also, the ability to introduce residues with particular spectral properties is allowing very detailed probing of phospholipases A_2. As will be described in Chapter 10, the study of specifically modified enzymes with metals and substrate analogs that can likewise be studied spectrally has allowed substantial progress in this field.

Chapter 10

Mechanism of Phospholipase A₂ Action

This chapter is a continuation of Chapter 9, which covered isolation and characterization of the structure of the pancreatic and venom phospholipases. In this chapter, emphasis shifts from protein structure to the interaction of the protein with water, metal ions, and substrates or their analogs. Since substrate–enzyme interaction is dependent on protein structure, frequent reference to Chapter 9 is essential. Also, where germane, references are made to the mechanism of action of other phospholipases.

10.1. General Considerations

From the many approaches taken to the study of phospholipase–substrate interaction, a picture is now beginning to clarify and some generalities can be looked for in this chapter. First, it appears that all phospholipases studied in detail thus far have two lipid binding regions. They are referred to by various names but basically one is the catalytic site and the other an activator site. Second, there appears to be some commonality in catalysis that involves Ca^{2+} and certain required features at the active site. Third, in all cases a water–lipid interface is required for optimal activity. Fourth, more evidence is accumulating that enzyme dimerization or polymerization accelerates hydrolysis by some but perhaps not all phospholipases A_2 of this class.

The interactions of phospholipases with their substrates serve as models for any enzyme that functions optimally at lipid–water interfaces. These studies, however, require a rather detailed knowledge of lipid–lipid, lipid–water, and protein–lipid interactions. In addition, rather innovative kinetic analyses have been required to determine the enzyme–substrate interaction and the mechanism of action of phospholipases. Considerable progress has been made with the venom and pancreatic phospholipases, to a great extent because these enzymes can be obtained in large quantities in homogeneous form. Studies done with these enzymes will undoubtedly serve as the basis for investigation on cellular mammalian, bacterial, and plant phospholipases in the future. While generalizations concerning all phospholipases are implied in this section, less is known about the phosphodiesterase phospholipases since their interaction with lipid has not been studied to the same extent as the acyl hydrolases. One might expect that even though the phosphodiesterases are activated by interfaces, they might have less hydrophobic character than the acyl hydrolases since their site of action is in the aqueous region of the substrate aggregate.

As indicated in the Introduction, two general types of phospholipase exist: those that are basically soluble and must first interact with or penetrate the organized lipid interface; and those that reside in membranes and, as a consequence, probably are not structured in such a way that penetration of a lipid interface is physiologically relevant. Out of necessity therefore, the purification of membrane-associated enzymes creates an artifactual condition in the study of their mechanism of action. This needs to be recognized in studies on their mechanisms of action and regulation.

Thus far, all phospholipases appear to have in common the feature of being more active on lipids at an interface than on lipids in a bulk three-dimensional solution. Three basic influences have been implicated as being responsible for this phenomenon, all of which probably contribute in varying degrees to this enhanced activity: (1) an effect on the enzyme and (2) and (3) an effect on the lipids (substrate and product).

The first of these is espoused by Verger and deHaas (1973, 1976) who originally postulated that a conformational change in the pancreatic phospholipase A_2 occurred following the initial binding (see Fig. 9-3). In this hypothesis, the penetration step is rate limiting and the induced conformational change in the enzyme causes hydrolytic rates orders of magnitude higher than the rate on monomers.

The second factor, the effect on the substrate, has two facets: (1) a concentration effect and (2) a conformational effect. The first is rather obvious since it is known from many studies that phospholipases bind to lipid interfaces and, as a consequence, are "afloat in a sea of lipid." Likewise, the hydrophobic environment provides a more thermodynamically favorable release of the lipid products than does an aqueous environment. More quantitatively, when the concentration of phospholipid exceeds the CMC, the effective concentration that the enzyme binds increases by orders of magnitude. For example, dihexanoyl phosphatidylcholine has a CMC of 10 mM (Wells, 1974) in the micellar form, a concentration that is several molar when calculated for the micellar surface. The enzyme molecules bound to the micelle therefore find a saturating environment. Once the substrate in a given aggregate is degraded, it is then essential that either the enzyme molecule dissociate and reassociate with new substrate or that substrate rapidly moves between aggregates. While the latter process is slow in bilayer systems, we have found that transfer of phospholipid molecules between micelles made of phospholipid in Triton is quite rapid with complete equilibrium being achieved in less than a second. The number of enzyme molecules capable of binding to the micelle and expressing catalytic activity can be limiting, however, so it is essential to determine both the saturation within the micelles and the number of micelles required to bind maximally all enzyme molecules. The work of Dennis (1983) and coworkers has established an elegant kinetic analysis of *N. naja naja* phos-

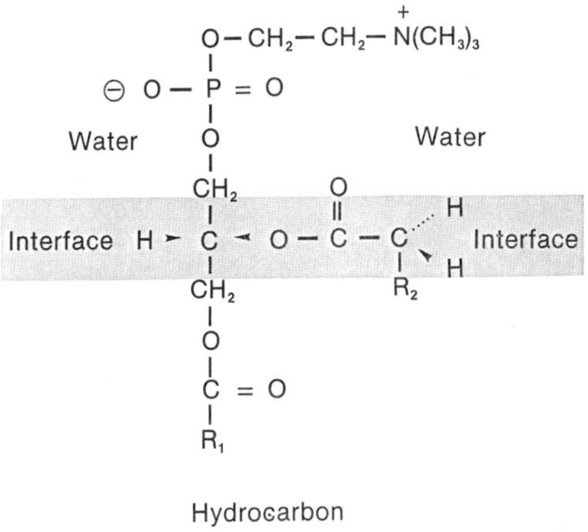

$$O-CH_2-CH_2-\overset{+}{N}(CH_3)_3$$

$$\ominus\ O-P=O$$

Water $\quad$ O $\quad$ Water

$$CH_2 \quad\quad O$$

Interface $\quad$ H ► C ◄ O — C — C $\quad$ H $\quad$ Interface

$$CH_2 \quad\quad R_2 \quad H$$

$$O$$

$$C=O$$

$$R_1$$

Hydrocarbon

Figure 10-1. The difference in the orientation of the acyl esters at positions 1 and 2 of the glycerol in phosphatidylcholine. (Adapted from Dennis, 1983.)

pholipase A₂ using a micellar suspension of phospholipid dissolved in Triton micelles.

A conformational change in the orientation of the glycerol backbone of the phospholipid, relative to the interface, is known to occur when the phospholipid is in an aggregated state (Roberts and Dennis, 1977; Roberts *et al.*, 1978; DeBony and Dennis, 1981). This change in orientation causes the two acyl esters to become nonequivalent; the ester group at position 1 is deeper in the hydrophobic region (Fig. 10-1). This dissimilar orientation of acyl esters was to be common for a wide range of phospholipids, regardless of their state of aggregation (Pluckthun *et al.*, 1986). As will be described later in this chapter, the orientations of the esters at positions 1 and 2 of the glycerol profoundly affect phospholipase interaction with substrate.

10.2. Metal Ion Binding to the Enzyme

10.2.1. Venom Phospholipases

It has long been recognized that Ca^{2+} is essential for the venom phospholipases and that the effect of the Ca^{2+} was on the enzyme and not on the substrate (Roholt and Schlamowitz, 1961). Wells (1973a) found that Ca^{2+}, as well as other alkaline earth cations, caused a spectral perturbation of the *C. adamanteus* enzyme in the region of 286 and 292 nm. Zn^{2+}, which does not support catalysis, caused a different shift (Fig. 10-2).

Wells concluded that this shift was a charge-induced rather than a solvent-induced perturbation of a tryptophan. Since the enzyme is a dimer, care was taken to rule out the possibility that the spectral change was the result of monomer formation. In agreement with equilibrium dialysis studies, 1 mole

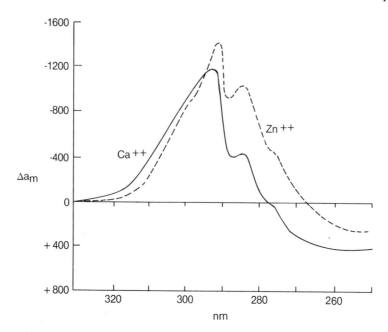

Figure 10-2. Spectral perturbations of phospholipase A_2 induced by divalent cations. Molar absorptivity changes (Δa_m) as a function of wavelength caused by 5×10^{-5} M Zn^{2+} or 1×10^{-3} M Ca^{2+} in 0.01 M Tris–0.05 M KCl (pH 8.0). The cell in the reference position contained the added cation. A change in absorbance of 0.035 = Δa_m of 1000. (From Wells, 1973a.)

of Ca^{2+} bound per monomer of enzyme. Comparison of kinetic, spectral, and dialysis measurements gave dissociation constants for Ca^{2+} of $4–8 \times 10^{-5}$M. The competitive inhibitor ions, Ba^{2+} and Zn^{2+}, had dissociation constants of $3–9 \times 10^{-5}$ M and $1–3 \times 10^{-6}$ M, respectively. The binding of Ca^{2+} was pH dependent and [H^+] was an uncompetitive inhibitor of Ca^{2+}.

The phospholipase A_2 from *C. atrox* has many properties in common with the enzyme from the closely related *C. adamanteus*. Spectral analysis of Ca^{2+} binding, however, failed to show any perturbation of the tryptophan absorption spectra of the *C. atrox* enzyme (Purdon *et al.*, 1976). On the other hand, a change in the circular dichroism spectrum of the *C. atrox* phospholipase suggested that 10^{-4} M Ca^{2+} induces a structural change in which tryptophan is exposed to the aqueous environment. From this, it was concluded that the K_D for Ca^{2+} was roughly 10^{-3} M, considerably higher than that found for the *C. adamanteus* enzyme. Binding of anilino-1-naphthalene sulfonate (ANS) was affected by Ca^{2+}; in the presence of Ca^{2+}, only one molecule was bound per dimer while in the absence of Ca^{2+} two ANS molecules were bound. ANS is thought to bind in the hydrophobic pocket of the enzyme occupied by the phospholipid substrate since it is protected against bromphenacyl bromide inhibition, at least in the case of the enzyme from *Hemachatus haemachatus* (Yang and King, 1980). With this enzyme, only one of four histidines interacted with ANS, His-48 (identified as His-47 in this paper). The cationic phospholipase A_2 (pI 10.6)from *N. nigricollis* was found by the same workers to have a number of similar properties. Both Ca^{2+} and ANS protected against inactivation by bromphenacyl bromide although the mechanism of protection by Ca^{2+} and ANS appeared to be different. This difference in protection was shown by the synergistic effect of ANS on Ca^{2+} protection. Also, Ca^{2+} appeared to alter the ANS binding site by creating a more hydrophobic environment as shown by an increase in the fluorescent emission quantum yield of the ANS.

The phospholipase A_2 from *Bitis gabonica* venom also binds two molecules of Ca^{2+} per dimer (Viljoen *et al.*, 1975). In this case, the nature of the tryptophan spectral changes suggested that both charge and solvent effects were involved. These workers concluded that a conformational change occurred which could have induced the solvent effects. In addition, this conformational change was proposed to enhance substrate binding. Since this enzyme acts *via* an ordered bi–ter mechanism in which Ca^{2+} adds first, such a conformational change might be expected. As with all other phospholipases with His-48 at the active site, the binding of Ca^{2+} was pH dependent and regulated by the residue with a pK of 6.4 (Viljoen and Botes, 1979). This value is somewhat higher than that found by van Eijk *et al.* (1984*b*), who showed that

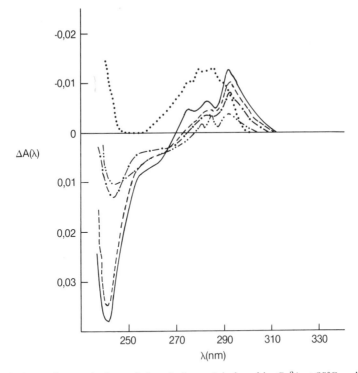

Figure 10-3. Spectral perturbations of phospholipase A induced by Ca^{2+} at 25°C and in 0.01 M Tris–0.05 M NaCl, pH 7.8. A protein concentration of 0.5 mg/ml was used in the experiment. · · · ·, 1×10^{-5} M Ca^{2+}; ------, 10×10^{-5} M Ca^{2+}; -·-·-, 25×10^{-5} M Ca^{2+}; ---, 50×10^{-5} Ca^{2+}; —, 100×10^{-5} M Ca^{2+}. The cell in the reference position contained the added cation. (From Viljoen *et al.*, 1975.)

Ca^{2+} binding perturbed the protonization of Asp-49 with a $pK_a = 5.2$. Unlike the enzyme from *C. adamanteus*, the *B. gabonica* enzyme also showed a major spectral change at 240–245 nm upon Ca^{2+} binding (Viljoen *et al.*, 1975) (Fig. 10-3). The N-terminal octapeptide of the *Trimeresurus flavoviridis* phospholipases was found to be responsible for increasing the pK_a for Ca^{2+} from 7.6 to 8.7 (Ohno *et al.*, 1984).

Modification of the *B. gabonica* enzyme by bromphenacylation abolished the shift at 242 nm but not at 279 nm induced by Ca^{2+}, which indicated that, although the enzyme still could bind Ca^{2+}, the binding region was altered. Three Ca^{2+} binding residues have now been proposed for this enzyme with pK values of 5.66, 6.75, and 9.15 involving Asp-46, His-45, and Tyr-25, respectively (Viljoen and Botes, 1979). Hydrogen ion was a noncompetitive inhibitor of Ca^{2+} and substrate binding, unlike the *C. adamanteus* enzyme that exhibited uncompetitive inhibition (Wells, 1974).

The neurotoxins from the venoms *Notechis scutatus scutatus* (notexin, Notechis II), *Oxyuranus scutellatus* (taipoxin γ chain), and *Bungarus multicinctus* (β-bungarotoxin, B chain) have subunits that bind Ca^{2+} in a manner similar to the porcine pancreatic phospholipase A (Verheij *et al.*, 1981b). The subunit that binds Ca^{2+} is catalytically active and involves Ca^{2+} binding at His-48 in a stoichiometric fashion. The enzyme from the first two venoms was measured

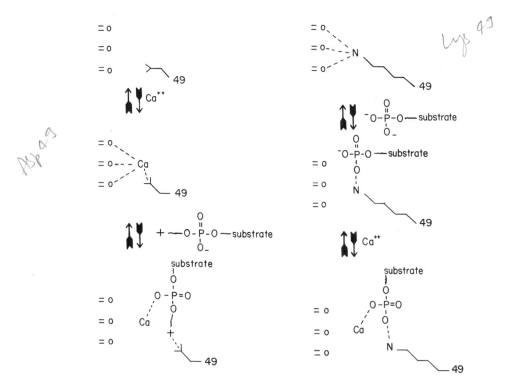

Figure 10-4. The proposed differences in the Ca^{2+} and substrate binding by the Asp-49 (*left*) and the Lys-49 (*right*) phospholipases A_2. (From Maraganore *et al.*, 1986a.)

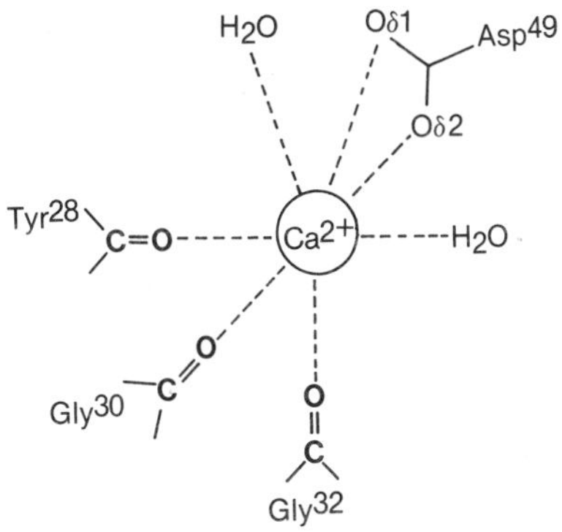

$Mechanism\ of\ Phospholipase\ A_2\ Action$

Figure 10-5. Schematic representation of the calcium ion and its ligands. (From Verheij *et al.*, 1981*b*.)

by spectral perturbations similar to the porcine enzyme while the binding of Ca^{2+} of the *B. multicinctus* enzyme was measured by ANS fluorescent changes (Abe *et al.*, 1977). The binding constant for Ca^{2+} by notexin ranged from 1.4×10^{-4} M for the native enzyme to 2.5×10^{-2} M for the phenacylated enzyme, at pH 7.4 (Halpert *et al.*, 1976).

Although some differences are noted between various species of snake venom and the pancreatic phospholipases, it is clear that Ca^{2+} binding involves His-48 that can be readily detected by spectral changes in most, but not all, instances. Also, it is clear from these various results that significant differences exist in the Ca^{2+} binding region of this class of phospholipases. These differences are the result of amino acid residue substitutions that can occur at a distance from the Ca^{2+} region (*loop*). Also, spectral and fluorescent studies show that there is an interplay between the binding of Ca^{2+} and the model substrate ANS. It will be of interest to use these techniques in concert with sequence and structural information to understand better the active site of venom and pancreatic phospholipases. Since most of the phospholipases A_2 described thus far have Asp-49 as an essential part of the Ca^{2+}-binding loop, special consideration must be given to its substitution by Lys. This replacement of an anionic group by the cationic Lys described by Maraganore *et al.* (1986*a*) accounts for the *A. p. piscivorus* phospholipase binding Ca^{2+} only after binding the substrate. Figure 10-4 shows the proposed differences in the Ca^{2+} and substrate binding by the Asp-49 (*left*) and Lys-49 (*right*) phospholipases A_2. Regardless of whether the initial binding of the enzyme is with Ca^{2+} (Asp-49) or with substrate (Lys-49), the same substrate–Ca^{2+} complex and mechanism of hydrolysis are proposed to occur. (Figure 10-5 shows the complete Ca^{2+}-binding loop.)

10.2.2. Pancreatic Phospholipases A₂

The porcine phospholipase A_2 as well as its precursor zymogen undergo spectral changes upon the binding of Ca^{2+}. Pieterson *et al.* (1974) found major peaks at 242 nm and smaller peaks at 282 and 288 nm. They attributed this change to a conformational rearrangement in the enzyme that resulted in a shift of a Tyr to a more polar environment. The pancreatic enzymes do show synergism between Ca^{2+} and substrate (or the ANS analog) binding (Pieterson *et al.*, 1974). As proposed for the venom enzymes, His-48 was also involved as well as carboxylate with an apparent pK = 5.2 tentatively assigned as Asp-49 (Fleer *et al.*, 1981). No fluorescent changes in the enzyme or its precursor were found upon Ca^{2+} binding. The absorption differences were pH dependent and allowed the estimation of Ca^{2+} dissociation constants; at pH 4.0, $K_D = 10^{-1}$ M while at pH 10.0, $K_D = 2 \times 10^{-4}$ M. The difference between the spectra of the pancreatic enzyme (or its precursor) and the venom enzymes is suggestive of a change in the charge in the region of a tryptophan residue which points out a fundamental difference in the architecture of the two classes of enzyme.

That a tyrosine undergoes a shift on Ca^{2+} binding by the pancreatic enzymes is supported by studies with Tb^{3+} as a luminescent probe. Brittain *et al.* (1976) found that the intensity of the green emission of Tb^{3+} at 545 nm was enhanced when the enzyme was illuminated in the tyrosine region, 280 nm (Tyr-28). They also concluded that tyrosine (Tyr-27) was involved in Tb^{3+} binding of the *C. adamanteus* but not the bee venom enzyme. In the latter case, a tryptophan residue (Trp-8) was thought to be involved. This conclusion is supported by the Ca^{2+}-*binding loop* proposed for the bee venom enzyme (Maraganore *et al.*, 1986*b*). The bovine enzyme that also has Tyr-28 had only 10% the intensity of the porcine or horse enzyme. Such studies should be applied further to phospholipases as the three-dimensional structures of the enzymes become better known.

The binding of $^{43}Ca^{2+}$ to the porcine pancreatic phospholipase was investigated using NMR measurements (Andersson *et al.*, 1981). The association constant found by these workers (2.5×10^{-3}) is in close agreement with values obtained by gel filtration and ultraviolet differential spectroscopy (2–4×10^{-3}M). Additional information was obtained that suggested that the Ca^{2+} binding site became more rigid and less accessible to income by a second incoming Ca^{2+}. The on-rate for the phospholipase A_2 was calculated to be at least 100-fold slower than that found for the Ca^{2+} on-rate to the regulatory sites of troponin C. Likewise, it was concluded that the bound Ca^{2+} has relatively low internal mobility. More recently, Drakenberg *et al.* (1984) extended these studies and proposed that the porcine prophospholipase A_2 had a more symmetrical Ca^{2+} binding site than that found with the activated phospholipase A_2.

A second Ca^{2+} binding site with a $K_D = 2 \times 10^{-2}$M (pH 7.5) was initially thought to be located in the N-terminal region involving Ala-1 (Slotboom *et al.*, 1978). Recent evidence suggests that a variable Glu-71 carboxylate of the pancreatic phospholipase A_2 is responsible for the binding of the second Ca^{2+} (Donne-Op den Kelder *et al.*, 1983). The binding of the second Ca^{2+} stabilizes interfacial binding at higher pH, presumably by shielding an internal salt bridge between the protonated amino terminus and a buried carboxylate. Spectral studies show that the binding of the second Ca^{2+} perturbs Trp-3. The low-affinity binding of Ca^{2+} at the amino terminus is independent of the high-affinity binding at the active site; blockage of the active site with 1-bromo-2-octane does not influence the low-affinity binding.

10.2.3. Model of the Ca²⁺-Binding Loop

The aforementioned studies, along with the x-ray structural analysis described in Chapter 9, have given rise to a model for Ca^{2+} binding in a region termed the *Ca²⁺-binding loop*, an octahedron with Ca^{2+} at the apex (Verheij *et al.*, 1981*b*) (Fig. 10-5).

In this model, the dihydrated Ca^{2+} is bound by the carbonyl oxygens of Tyr-28, Gly-30, and Gly-32 and the oxygens of the carboxyl group of Asp-49. While this proposal appears to be valid for all enzymes with Asp-49, it is clear that a different structure exists for the Lys-49 phospholipases A₂.

A proposal that relates the protein structure to ordered sequence of

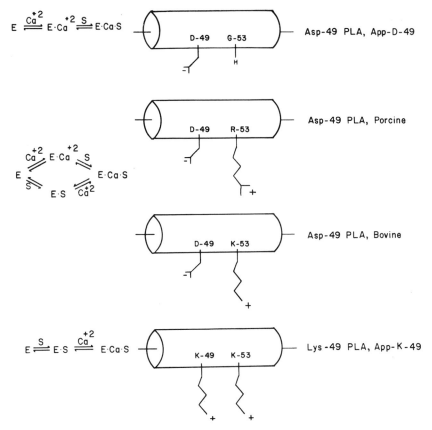

Figure 10-6. Schematic representation of the proposed relation between cationic and/or anionic side chains at positions 49 and 53 and the ordered or random formation of the catalytic complex. (From Maraganore and Heinrikson, 1985.)

substrate and Ca^{2+} was made by Maraganore and Heinrikson (1985). In addition to the *Ca^{2+}-binding loop* and its modification in the Lys-49 enzyme, this proposal also postulates a role for the variable residue at position 53 which, in part, accounts for the order of addition (Fig. 10-6). Three lines of evidence were cited that implicate the residue at position 53 in the active site. First, it is one turn of the α helix from position 49 and therefore oriented properly. Second, monomolecular substrate enhanced the reaction of trinitrobenzene sulfonate with Lys-53 of the *A. p. piscivorus* Lys-49 enzyme. Third, an essential Arg was implicated in activity of the porcine phospholipase A_2 by Vensel and Kantrowitz (1980). This enzyme, which has Arg at position 53, has the same random-order mechanism as the bovine pancreatic phospholipase A_2. As shown in Fig. 10-6 the phospholipase A_2 with an anionic site, *A. p. piscivorus* and *C. adamanteus* with Asp-49 and Gly-53, must first bind Ca^{2+} (Wells, 1972), the zwitterionic pancreatic enzymes have a random order (deHaas *et al.*, 1971), and the cationic Lys-49 and Lys-53 must bind substrate before Ca^{2+} can enter. Therefore, charge of this expanded *Ca^{2+}-binding loop* that includes the residue at position 53 is thought to regulate the order of addition of substrate and Ca^{2+} to the enzyme.

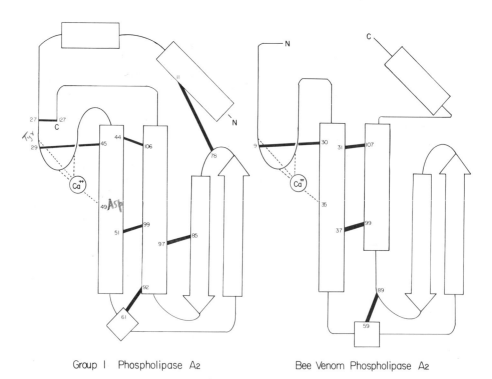

Group I   Phospholipase A_2 Bee Venom Phospholipase A_2

Figure 10-7. Comparison of the Ca^{2+} binding regions of group 1 and bee-venom phospholipases A_2. (From Maraganore *et al.*, 1986*a*.)

10.2.4. Bee Venom Phospholipase A₂

Little is known about the three-dimensional structure of the bee venom phospholipase A_2. Indeed, as pointed out in Chapter 9, assignment of the disulfide bridges is not certain. Nonetheless, Maraganore *et al.* (1986*a*) have proposed the following model for the Ca^{2+} binding region, based on their assignment of disulfide bridges (Fig. 10-7). In this proposal, the binding is to the carbonyl oxygens of Trp-8, Gly-10, Gly-12, and the free carboxyl of Asp-49.

10.3. Phospholipase Substrate Binding

10.3.1. General Considerations

Clearly, the binding of an amphipathic molecule such as a phospholipid in the active site of an acyl hydrolase phospholipase A must accommodate insertion of the polar phosphoryl base group into the active pocket through the *hydrophobic wall*, which is closely associated with the active site. Accordingly, a number of changes in the structure of the enzyme molecule would be expected upon substrate binding that are monitored by the aromatic amino acid residues in this hydrophobic region. With the recognition that two types of binding of substrate to phospholipases can occur, spectral changes induced by monomeric versus aggregated lipid can be used to probe the microenvironment of the two types of site. Coupling of this approach to chemical modification or substitution of particular residues yields additional information on the interaction of various residues with one another and with substrate. Often, nonhydrolyzable substrate analogs are used in studies of substrate binding, provided that they behave as competitive inhibitors. Furthermore, monomer binding studies require that the dissociation constant be below its CMC (Verheij *et al.*, 1981*a*) although evidence exists that true monomer–enzyme interaction is complex (see Fig. 10-11). Some compounds that have been used successfully include the D isomers of short-chain phosphatidylcholines, 1-acyl lysophosphatidylcholines, and *N*-alkyl phosphocholines.

10.3.2. Interaction with Monomers

10.3.2.1. Venom Phospholipase A₂

Roholt and Schlamowitz (1961) first described kinetically a venom (*C. durissus terrificus*) phospholipase A_2 using monomers of dihexanoyl phosphatidylcholine. Most important in this study was the observation that the hydrolysis of dihexanoyl phosphatidylcholine was enhanced severalfold by the addition of micellar lysophosphatidylcholines and that emulsions of dioctanoyl phosphatidylcholine were degraded 70 times faster than soluble dihexanoyl phosphatidylcholine.

Dennis and coworkers have shown that the *N. naja naja* phospholipase A$_2$, while not particularly active on monomers of short-chain phosphatidylcholines, can be activated by these substrates to hydrolyze phosphatidylethanolamine present at interfaces. Likewise, the Lys-49 phospholipase A$_2$ from *A. p. piscivorus* could degrade monomers; however, it required an anionic substrate such as phosphatidic acid (Maraganore *et al.*, 1986*a*). By comparison, the Asp-49 enzyme from the same species could degrade monomers of phosphatidylcholine but not phosphatidate. Both enzymes degraded dihexanoyl phosphatidylcholine optimally when present in an interface. While the differences between activity on monomers and micelles by the Lys-49 enzyme are not resolved as yet, it is clear that charge interactions are important in substrate binding at the active site.

10.3.2.2. Pancreatic Phospholipase A$_2$

Binding of monomers of D-diheptanoyl phosphatidylcholine to the porcine pancreatic phospholipase A$_2$ and its zymogen induced spectral change with peaks at 282 and 288 nm assigned to a Tyr near the active site, probably Tyr-52 (Pieterson *et al.*, 1974) (Fig. 10-8). From these data as well as kinetic and gel filtration studies, the $K_{S(m)}$ for the monomer substrate was calculated to be 0.4–0.9 × 10^{-3} for the activated enzyme and 0.7–1.5 × 10^{-3} for the zymogen. The binding of monomers is primarily due to hydrophobic bonding; with increasing acyl chain length of 1-acyl lysolecithin from 7 to 14 carbons, the K_D decreased from 43 to 0.06 mM (Verheij *et al.*, 1981*a*). Inactivation of the porcine and horse pancreatic phospholipases with bromphenacyl bromide blocked binding of the monomers. The affinity of the porcine enzyme for monomers was independent of Ca^{2+} between pH 4.0 and pH 7.0 but decreased with pH values above 7.0 in the absence of Ca^{2+}. Nitration of the enzyme with tetranitromethane modified Tyr-69 and in some cases Tyr-19

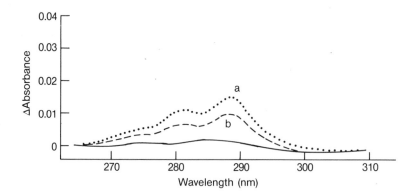

Figure 10-8. UV absorption difference spectrum of phospholipase A (curve a) and the zymogen (curve b) induced by D-diheptanoyllecithin monomers. Buffer, 0.05 M NaAc–0.1 M NaCl-10^{-3} M ethylenediaminetetraacetate (pH 6.0); protein concentration, 60 μM; D-diheptanoyllecithin concentration, 1.2 mM. (From Pieterson *et al.*, 1974.)

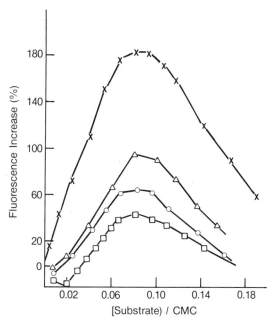

Figure 10-9. 8-Anilino-1-naphthalene sulfonate (ANS) fluorescence at various substrate : CMC ratios in the presence of 7.7 μM phospholipase A₂. Conditions: 10 mM 1,4-piperazinedithane sulfonic acid buffer, 25 mM SrCl₂, $T = 25°C$, [ANS] = 16.1 μM. □, (*n*-Heptanoylthio)glycol sulfate, R = 6; ○, (*n*-octanoylthio)glycol sulfate, R = 7; △, (*n*-nonanoylthio)glycol sulfate, R = 8; ×, (*n*-decanoylthio)glycol sulfate, R = 9. (From van Oort *et al.*, 1985*a*.)

and Tyr-124 but did not attack Tyr-52. Monomer binding at pH 6.0 produced spectral changes consistent with perturbations at NO₂–Tyr-19 and NO₂–Tyr-69. Dansylation of Tyr-19 and Tyr-69 increased the affinity for lipid which could be monitored by enhanced fluorescence of the dansyl groups. At higher pH values, binding of the NO₂-modified enzymes decreased, presumably the result of charge repulsion between NO₂–Tyr-69 and the phosphate group of the substrate. This demonstrates that while hydrophobic interaction is a dominant factor in substrate binding, charge requirements of the amino acid residues must be met.

In addition to the spectral changes noted, hydrophobic binding of ANS to the enzyme was enhanced by monomers of single-chain substrate analogs (van Oort *et al.*, 1985*a*). As shown in Fig. 10-9, this enhancement was dependent on the ratio of the substrate used to the CMC of the lipid. The acyl chain of the analog also determined the magnitude of the ANS fluorescent change. As described later (see Fig. 10-11), this effect is directly related to aggregation of the enzyme with monomers yielding a complex with molecular weight ranging up to 90,000 daltons. It would therefore appear that interpretation of studies using substrates below their CMC must take into account a *seeding* effect that results in enzyme–substrate aggregation.

10.3.3. Interaction with Substrate Micelles

10.3.3.1. Micelle Formation and Influence of Pancreatic Phospholipase

A number of studies have shown that there is not a clear distinction between the interaction of the phospholipases A with monomers below the

CMC and monomers plus aggregates above the CMC when a lipid interactive protein such as phospholipase is studied. One way in which this has been determined is by Gd^+–phospholipase–alkyl phosphatidylcholine interaction using [^{1}H] NMR relaxation studies (Hershberg *et al.*, 1976). They concluded "that binding of monomeric *N*-alkyl phosphatidylcholines to the enzyme completes effectively aggregation equilibrium and increases the apparent CMC of the amphiphile." It therefore appears that a phospholipase can bind a number of molecules of amphiphile below the CMC of pure amphiphile. These workers found that the spectra observed for enzyme–Gd^{3+}–micelle formation shifted with increasing enzyme concentration, consistent with monomer amphiphile–enzyme interaction, effectively lowering the monomer concentration. The binding of more than one molecule of amphiphile per enzyme molecule did not appear to induce micelle *seeding* since the spectral shift did not occur until higher concentrations of amphiphile were reached. The binding of micellar substrate induced a major enhancement in proton relaxation rate that indicated binding of the micelle induced a conformational change in the enzyme that was transduced to the active (Gd^{3+}) site. The nature of micelles was dependent on the structures of the lipid as well as the solute composition (Tanford, 1976; Small, 1986). For example, under comparable conditions dihexanoyl phosphatidylcholine forms micelles with molecular weights of 15,000–20,000 daltons, whereas diheptanoyl phosphatidylcholine forms micelles of 20,000–100,000 daltons (Tausk *et al.*, 1974). The former contains about 34 molecules per micelle. Increasing salt concentration from 0 to 3 M increases the ideal molecular weights of the micelles some 25–30%. Based on the dimensions of the molecule and the size of the micelle, Tausk *et al.* (1974) suggested that the structure of dihexanoyl phosphatidylcholine might be spherocylindrical. Furthermore, the formation of micelles is not a simple process. The nature of the structures formed as the CMC is approached varies although the precise nature of the different micelles existing at lipid concentrations around the CMC is not yet clear.

Proton NMR studies have shed some light on the nature of micelles formed at the CMC (Hershberg *et al.*, 1976). At concentrations of dihexanoyl phosphatidylcholine one to four times the CMC (10–40 mM), rapid exchange occurs between the individual molecules in the monomeric and micellar states. Above this range, however, the signals for the α-methylene protons of the acyl chains indicate a second, slow-exchanging environment. The structure formed at high lipid concentration was termed the second CMC. Based on these observations, it would appear that the phospholipases A_2 studied thus far attacked the freely exchanging phospholipid molecules more rapidly than monomers but less well than molecules in micelles with restricted exchanges. It is possible that the high activity on the poorly exchanging substrate micelles is due to a change in the orientation of the acyl–ester bonds described in the beginning of this chapter.

Anomalous kinetic behavior was observed for the porcine phospholipase A_2 around the CMC of the substrate (Fig. 10-10). In this case, the substrate thioester analogs of phosphatidylcholine were used as the substrates (Volwerk *et al.*, 1979). The thiol produced by hydrolysis was coupled with Elman's

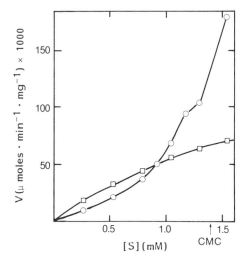

Figure 10-10. Hydrolysis rates of 2-(decanoylthio)ethyl phosphatidylcholine by phospholipase A₂ (○) and by prophospholipase A₂ (□) at pH 6.0 in the presence of 10 mM CaCl₂, 100 mM NaCl, and 200 mM sodium acetate at 25°C. (From Volwerk *et al.*, 1979.)

reagent, permitting a spectrophotomeric assay. Hydrolysis of 2-(decanoyl-thio)ethyl phosphatidylcholine exhibited a preactivation just below its CMC (1.3 mM). The zymogen proenzyme, while degrading monomers of this substrate, did not show interfacial recognition for any aggregated form of substrate. The use of the thiol-containing substrate yielded some additional new results as well. For example, the product of hydrolysis, 2-mercaptoethyl ly-sophosphatidylcholine, is water soluble and therefore has no hydrophobic affinity for the enzyme. Since interfacial activation of the processed enzyme was still observed, it would appear in this case that enhanced product diffusion at the interface did not account for interfacial activation. Also, the proenzyme had a rather broad pH optimum, compared with the activated enzyme. Based on differences in rates and pH activity profiles, it was suggested that the formation of the enzyme–product complex could be rate limiting for the proenzyme while product release is rate limiting for the processed enzyme.

The model lipids *n*-octadecyl phosphocholine and *n*-hexadecyl phospho-choline have 200 and 155 molecules per micelle, respectively (deAraujo *et al.*, 1979; Hille *et al.*, 1981). The studies with this and dihexanoyl and diheptanoyl phosphatidylcholine demonstrate the marked change in micellar structure that results from the addition of a single methylene group to the acyl chain, whether it be a monoacyl or diacyl phospholipid. As described in the preceding section of this chapter, the presence of the porcine pancreatic phospholipase A₂ drastically altered the micellar structure of the two model lipids. In both cases, the number of lipids in the micelle decreased by 50% and two enzyme molecules bound per micelle (deAraujo *et al.*, 1979) (Fig. 10-11).

The mechanism of enzyme–lipid interaction is thought to proceed via pathway B; that is, direct insertion of the two enzyme molecules (indicated

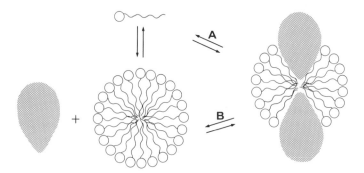

Figure 10-11. Schematic view of the pathways for the formation of a complex between phospholipase A_2 and micelles of *n*-hexadecanyl phosphatidylcholine. (From Slotboom *et al.*, 1982.)

by stippled figure on left) into the micelle occurs rather than an association of enzyme with soluble monomers. The arguments of deAraujo *et al.* (1979) were based primarily on studies in which the proenzyme and activated pancreatic enzyme were compared. However, a number of studies have now shown that both pancreatic and venom phospholipases A_2 can be aggregated and are activated by substrates below the CMC of the lipid. The results of various studies on the aggregation of phospholipases suggest that the activation of the enzymes can occur under conditions where aggregation of the enzyme occurs below the CMC of the substrate. This apparently does not occur when the pancreatic phospholipase A_2 interacts with zwitterionic substrate. However, as described below, this is not the case when anionic substrate is used and aggregation does occur.

The protein–lipid association is characterized by low temperature-dependent enthalpy and positive temperature-independent entropy change characteristic of hydrophobic bonding. Since the enthalpy change of demicellarization is low, the major enthalpy change was attributed to lipid–protein interaction; however, the contribution of protein conformational changes could not be assessed. A protein conformational change would be expected to be enhanced by hydrophobic interaction with lipid.

Since it was concluded that the lipid–protein interaction was hydrophobic primarily, it would appear that the enzyme binding perturbs only a limited region of the lipid structure, allowing for variation in the amount of lipid present in the complex. Hille *et al.* (1981) cited unpublished experiments that the overall environment of the acyl chains is not changed significantly upon complex formation, as determined by [^{1}H] NMR studies.

More recent studies by Hille *et al.* (1983*a*, 1983*b*) showed that the anionic substrate analog S-*n*-alkanoylthioglycol sulfate (or *n*-alkyl sulfate) causes aggregation of the porcine pancreatic phospholipase A_2. At concentrations equal to 0.07 times the CMC, a complex was formed with six enzymes and 40 detergent molecules. These results give further evidence that pathway A (Fig. 10-11) is important. It was concluded that the initial event involved two mono-

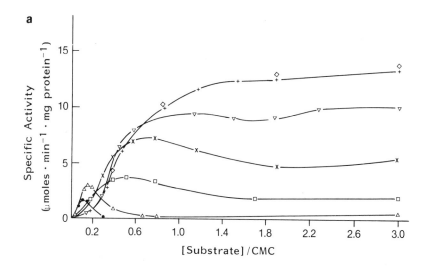

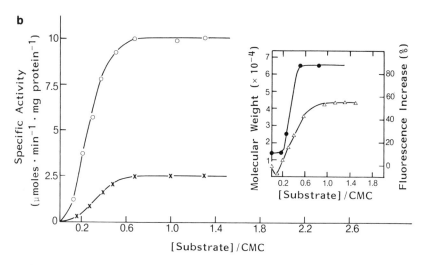

Figure 10-12. (a) Specific activity of porcine pancreatic phospholipase A₂ as function of (*n*-decanoylthio)glycol sulfate concentration between pH 6 and pH 10. Conditions: (pH 6.0) 1 mM 1,4-piperazinediethane sulfonic acid, 25 mM CaCl₂; (pH 7.0) 1 mM 4-(2-hydroxyethyl)-1-piper-azineethane sulfonic acid, 25 mM CaCl₂; (pH 8.0–10.0) 1 mM borate, 25 mM CaCl₂; $T = 25°C$. ●, pH 6.0; △, pH 7.0; □, pH 8.0; ×, pH 8.5; ▽, pH 9.0; +, pH 9.5; (◇), pH 10.0. (b) Specific activity as function of substrate: CMC ratio at pH 9.0: (*upper curve*) (*n*-decanoylthio)glycol sulfate (○); (*lower curve*) *n*-decanoylglycol sulfate (×). Conditions: 1 mM borate buffer, 25 mM CaCl₂, pH 9.0, $T = 25°C$. (*Insert*) Molecular weight of the phospholipase A₂ substrate complex as a function of (*n*-decanoylthio)glycol sulfate concentration [left ordinate (●)]. The right ordinate gives the 8-anilino-1-naphthalene sulfonate (ANS) fluorescence increase as a function of (*n*-decanoylthio)glycol sulfate concentration (△). Conditions: 10 mM borate buffer, 25 mM SrCl₂, pH 9.0, $T = 25°C$, [ANS] = 16.1 μM. (From van Oort *et al.*, 1985a.)

mers of detergent bound to the *hydrophobic surface region* (interfacial recognition site), leading to the polymerization. Below this concentration of detergent (0.07 times the CMC), only low activity was noted whereas with an increase to only 0.08 times the CMC there was a marked activation of hydrolysis. In many respects, this is similar to the polymerization and activation of venom phospholipases A_2. The authors left the interpretation of the significance of polymerization open and pointed out that the proenzyme, while aggregating, was not activated. This indicated that a conformational change in the protein was still required for full hydrolytic activity. It is of interest that neutral substrates led to dimerization of the pancreatic phospholipase A_2 (deAraujo *et al.*, 1979) whereas the anionic substrate caused hexamer formation (Hille *et al.*, 1983*b*). The pH was crucial to the aggregation when anionic substrates, *N*-acylglycol sulfates, were used (van Oort *et al.*, 1985*a*). It was concluded that charge–charge repulsion occurred when an excess of anionic substrate was present that could be overcome at higher pH values. Indeed, Fig. 10-12a shows that aggregate molecular weight, ANS fluorescence, and specific activity remained high when studied at pH 9.0, relative to that observed at lower pH values (Fig. 10-12b). Although not proven, these authors suggested that at high pH values charge–charge repulsion of enzyme and substrate prevented excess substrate binding and permitted the maintenance of a more stable enzyme–lipid aggregate. This conclusion was supported by the finding that the aggregate molecular weight of the anionic substrate–enzyme complex was about 50% lower than that of the zwitterionic substrate–enzyme complex (*n*-acylglycolphosphocholines). Using the diacyl analogs of these substrates, similar aggregations were observed, including aggregation of the prophospholipase A_2 (van Oort *et al.*, 1985*b*). The proenzyme, however, was not activated by aggregation.

The spectral changes induced by micelle binding to the interfacial recognition site of the porcine pancreatic phospholipase has been assigned to Trp-3. This is possible since monomer binding had little effect on the Trp spectra (Slotboom and deHaas, 1975). When the CMC of the lipid was exceeded, the spectra suddenly changed, characterized by more pronounced absorption at 300 nm and an increased trough at 260 nm (van Dam-Mieras *et al.*, 1975). The binding of micelles produced spectral shifts that were dependent on pH. Deprotonation of the α-amino group led to a loss in Trp absorbance, and Ca^{2+} helped to stabilize the bridge between substrate and enzyme. Two observations permitted the assignment of these shifts to Tyr-3: First, the semisynthetic modified enzyme, amidated [Ala[1], Leu[2], Phe[3]], did not show any spectral shift in the presence of micelles; and second, the zymogen that did not bind micelles and the Trp did not show a spectral shift. The fluorescent properties of Trp also monitored micellar binding and actually provided an easier method for studying this interaction. This was characterized by a shift in the emission wavelength from 340 to 330 nm and a marked increase in the fluorescent intensity. However, it was not certain that this fluorescent shift, characteristic of a hydrophobic environment, was the result of Trp-3 interaction with lipid acyl chains or an interaction with the apolar residues of the *hydrophobic wall* within the enzyme.

10.3.3.2. Micelles as Substrates for the Venom Phospholipases

There are many similarities between the action of the pancreatic and venom phospholipases on micelles, especially in the range of the CMC. Wells found that below the CMC of dihexanoyl phosphatidylcholine, a K_M value of 4 mM and V_{max} of 6.9 μmol min^{-1} mg^{-1} were obtained. However, when the CMC was passed (about 10 mM) but at a concentration below 25 mM, anomalous results were obtained resulting from a concentration-dependent change in the nature of the micelle formed (Wells, 1974) (Fig. 10-13).

This anomalous region is characterized by parabolic rather than hyperbolic kinetics. The molecular basis of the changes in organization has not yet been determined although the range of micellar size changes has been established (Allgyer and Wells, 1978). Above 25 mM the plot of [S]/V versus [S] was again linear. The addition of 2 M KCl both lowered the CMC to 3 mM and shifted the concentration at which the anomalous behavior was observed downward. Since the activity on monomers was not affected by KCl, it appeared that KCl had no direct effect on the enzyme.

A similar study was done on the phospholipase D purified from savoy cabbage. This enzyme has kinetic properties quite different from venom phospholipase A₂ when assayed using monomers of dihexanoyl phosphatidylcho-

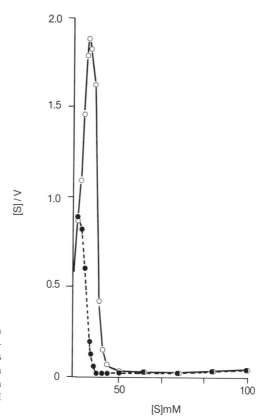

Figure 10-13. Effect of the concentration of dihexanoyllecithin on the rate of hydrolysis by phospholipase A₂. Reaction was carried out at 45°C and pH 8.0; ○, reaction in the presence of 1 mM Ca^{2+}; ●, reaction in the presence of 1 mM Ca^{2+} and 2 M KCl. (From Wells, 1974.)

line (Allgyer and Wells, 1979). In this case, the reaction velocity was a linear function of enzyme concentration at different concentrations of monomeric substrate, rather than the expected hyperbolic relation. The dependency of activity on substrate concentration was parabolic, similar to the effect seen with phospholipase A_2 at the CMC of the substrate. The phospholipase D, however, showed enhancement in activity at 4.3 mM dihexanoyl phosphatidylcholine where only monomers were present. No change in rate was noted above the CMC. While this would be expected of a multisited enzyme exhibiting cooperativity, Hill plots were not linear, leaving the authors with the conclusion that cooperativity was not responsible.

van Eijk *et al.* (1983) compared the kinetics of the phospholipases A_2 from *N. melanoleuca* and porcine pancreas using micelles of diheptanoyl phosphatidylcholine (Fig. 10-14). The venom enzyme does not have biphasic kinetics as found with the pancreatic enzyme. As shown with a number of other phospholipases A_2, polymerization (in this case tetramers) occurred below the CMC of the substrate. Unique to this study was the demonstration that polymerization required the presence of Ca^{2+} and that this interaction caused a spectral shift at 293 nm indicative of a cooperative formation of the complex.

Chemical conversion of the α-amino to a keto group lowers the affinity of three venom (*N. melanoleuca*, *C. atrox*, and *V. berus*) phospholipases for micelles of *n*-hexadecyl phosphatidylcholine (Verheij *et al.*, 1981a). This change was quantitated by lipid-induced shifts in the ultraviolet absorption at 277 and 290 nm. The shape of the spectra indicated that two tyrosines in the *Viberus* enzyme (no tryptophan) are perturbed by lipid binding whereas only one is involved in the α-amino modified enzyme. The modified enzyme is catalytically active on monomers and monomolecular films at low surface pressures but is unable to degrade micelles of lipid because of the low binding

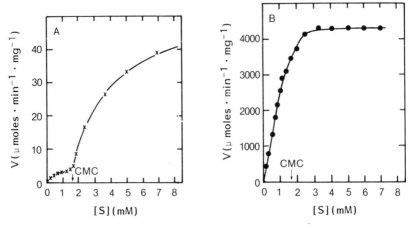

Figure 10-14. Hydrolysis of diheptanoyl glycerophosphatidylcholine catalyzed by (A) porcine pancreatic or (B) *N. melanoleuca* phospholipase A_2. Experimental conditions: (A) pH 7.0, 40°C, 1 mM $CaCl_2$, 0.5 mM Tris, 100 mM NaCl, 1 μg protein/ml; (B) pH 8.0, 25°C, 50 mM $CaCl_2$, 0.5 mM Tris, 100 mM NaCl, 10 ng protein/ml. (From van Eijk *et al.*, 1983.)

capacity induced by modification of the hydrophobic surface region. The change in binding capacity was the result of a decrease in affinity for lipid rather than a change in the stoichiometry of binding. Two possible explanations were offered for the difference in the affinity of the modified enzymes for lipid: (1) Modification induces a structural change that does not permit proper orientation of the active site and substrate, or (2) modification prevents the N-terminus α-amino group from stabilizing a conformational state with a maximal number of binding residues. In either event, the structural changes lead to a greater susceptibility of denaturation by micellar substrate.

An interesting micellar system was investigated by Wells' group, which studied the hydrolysis of reversed micelles of phosphatidylcholine in ether solutions. In this system, limiting amounts of water were used so that the enzyme was trapped in the internal water space surrounded by the polar head groups of the substrate. Their study was prompted by the early observation that the phospholipase A_2 from *C. adamanteus* venom was extremely active in the presence of ether (Hanahan, 1952). They found that phosphatidylcholine could exist in four states in ether solutions: (1) an anhydrous species, (2) a species of limited hydration with bound water only (6–7 water molecules per molecule of phosphatidylcholine), (3) a moderately hydrated state with 32–37 water molecules per molecule of phosphatidylcholine, and (4) a completely water-saturated system with up to 60 water molecules per molecule of phosphatidylcholine (Misiorowski and Wells, 1974; Poon and Wells, 1974). The *C. adamanteus* enzyme was only active in system 3 and apparently was unable to enter into system 4. It was calculated that the fully hydrated enzyme contained about 350 molecules of water and had a hydrate radius of 20 Å. The core of the micelle with a radius of 20 Å contained about 1100 molecules of water, the minimal amount of water supporting hydrolysis. It was concluded that the enzyme must be able to fit into the hydrated core of 20-Å diameter for activity. It is not clear, however, why the more fully hydrated micelles that had the size to accommodate the enzyme could not support activity. Ca^{2+}, although required for activity, acted as a competitive inhibitor with water in the reaction, presumably by forming a complex with the substrate and water that did not allow interaction with the enzyme. Likewise, the enzyme was not active in system 2 since no free water was available. These studies gave good insight into the nature of the water that participates in hydrolysis and the interactive role Ca^{2+} plays at the interface. Unfortunately, as pointed out by these workers, no information is available on the conformation of the enzyme under these conditions since such a small amount is absorbed into the aqueous compartment. However, the reversed micellar system offers an interesting approach to the study of enzymes that are stable in organic solvents.

10.3.4. Substrates in Mixed Micelles

10.3.4.1. Hydrolysis of Triton X-100–Phospholipid Micelles; N. naja naja Phospholipase A₂

There are a number of advantages in using phospholipid substrates dissolved in a detergent (mixed micelles) for kinetic analysis of phospholipases.

Perhaps most important, mixed micellar systems permit a more realistic expression of concentration than that obtained with bilayer liposomes. In the case of mixed micelles, the concentration is expressed in two rather than three dimensions (Deems *et al.*, 1975) and the concentration can be regulated by the ratio of detergent to phospholipid (Fig. 10-15).

The detergent that has been used primarily in mixed micelles is Triton X-100 (Dennis, 1983). While charged detergents such as deoxycholate have been used successfully in the study of a number of phospholipases, two problems are encountered that make those detergents less attractive: (1) The effect of charge on enzyme activity can add another variable, and (2) those detergents usually have high CMC (4–6 mM) (Mukerjee and Mysels, 1971).

The concentration expression of pure lipid aggregates defined as normal solutions has the limitation of not including the concentration of the lipid at the interface which is in the range of several molar. Thus, a difference in the concentration of substrate at the interface is crucial in the kinetic analysis of phospholipases since the enzyme is active at the interface. Likewise, in the study of substrate specificity, the mixed micellar system has the advantage of providing a more uniform dispersion of the phospholipids. Pure phospholipids assume different states of aggregation depending on their geometry and on the composition of the aqueous solution.

There are, however, limitations to the use of Triton mixed micelles. First, some phospholipases do not act optimally on the mixed micellar substrate. For example, Upreti and Jain (1978) found that the bee-venom phospholipase more rapidly attacked osmotically shocked liposomes in the presence of a large excess of hexanol (30–50 mM) than mixed micelles. While the molecular basis for this difference in hydrolysis was not established, it is possible that the enzyme bound the Triton. Such was found to be the case for the phospholipase C from *B. cereus* (Burns *et al.*, 1982). As pointed out by Burns and coworkers, the problem potentially was compounded by the enzyme binding to both micelles and monomers of Triton (CMC = 0.3 mM). In addition,

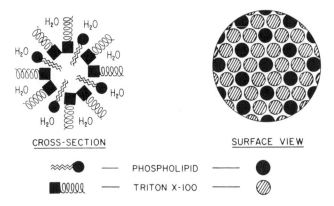

Figure 10-15. Schematic diagram of the average structure of mixed micelles of phosphatidylcholine and Triton X-100 at a stoichiometry of 2:1 Triton X-100 : phospholipid. (From Deems *et al.*, 1975.)

Zn^{2+}, which is required by this enzyme, influences the binding of Triton by phospholipase C. In an attempt to overcome these problems, detergent binding constants were estimated by gel filtration analysis and the derived dissociation constant used in the kinetic analysis of the hydrolysis of dimyristoyl phosphatidylcholine. A similar problem was encountered by Sundler *et al.* (1978) who found surface dilution inhibition when Triton X-100–phosphatidylinositol mixed micelles were attacked by a phosphatidylinositol-specific phospholipase C. On the other hand, this type of inhibition was not observed when the hydrolysis of phosphatidylinositol was measured in vesicles made with a mixture of phosphatidylcholine. This suggested that the enzyme recognized either the hydroxyl or the oxyethylene moieties of the Triton.

Another potential problem with using the Triton X-100–phospholipid mixed micelle is the heterogeneity of the commercially available Triton X-100. As shown by Robson and Dennis (1978), Triton X-100 contains a wide range of components that vary in the number of oxyethylenic units with an average of 9.5 units per monomer.

$$—C_6H_4—$$
$$(CH_3)_3C\ CH_2C(CH_3)_2 \longleftarrow\hspace{-0.2cm}\bigcirc\hspace{-0.2cm}\longrightarrow O(CH_2CH_2O)_{9.5}H$$
$$(OPE_{9.5})$$

These workers also found that the micelles of Triton X-100 were heterogeneous in size when compared with Triton with only nine oxyethylenic units, OPE_9, as measured by gel filtration. Likewise, the size of the micelle was dependent on the ratio of phospholipid to OPE_9; the higher the percentage of phospholipid to OPE_9, the larger the size. The micelles of pure OPE_9 contained about 170 OPE_9 monomers and were thought to be in the form of an oblate ellipsoid. Mixed micelles that contained dipalmitoyl phosphatidylcholine and OPE_9 (1 : 2) had an apparent molecular weight of 1.2×10^5 daltons. Although these differences in size theoretically could influence hydrolysis, studies thus far have not suggested that the size difference of micelles is a significant problem.

The first detailed kinetic study of mixed micelles was in 1973 (Dennis, 1973*b*) when Dennis found that above a phosphatidylcholine : Triton ratio of 1 : 2, surface dilution occurred that decreased the activity of *N. naja naja* phospholipase A_2. Shortly thereafter, Deems *et al.* (1975) described a kinetic approach to the study of phospholipases that assumed a two-step process. First, the enzyme interfacially adsorbed to the micelle with subsequent binding of the substrate phospholipid molecule to the active site of the enzyme. (As will be described later in this section, the initial binding is now thought to be specific for the phospholipid in the micelle rather than a nonspecific interaction.) It was possible using this approach to determine the first step independent of the second step by varying the total number of micelles at a fixed phospholipid : Triton ratio. This constant, K_S^A, was expressed as a molar concentration based on the total Triton and phospholipid present. The binding of the phospholipid to the active site was determined by varying the ratio of phospholipid to Triton while holding the total phospholipid and Triton constant. (As previously mentioned, the number and size of the micelles will vary since the size is dependent on the ratio of phospholipid to Triton.) From this analysis, the *surface* K_m or K_m^B can be determined and is expressed in moles/cm². V is comparable to the usual $V_{\max}$. This kinetic scheme was proposed as follows:

$$E + A \underset{k_{-1}}{\overset{k_1}{\rightleftharpoons}} EA$$

$$EA + B \underset{k_{-2}}{\overset{k_2}{\rightleftharpoons}} EAB \underset{k_{-3}}{\overset{k_3}{\rightleftharpoons}} EA + Q$$

In this scheme, E is the enzyme, A is the mixed micelle, B is the phospholipid molecule (dipalmitoyl phosphatidylcholine), and Q is the product. Deems *et al.* (1975), using this kinetic approach, determined K_S^A to be about 5×10^{-4} M, K_m^B to be $1-2 \times 10^{-10}$ moles/cm², and V to be 4 mmoles/min/mg of protein of the phospholipase A_2 from the venom of *N. naja naja*. This yields a turnover number of about 1000 moles of phosphatidylcholine per second per mole of enzyme. The major problem encountered in this approach was the high affinity of the enzyme for the phospholipid molecule in the micelle; the expression derived for this affinity for phospholipid in the micelle was above the maximal amount of phosphatidylcholine that can be dispersed in Triton as a mixed micelle. For this reason, data could not be obtained in the region of the K_m^B and the authors felt that the K_m^B was only valid within an order of magnitude.

An inherent difficulty is faced in the study of phospholipase hydrolysis of any aggregated lipid. Once the lipid in the aggregate has been hydrolyzed, either the enzyme must desorb from the micelle and readsorb to a new micelle or substrate and/or product must move freely between micelles. It was established that phospholipid does not move freely between bilayer liposomes (Kornberg and McConnell, 1971) which could be one reason that phospholipases do not optimally attack stable bilayer systems. However, we have found that phospholipid rapidly moves between mixed micelles of Triton X-100 and phosphatidylcholine. An electron spin resonance study of the relief of spin-label quenching of Triton-2 (12 doxyl) distearoyl phosphatidylcholine by non-spin-labeled mixed micelles showed that complete equilibrium of phosphatidylcholine between micelles occurred in less than 1 sec; [¹H] NMR studies extended this value to the millisecond range for exchange between mixed micelles (unpublished data). This minimal exchange rate was compared with the catalytic rate of the rat liver plasmalemma phospholipases A_1. We estimated that a molecule of this enzyme could completely degrade the phosphatidylethanolamine of 10 micelles in 1 sec. Compared with the turnover number of the *N. naja naja* phospholipase A_2, it can be estimated that a molecule of the venom enzyme could degrade the phospholipid in 100–500 micelles in 1 sec. Clearly, more information is needed to determine the limiting step in hydrolysis, the availability of substrate in a micelle, or the catalytic rate. (This general problem will be discussed in the next section on the hydrolysis of bilayer structures.)

More recently, Dennis and coworkers found that the *N. naja naja* phospholipases A_2 exhibited a phenomenon they termed *substrate specificity reversal* (Roberts *et al.*, 1979). The *N. naja naja* enzyme hydrolyzed phosphatidylcholine 10 times more rapidly than phosphatidylethanolamine when assayed separately. However, when present in the same mixed micelle, phosphatidylethanolamine was preferentially hydrolyzed at a rate 1.5–3 times higher than phosphatidylcholine, depending on the ratio of the phospholipids and Triton. Furthermore, soluble dibutyryl phosphatidylcholine and sphingomyelin, which were not substrates, stimulated the hydrolysis of phosphatidylethanolamine (Pluckthun and Dennis, 1982*b*). Based on these observations, the proposal was made that the enzyme has two types of functional site: one that is an activator and is specific for the phosphocholine moiety, and the other that is the catalytic site that has a low degree of specificity. Such an interpretation accounts for the apparent specificity when substrates are assayed singly. Since phosphatidylethanolamine cannot bind to the activator site, catalysis cannot occur even though it is readily accommodated in the catalytic site. This approach also helped to resolve the question of the requirements for the initial binding step. Since soluble dibutyryl phosphatidylcholine was an effective activator and eliminated the surface dilution effect, it was proposed that the enzyme specifically binds a phospholipid molecule in the initial step rather than a nonspecific binding to the micelle (Dennis, 1983). Although high concentrations of dibutyryl phosphatidylcholine were required to eliminate surface dilution (60 mM), the fact that no decrease in activity was noted when the Triton X-100 : phosphatidylethanolamine ratio increased from 4 : 1 to 16 : 1 indicated that once the enzyme is activated the binding and hydrolysis of substrate is not limiting. This would also suggest that the movement of phospholipid and/or enzyme between micelles is not limiting, assuming that soluble phosphatidylcholine does not favor desorption and resorption of phospholipid in the micelle.

Two recent studies further supported the hypothesis that both an activator molecule and an interface are required for optimal activity of the *N. naja naja* phospholipase A_2 (Pluckthun and Dennis, 1985; Pluckthun *et al.*, 1985). Low activity was found when dihexanoyl phosphatidylethanolamine was used as monomers or when micelles of Triton were present (Table 10-1). Furthermore, the addition of an activator choline-containing lipid did not increase activity unless Triton was added to provide the required interface. This activation was not seen when the phospholipases A_2 from porcine pancreas, *C. adamanteus*, and bee venom were studied (Pluckthun and Dennis, 1985). Importantly, only the *N. naja naja* enzyme was found to polymerize in the presence of the choline activator lipid. (It should be noted, however, that the *C. adamanteus* phospholipase A_2 is isolated as a dimer.) The relation be-

Table 10-1. Rate of Dihexanoyl Phosphatidylethanolamine[a] Hydrolysis by Cobra Venom Phospholipase A_2 in the Presence of Various Phosphocholine-Containing Compounds[b]

Addition	Concentration (mM)	Triton X-100 (8 mM)	Specific activity (μmoles/min/mg)
None	—	−	20 ± 3
None	—	+	40 ± 4
1,2-Bis[(butylcarbamyl)-oxy]-sn-glycero-3-phosphatidylcholine	5.5	−	25 ± 0
1,2-Bis[(butylcarbamyl)-oxy]-sn-glycero-3-phosphatidylcholine	5.5	+	283 ± 3
Sphingomyelin	1.0	+	866 ± 12

[a] Each assay included 1 mM dihexanoyl phosphatidylcholine and 1 mM Ca^{2+}.
[b] From Pluckthun *et al.* (1985).

tween aggregation of *N. naja naja* enzyme and activation was further strengthened by the finding that the enzyme, when immobilized on agarose beads, could not be activated by choline lipids (Lombardo and Dennis, 1985*b*).

Several examples have now been cited that implicate polymerization of phospholipases in an activation process. Some of these are summarized in Table 10-2 and show the complexity of the process. Some generalities might be suggested despite the incomplete picture. It is quite clear that for some phospholipases A_2 polymerization is required for optimal activity. In cases where this has been investigated and not yet observed, it is possible that the substrate is the activator (e.g., *N. naja naja*) or that the proper activator has not yet been found. Historically, the studies on the porcine pancreatic phospholipase A_2 are an excellent example. Under a wide variety of conditions, the enzyme was found to act as a monomer. However, it was known that the enzyme was activated by anionic bile salts and was autocatalytic. It has now been shown to be activated by anionic substrates and fatty acids and that the

Table 10-2. Activation and Polymerization of Phospholipases A_2 by Lipids

Source of phospholipase A_2	Native state	Effect of lipid addition		
		Lipid added	Polymerization	Activation
N. naja naja	Monomer	Choline-containing	+	+
		Fatty acid	?	+
Porcine pancreas	Monomer	Choline-containing (below CMC)	−	−
		Choline-containing (above CMC)	+	−
		Anionic	+	+
Bee venom	Monomer	Choline-containing	−	−
C. adamanteus	Dimer	Choline-containing	−	−
N. melanoleuca	Monomer	Choline-containing (below CMC)	+	+

activation by anionic substrates (or their analog nonsubstrates) is accompanied by polymerization (Hille *et al.*, 1983*a*, 1983*b*). Such a correlation would suggest that the polymerized enzyme might be the form active in a bilayer. However, forthcoming reports from the work of Jain and coworkers do not support this possibility.

The kinetic model for the *N. naja naja* phospholipase A₂ has now been modified.

$$\text{E} + \text{A} \underset{k_{-1}}{\overset{k_1}{\rightleftharpoons}} \text{EA} + \text{S} \underset{k_{-2}}{\overset{k_2}{\rightleftharpoons}} \text{EAS} \overset{k_3}{\rightarrow} \text{EAS} + \text{product}$$

[This scheme is a modification of the one by Hendrickson and Dennis (1984). They specified E + S in the first step. Since the nonsubstrate sphingomyelin can activate the enzyme, A is used for activator rather than specifying S, the substrate. The activator and substrate molecules are the same when a choline-containing phospholipid is the substrate.]

Thioester substrates were used to explore this kinetic model (see Chapter 1), providing a much more sensitivity analysis (Hendrickson and Dennis, 1984). The kinetics are consistent with the modified model in which the enzyme initially can bind to a phospholipid molecule, either at the catalytic or activator site. The initial binding ($K_S = k_1/k_{-1}$) and Michaelis–Menten constants ($K_m = k_{-2} + k_3/k_2$) were similar for phosphatidylcholine and phosphatidyl-ethanolamine thioesters when assayed separately. Since the V values were about fivefold higher for phosphatidylcholine thioester (440 vs. 89 μmoles/min/mg), it was suggested that the catalytic rate (k_3) rather than either of the binding steps was limited in hydrolysis. It is of interest to note that the initial binding of the thio compound was about 10-fold greater than that of the oxy compound, perhaps because of the greater hydrophobicity of the thio compound. Similar results were obtained earlier with the pancreatic enzyme (Volwerk *et al.*, 1979). If such is the case, it might be expected that dissociation of the product had been reduced as the result of the increased hydrophobicity, causing a decrease in velocity.

These authors also raise the possibility that the dual binding capacity of the enzyme could induce a conformational shift in the enzyme that permitted more efficient catalysis, comparable to the mechanism proposed for the pancreatic phospholipase A₂ (Verheij *et al.*, 1981*b*). If such were the case, the activation of the venom enzyme should be more rapid for the *N. naja naja* than the pancreatic enzyme since a lag phase is observed with the pancreatic but not a bee-venom phospholipase (Verger *et al.*, 1973), and a minimal lag in the action of the *N. naja naja* (Pluckthun and Dennis, 1985).

Four possible models have been discussed by Dennis and Pluckthun (1985) for the action of *N. naja naja* phospholipase A₂ (Fig. 10-16). Briefly, model A predicts that the activator increases the affinity of the enzyme for the substrate but not the V_{max} of the reaction, whereas model B proposes that the activator stimulates the release of product and increases the V_{max}. Both models propose that the monomeric enzyme is the activated species, as does model C. In model C, however, it is proposed that the activator, binding in

A

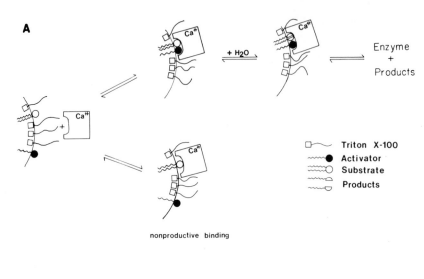

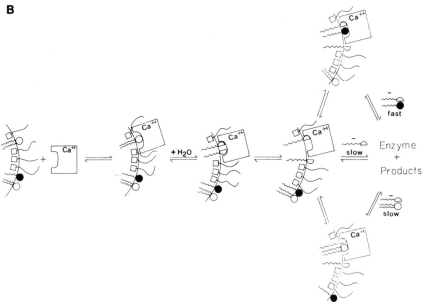

Figure 10-16. Four possible models for the action of *N. naja naja* phospholipase A_2. (A) Productive binding model; (B) product release model; (C) two-site, single-subunit model; (D), two-site dimer model. (From Dennis and Pluckthun, 1985.)

a distinct site, causes a conformational change (square) that in some fashion enhances activity. Model D differs by requiring that a dimer be formed and, as shown here, the activator molecule binds one monomer of enzyme while the substrate binds the other. As thoroughly discussed in this review by Dennis and Pluckthun, a final conclusion about which, if any, of these models is correct

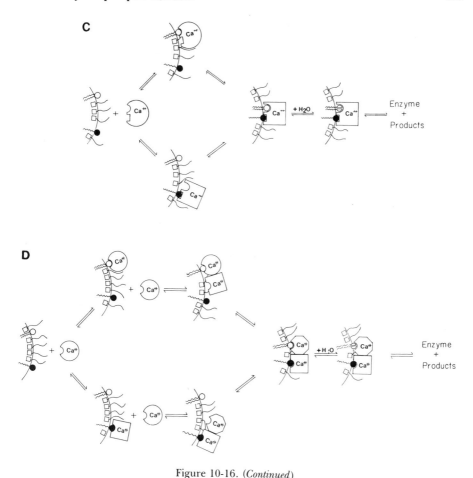

Figure 10-16. (*Continued*)

cannot yet be drawn. However, most evidence favors model D or some variation thereof, possibly a combination of models C and D.

10.3.4.2. Triton X-100 Phospholipid Micelles: Bee-Venom Phospholipase A$_2$

Chemically pure phosphatidylcholines with a sulfur present on the phosphate were used in a mixed micellar system to probe the interaction of Ca^{2+} and substrate with this phospholipase A$_2$ (Tsai *et al.*, 1985). It was concluded that the enzyme was stereospecific for the sulfur (oxygen) of the phosphate and that the Ca^{2+} was bridged between the *pro-S* oxygen of the phosphate and the carbonyl oxygen of the acyl group at position 2 of the glycerol backbone (Fig. 10-17). It will be of considerable interest to use this elegant approach with phospholipases A$_2$ that have different active-site structures such as the pancreatic and venom Asp-49 and Lys-49 enzymes.

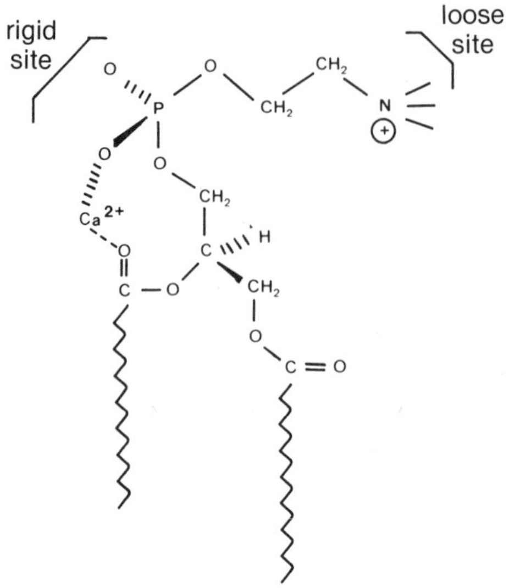

Figure 10-17. Structure showing binding of Ca^{2+} with the *pro-S* oxygen and possibly with the 2-carbonyl oxygen of phospholipid substrate at the active site of phospholipase A_2. (From Tsai *et al.*, 1985.)

10.3.4.3. Studies with Other Detergent–Phospholipid Micelles

Mixed micelles of natural detergents such as cholate and deoxycholate have been used successfully in the study of a variety of phospholipases. Most studies, however, have used bile salts empirically without fully exploiting the nature of the mixed micelle formed in the kinetic analysis. [The nature of phospholipid–bile salt complexes has been well studied and is detailed in Small's book in this series on the physical properties of lipids (Small, 1986).] Also, as mentioned before, cholate and deoxycholate have rather high CMCs so the possibility of enzyme stimulation by monomers cannot be ruled out. An example of a possibility was raised by Kannagi and Koizumi (1979) who found that a phospholipase A_2 partially purified from rabbit platelets was stimulated by bile salts far below their CMC. Triton X-100, on the other hand, stimulated hydrolysis only above its CMC. Likewise, deoxycholate had no effect on the hydrolysis of short-chained phosphatidylcholine by the phospholipase C from *B. cereus* but was essential when longer acyl chains were used above their CMC (Little, 1977*b*). This indicates that deoxycholate did not act directly on the phospholipases but acted on the substrate, presumably by converting the bilayer liposome to a mixed micelle.

The stimulation of hydrolysis by bile salts is highly dependent on the bile salt : phospholipid ratio and on the origin of the phosphatidylcholine (Nalbone *et al.*, 1980). For example, mixed micelles of bile or egg phosphatidylcholine and bile salts were optimally attacked at a ratio of 1 : 2 with little or no hydrolysis occurring in the absence of bile salt or at phosphatidylcholine : bile salt ratios of 1 : 6. Native bile was poorly degraded, presumably because of the presence of two pools of phosphatidylcholine, only one of which can be attacked by the pancreatic phospholipase A_2. The neutrophil phospholipase C is likewise stimulated by dexoycholate at a phosphatidylinositol : deoxycholate ratio of 1 : 2. Ratios as high as 1 : 4 completely block hydrolysis (Smith and Waite, 1986). Clearly, much can be learned from the study of activation by bile salts if physical studies are done on the mixed micelles used in the phospholipase assay mixtures.

To examine the question of enzyme–substrate interaction, two enantiomers of phosphatidylcholine have been studied using mixed micelles. This approach was used because the binding of pancreatic phospholipase A_2 to monolayers is rather inefficient and therefore it is difficult to determine if any differences in activity noted between its attack on different substrates is a difference in enzyme penetration and activation, $E \rightarrow E^*$ (K^*_m), or the actual hydrolytic rate, k_{cat} (see Fig. 10-24 and Section 10.3.6). Therefore, rather than pursuing the problem with the monolayer system, mixed micelles were used (Slotboom *et al.*, 1976). The very low affinity of the pancreatic phospholipase A_2 for mixed micelles of phosphatidylcholine and Triton X-100 led Slotboom *et al.* (1976) to use mixed micelles that employed the substrate analog *n*-tetradecyl phosphocholine rather than Triton X-100. This approach was similar to that used by Dennis and coworkers previously described; the concentration of the substrate could be regulated by the substrate : analog ratio. It was assumed that the analog would not change the *interfacial quality* as Triton X-100 might. However, like Triton X-100, the binding of enzyme to the analog had to be established to obtain accurate kinetic analysis. It was found that 1-hexanoyl-3-dodecanoyl-*sn*-glycerol-2-phosphocholine (6/12 substrate) was degraded at only 10% the rate of substrate that had the dodecanoyl group in position 1 and the hexanoyl group in position 3 (12/6 substrate). Since the quality of the interface was found to remain constant kinetically when the analog : substrate ratio was varied, it was concluded that the binding properties were constant for the different micelles. Also, these micelles had the advantageous property of binding essentially all the enzyme so that the hydrolysis followed a hyperbolic saturation curve as the substrate increased proportionally in the mixed micelle. It was found that the K^*_m (K^B_m) was 2.2×10^{14} for the C_6/C_{12} substrate and 0.9×10^{14} for the C_{12}/C_6 substrate, while the k_{cat} (V) was 1456 and 112, respectively. It is obvious that even though the binding of the enzyme to the 6/12 substrate is weaker than to 12/6 substrate, the hydrolytic rate is much higher. It was suggested from this study that the rate-limiting step might be deacylation of an enzyme–acyl intermediate. Since it is now thought that an enzyme–acyl intermediate does not exist, this difference could be the result of product dissociation from the enzyme. However, this study does show very nicely that when the initial steps of interfacial binding are held constant, it is possible to dissect the kinetics of the various steps in the reactions.

10.3.5. Substrates in Aggregates (Nonmicelle)

As described in Chapter 1, liposomes are sealed bilayer vesicles that can be single- or multishelled. Furthermore, they can vary considerably in size, depending on the method of preparation and the chemical nature of the phospholipid. In general, these are prepared using pure phospholipids or mixtures of phospholipids with other components such as cholesterol. Some phospholipids, however, either pure or in mixtures with other amphipaths, do not form liposomes in aqueous dispersions; rather, they form hexagonal arrays or related structures. It is important to appreciate that the composition of the solute can influence the nature of the aggregate. This is particularly important when studying the effect of Ca^{2+} or pH on phospholipases since they can cause transitions between bilayer structures and hexagonal arrays. These considerations have not always been appreciated and, as a consequence, often the term liposome has been used improperly.

10.3.5.1. Bilayer Liposomes

By and large, liposomes are poor substrates for phospholipases. On the other hand, they have been useful in studies on the enzyme's interaction with organized lipid interfaces. As described in the section on mixed micelles, pure lipids are often more readily attacked when detergents or other agents are present that facilitate the enzyme's capacity to interact with the interface. Various studies have addressed this problem and, in general, it is thought that the initial binding to the liposome is limiting and that conditions that favor the initial binding enhance the overall rate of hydrolysis. Studies on monomolecular films described in the next section support this interpretation. When physical conditions are employed that produce surface discontinuities, hydrolysis proceeds at a much higher rate. Perhaps the first example of this was the study of op den Kamp *et al.* (1974) who demonstrated that the pancreatic phospholipase A_2 readily attacked liposomes comprised of saturated phosphatidylcholines at the phase transition of the phospholipid. At the phase transition, domains of gel and liquid crystalline phospholipid coexist and it is thought that the enzyme penetrates the bilayer at the interface of the two domains initiating catalysis. Looser packing occurs with phospholipids containing unsaturated fatty acids and the essentiality of surface discontinuities is less pronounced. Likewise, the enhanced activity at the phase transition is most pronounced for phospholipases with weak penetrating power such as the pancreatic phospholipases (Verheij *et al.*, 1981*b*). This work has been confirmed and expanded in studies on pancreatic phospholipase A_2 interaction with liposomes of dipalmitoyl phosphatidylcholine (Lichtenberg *et al.*, 1986; Menashe *et al.*, 1986). These workers concluded that a Ca^{2+}-independent binding of enzyme to the liposomes occurred best when the lipid was in the gel phase. Before hydrolysis could occur, however, a Ca^{2+}-dependent activation must occur at the phase transition. They postulated further that this activation involved enzyme–enzyme interaction, an event that may be comparable to the previously described polymerization of the enzyme.

A different approach to the problem of enzyme–liposome interaction was used by Upreti and Jain (1980) and Upreti *et al.* (1980). They osmotically shocked multilayered liposomes to show that discontinuities were important in hydrolysis. Under the conditions employed, the bee venom phospholipase could attack the substrate for a period of several minutes following osmotic shock or the addition of alkanols that allowed penetration of the enzyme.

Indeed, the addition of hexanol to osmotically shocked liposomes of egg phosphatidylcholine gave a maximal hydrolysis (Upreti and Jain, 1978). A second phase of hydrolysis occurred when sufficient product was present that reintroduced surface defects. This could occur as the result of hydrolysis or by the addition of reaction products to the reaction mixture (Apitz-Castro *et al.*, 1982). They dissected the course of hydrolysis into three phases: the initial burst, the latency period, and the final burst (Fig. 10-18).

The latency was inversely proportional to the enzyme concentration. The authors suggest that catalytic activity is required for incorporation of the enzyme into the liposomes. If such is the case, only limited amounts of enzyme can bind initially, even at the phase transition. The *second wave* of binding is dependent on product formation. More recently, however, Lichtenberg *et al.* (1986) have concluded that under the appropriate conditions product formation is not essential for activation of the pancreatic phospholipases A₂. The apparent difference between these two reports might reside in the difference between binding and activation of the enzyme. The steady-state rate is reached when 5% of the substrate is hydrolyzed. It was possible to quantitate the binding of the pancreatic phospholipase A₂ to liposomes by measuring the fluorescent enhancement of the tryptophan of the phospholipase when the active Michaelis–Menten complex was formed (Jain and deHaas, 1983). An

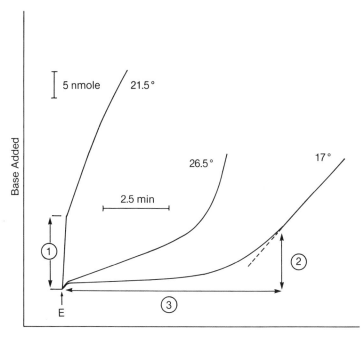

Figure 10-18. The reaction progress curves for the hydrolysis of sonicated dimyristoylphosphatidylcholine (250 μM) vesicles by pig pancreas phospholipase A₂ (0.25 μg/5 ml). All conditions were identical except for the temperatures indicated in the graph. Reaction medium contained 100 mM KCl/10 mM CaCl₂ at pH 8.0. Initial burst (burst at $t = 0$, ①), final or total burst (burst at $t = \tau$, ②), and the latency period (③) are defined as indicated. (From Apitz-Castro *et al.*, 1982.)

increase in hydrolysis was observed when lysophospholipid was added to preexisting liposomes. However, lysophospholipid did not stimulate hydrolysis when equilibrated on both sides of the bilayer. This demonstrated that the enhancement of binding and hydrolysis is the result of the discontinuity introduced rather than to binding of the enzyme to its product, the lysophospholipid. This would be expected since many agents including detergents and alkanols can enhance activity. Recent work by Jain and coworkers indicated that in addition to increasing enzyme binding, surface discontinuities enhance the catalytic turnover of the bound enzyme.

Jain's group has used a variety of membrane-perturbing agents to probe the effect of surface defects on hydrolysis. Through the extensive use of alkanols, several phospholipases were shown to have biphasic activation (Jain and Cordes, 1973a, 1973b; Jain and Apitz-Castro, 1978; Upreti and Jain, 1978; Upreti et al., 1980; Apitz-Castro et al., 1981). The concentration of alkanol required for activation decreased with increasing chain length but decreased with increasing unsaturation. However, these effects were thought to be on the microscopic rather than macroscopic structure of the liposome (M. K. Jain, personal communication). Regulation of vesicle fusion was postulated also to regulate activity of the pancreatic phospholipase. It therefore appears that factors capable of promoting greater access of substrate to the enzyme such as vesicle fusion or exchange enhance activity. This is particularly true of the pancreatic enzyme that does not readily dissociate from the phospholipid interface.

Some studies have indicated that binding of phospholipases A_2 is more efficient below than above the phase transition of the phospholipid when small or large vesicles are studied (Smith et al., 1972; Kensil and Dennis, 1979, 1985; Menashe et al., 1981, 1986). The pancreatic phospholipase A_2 did not exhibit a lag phase when assayed on dipalmitoyl phosphatidylcholine at 23° C, whereas a distinct lag phase occurred slightly above the phase transition, 39°C, and no activity was observed at 42°C (Menashe et al., 1981). However, if the enzyme was preincubated with the substrate vesicles for as little as 2 min at 23°C and then assayed at or above the phase transition, no lag period was observed. Also, the initial rate observed was considerably higher than that observed at 23°C. It was suggested that the greater tilt of the acyl chains in the gel state favored penetration and interfacial activation of the enzyme, whereas hydrolysis occurred maximally in the liquid crystalline state that permitted greater motion of the phospholipid in the bilayer (Kensil and Dennis, 1979). The work of Thuren et al. (1984) further emphasizes the importance of the orientation of the acyl esters in hydrolysis.

The nature of the bilayer interface, as determined by the radius of curvature, will also regulate attack by the phospholipase A_2 from N. naja naja. Kensil and Dennis (1985) found that small unilamellar vesicles of dipalmitoyl phosphatidylcholine were more rapidly hydrolyzed than large unilamellar or multilamellar vesicles, in confirmation of the earlier work of Wilschut et al. (1978) who studied the porcine pancreatic and bee venom phospholipases A_2 (Fig. 10-19). On the other hand, Lichtenberg et al. (1986) found that the presence of large unilamellar liposomes inhibited the hydrolysis of small unilamellar liposomes.

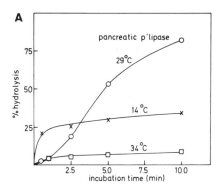
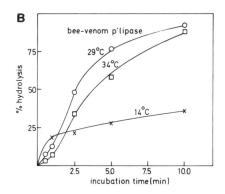

Figure 10-19. Time course of the degradation of a myristoylphosphatidylcholine sonicate at 14°C (×), 29°C (○), and 34°C (□) by pancreatic (A) and bee-venom (B) phospholipase A₂. Incubations were carried out in 0.05 M Tris-HCl (pH 7.5) in a volume of 1.0 ml. Concentrations: lipid, 1.0 mM; CaCl₂, 2.5 mM; pancreatic phospholipase, 5.0 μg ml; bee-venom phospholipase, 2.5 μg/ml. (From Wilschut *et al.*, 1978.)

The pancreatic phospholipase A₂ was shown not to transfer readily between liposomes (Wilschut *et al.*, 1978). In their studies, about 9×10^{12} molecules (0.175 μg) of enzyme were used with 1.3 mmoles of substrate. Although no data were given on the size of the liposomes used, if one were to assume that each liposome contained 1000 molecules of phosphatidylcholine, then the reaction mixture would contain 8×10^{17} liposomes—about 1×10^{5} times the number of enzyme molecules. This raises the possibility that in addition to the products stimulating binding of the enzyme, product formation might promote substrate and/or product transfer between liposomes (without specifying mechanism). Such a possibility was raised by the authors. These conclusions are in contradistinction to those of Tinker *et al.* (1980) who concluded that surface discontinuities and reaction products favored desorption of the enzyme from the micelle, permitting readsorption to a new micelle rich in substrate. Figure 10-20 depicts the various models accounting for enzyme desorption. Briefly, it was proposed in model A that desorption of the snake venom enzyme is the rate-limiting step, while model B proposes that adsorption of the pancreatic phospholipases is limiting. In both cases, product increases the rate-limiting step. A modification of model A, model C, proposes that reversible adsorption of the venom phospholipase can occur as a side reaction. While not yet resolved, consideration should be given to the assessment of transfer of the components of the reaction between aggregates, be they mixed micelles or liposomes.

10.3.5.2. Hexagonal Arrays

One might expect that structures such as hexagonal arrays might be resistant to attack by all classes of phospholipases since the esters would be shielded from the external environment by the acyl chains. It has been known, however, that many cellular phospholipases A₁ and A₂ attack aggregates of pure phosphatidylethanolamine more rapidly than other phospholipids (Waite

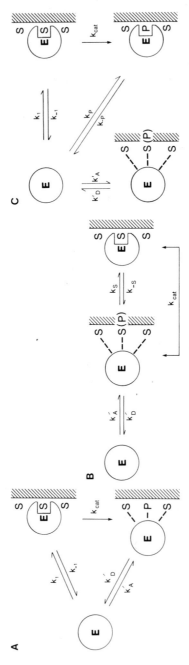

Figure 10-20. Kinetic models for hydrolysis of lipid (S) at the surface of an insoluble-phase catalyzed by a water-soluble enzyme; P represents product(s) of the reaction. Model A is the Path 1 model of Tinker *et al.* (1980); model B is the mechanism of Verger and deHaas (1976). Model C is a modification of model A that allows reversible adsorption of the enzyme to the surface as a side reaction. (From Tinker *et al.*, 1980.)

and van Deenen, 1967; van den Bosch, 1980). The conditions employed would suggest in many cases that hexagonal arrays were used even though it was not recognized at that time. More direct evidence for the ability of phospholipases to attack hexagonal arrays has been obtained by two groups, although direct studies showing the structure of the substrate used are still lacking. Dawson *et al.* (1983*a*, 1983*b*) showed that a rat liver cytosolic phospholipase A_1 most rapidly attacked phospholipids that are nonbilayer in structure. This conclusion was reached in a study in which an excess of phosphatidylethanolamine stimulated hydrolysis of phosphatidylcholine, normally a poor substrate. Likewise, this group demonstrated that diacylglycerol stimulated the hydrolysis of phosphatidylcholine and phosphatidylinositol by rat mucosal phospholipase A_2 and platelet phospholipase C under conditions thought to disrupt the bilayer structure. Key to the interpretation that hexagonal arrays were present was the finding that hydrolysis of phosphatidylethanolamine was not stimulated by diacylglycerol. When phosphatidylcholine was added to phosphatidylethanolamine, an inhibition occurred that could be overcome partially by the addition of diacylglycerol. It was concluded that the bilayer structure formed by the mixture of phosphatidylcholine and phosphatidylethanolamine was disrupted by the addition of diacylglycerol.

Our group reached a similar conclusion from studies on the rat liver lysosomal phospholipase A_1 that is optimally active at pH 4.0 (Robinson and Waite, 1983). Without Triton, phosphatidylethanolamine was most readily attacked, whereas phosphatidylcholine and strongly acidic phospholipids such as phosphatidylglycerol were readily attacked when present in mixed micelles (See Fig. 6-3). The results suggested that the mixed micelle was degraded more rapidly than the hexagonal array, but the bilayer system is a poorer substrate than the hexagonal array.

The complete understanding of these studies is far from certain and awaits more detailed physical study of the substrate structures. While a number of possibilities exist to account for the preference of these enzymes for nonbilayer structures, three conditions may exist that could help further interpretation. First, it is possible that the conditions employed produced heterogeneous structures presenting the enzyme with surface discontinuities that facilitated access to the enzyme to the ester bonds. Indeed, the [31P] NMR spectra indicate that some degree of heterogeneity may exist with pure phosphatidylethanolamine samples (Cullis and deKruijff, 1978). Second, hexagonal arrays may be formed with an outer shell or monolayer of phospholipid coating the outer acyl chains. Such a layer would have a more hydrated and accessible region about the interface and polar esters. Third, the enzyme may enter the aqueous phase within the hexagonal array and hydrolyze from within.

10.3.5.3. Surface Charge

Liposomes, and in all probability hexagonal arrays, have been useful in determining the effect of surface charge on the attack of phospholipases on aggregated lipids. The classic works of Bangham and Dawson beginning in

1959 showed that the electrostatic nature of the phospholipid–water interface has a marked influence on the hydrolysis of the lipid (Bangham and Dawson, 1959; for review of model membranes, see Bangham, 1968). Both the nature of the polar head group of the phospholipid and the ionic composition of the media can regulate the apparent charge at the interface (Dawson, 1966) (Fig. 10-21).

The fixed charge at the interface is regulated by the concentration of net charge moieties present. The surface charge is also regulated by the phospholipids present or by the addition of nonsubstrate charged amphipaths such as dicetylphosphate and cetyltrimethylammonium ions. Counterions are attracted to the interface and tend to balance the surface charge. Experimentally, the determination of surface charge has been done using a microelectrophoretic apparatus that measures the mobility of vesicles in an electrical field (Bangham and Dawson, 1962). This in turn can be used to derive the Zeta potential and surface-charge density. On the average, a number of fixed counterions are mobile with the vesicles within the plane of shear. When polyvalent cations are present in sufficient concentrations, negatively charged phospholipid aggregates actually have an apparent net positive charge.

Another approach to the study of surface-charge influence on a phospholipase was used by Wells (1974). *C. adamanteus* phospholipase A_2 that requires Ca^{2+} was stimulated by Mg^{2+} but inhibited by K^+. This suggested that these two ions have different influences on the orientation of the polar head group. It was found that the surface potential at the polar head group region, as determined by the titration of a dye, was significantly different from the potential. The latter was measured by the base-catalyzed hydrolysis of the phosphlipid in the presence of different ions. Wells concluded that K^+ induced one type of shift in the dipole between $P—O^-$ whereas divalent Mg^{2+} and Ca^{2+} induced a different type of orientation that favored hydrolysis

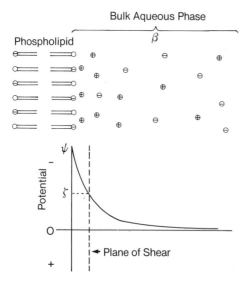

Figure 10-21. Nature of the polar head group of the phospholipid and the ionic composition of the media can regulate the apparent charge at the interface. (From Dawson, 1966.)

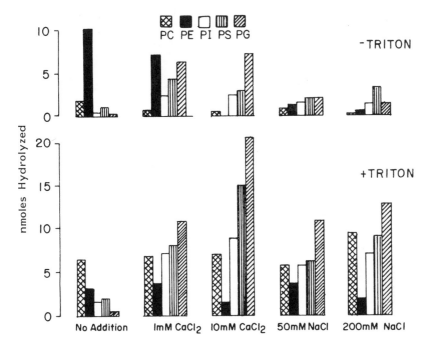

Figure 10-22. The effect of Ca^{2+} and Na^+ on the activity of phospholipase A_1. Each assay contained 100 nmoles of each phospholipid substrate in the absence (*top*) or presence (*bottom*) of 0.5 μmole of Triton WR1339 and 0.37 μg of lysosomal protein. (From Robinson and Waite, 1983.)

(perhaps an inward movement of the nitrogen from the surface). The conclusion was drawn that the potential effects on polar head group orientation are important in understanding phospholipases "and show that pH studies with aggregate substrates are essentially uninterpretable." While such a statement reflects the difficulty in understanding the effect of bulk pH on the charge at a given distance from the ester bond under attack, surface-charge measurements made using microelectrophoresis in addition to defining bulk pH can be of use in describing the properties of phospholipases.

In general, those phospholipases that have been studied tend to favor lipid aggregates with a slight negative charge. A notable exception is the phospholipase C from *C. perfringens* that requires a slight positive charge (Bangham and Dawson, 1962). We found that the lysosomal phospholipase A_1 has optimal activity on phosphatidylcholine or phosphatidylethanolamine at a net negative charge of about 2 units $(\mu m/sec^{-1} \times volt^{-1} \times cm^{-1})$ (Robinson and Waite, 1983). This enzyme does not require but is inhibited by Ca^{2+} when phosphatidylethanolamine was the substrate. Additional evidence that the effect of Ca^{2+} on hydrolysis was mediated through charge effects was shown by Ca^{2+} stimulating activity when strongly acidic substrate was used such as phosphatidylglycerol. The effect of adding Ca^{2+} or charged amphipaths to regulate the apparent mobility was equally effective in regulating activity (Fig. 10-22).

Clearly, charge of the phospholipid is extremely important in regulating activity since it determines both bulk interaction of the enzyme with substrate and the formation of the Michaelis–Menten complex. As pointed out by Wells (1974), the orientation of the polar head group and the lack of a fixed point charge add to the complexity of the system and a better understanding of the aggregated substrate is needed at this time.

10.3.6. Monomolecular Films of Phospholipid as Substrate

One of the most powerful approaches to study the interaction of phospholipases at interfaces has been with the use of monolayers of phospholipid at the air–water interface. These approaches have been reviewed by Verger (1980), which is recommended to the reader. Likewise, the physical–chemical aspects of monolayers are covered in Small's treatise in this series (Small, 1986) as well as in *Interfacial Phenomena* by Davies and Rideal (1961). This technique has the advantage of determining the effect of molecular packing on the penetration of the enzyme into the lipid interface and on hydrolysis without the use of detergents or other agents. The main limitation to this technique, however, has been the technical difficulties associated with its use. As a consequence, relatively few groups have used it consistently.

Key to the successful use of monolayers of phospholipid as substrate for the study of phospholipases was the development of a modified Langmuir–Adams trough that could maintain a constant surface pressure of the film (Gaines, 1966). The first approach using the *zero-order* rather than the *first-order* trough was in 1973 when Verger and deHaas (1973) investigated the kinetics of pancreatic phospholipase A_2 catalyzed degradation of didecanoyl phosphatidylcholine (Fig. 10-23).

In this system, the enzyme is injected into the oval left-hand chamber that is connected with the right-hand reservoir in the surface region only. Activity is measured by constant titration of the proton release. As the reaction proceeds and the products diffuse into the subphase, replenishing substrate is transported into the reaction chamber by the means of a mechanical bar connected to the microbalance. Continual stirring of the subphase is essential to ensure movement of products away from the interface. In contrast, the first-order trough has enzyme in the entire subphase volume and, as a consequence, has a decreasing amount of substrate film as the reaction proceeds.

It was recognized in the development of the kinetics of hydrolysis that only a limited amount of enzyme actually penetrated the monolayer since the ratio of the monolayer to the bulk volume is small (Verger *et al.*, 1973). Also, enzyme binding did not alter the surface pressure of the monolayer. The velocity of hydrolysis, however, was still found to be linearly dependent on the concentration of the pancreatic enzyme in the subphase. The phospholipase A_2 from *C. adamanteus*, which is active as a dimer, differs in that respect from the pancreatic enzyme (Shen *et al.*, 1975). At low enzyme concentrations, $1–10 \times 10^{-10}$ M, the hydrolysis is not proportional to the concentration of the enzyme—the result of dissociation of the enzyme into inactive monomers. This monolayer study established the K_D for the *C. adamanteus* enzyme to be

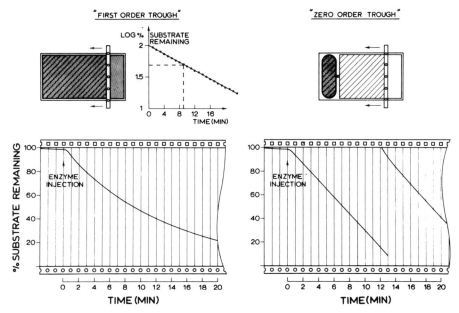

Figure 10-23. Comparison between the recorded kinetic plots obtained with the first-order trough (*left side*) and with the zero-order trough (*right side*). (From Verger and deHaas, 1973.)

2×10^{-9} M. No deviation from a linear dependence of the reaction was found with the pancreatic phospholipase A_2, which is consistent with its functioning as a monomer. The use of $[^{125}I]$ or $[^{3}H]$ ε-amidated porcine phospholipase A_2 allowed direct measurement of the binding of the pancreatic enzyme to the monolayer (Pattus *et al.*, 1979*a*, 1979*b*, 1979*c*). Also, it was recognized that at all working surface concentrations of substrate, compression of the film to higher pressures not only changed S, the surface concentration of substrate, but also the *interfacial quality* and the binding of the enzyme to the monolayer. As a consequence, V_{max} is not directly proportional to S. In the model proposed, two induction rates, τ, are presumed: $τ_1$ reflects the equilibrium of penetration–desorption of the enzyme to the monolayer, and $τ_2$ represents the formation of the active Michaelis–Menten equilibrium (Verger *et al.*, 1973) (Fig. 10-24).

It was found that $τ_1$ was much greater than $τ_2$, as shown by the long time required for the pancreatic phospholipase to reach maximal velocity. As was shown in Fig. 9-2, the penetration and initiation of hydrolysis by the bee-venom enzyme are very rapid and no lag period is observed (Verger *et al.*, 1973).

As predicted by the kinetic model, the steady-state rate of hydrolysis, but not the induction time, was dependent on enzyme concentration. Calcium, which causes a conformational change in the pancreatic phospholipase A_2, was also required for binding of the enzyme to the monolayer. This effect of Ca^{2+} was different from its effect when liposomes were used (Lichtenberg *et al.*, 1986). In that study, Ca^{2+} was required for activation of the enzyme but

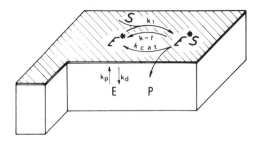

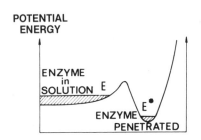

Figure 10-24. Proposed model for the action of a soluble enzyme at an interface. (From Verger *et al.*, 1973.)

not its binding to lipid. The induction period for hydrolysis was the same for enzyme in the subphase prior to the addition of Ca^{2+} as enzyme injected into the subphase containing Ca^{2+}. Apparently, the conformational change induced by Ca^{2+} is required for binding of the pancreatic enzyme to substrate. The Ca^{2+} concentration influences the pH dependency of both the induction time as well as the steady-state velocity.

A term has been coined to describe the interaction and regulation of phospholipases with interfaces, namely, the *quality of the interface*. However, thus far it has not been possible to define physically the meaning of *quality of the interface*. Kinetically, it can be expressed in terms of velocity, substrate concentration, and total enzyme concentration (V_{max}, K_m, and E in bulk hydrolysis studies). Most important to this concept is the finding that the interfacial quality for bulk hydrolysis of diheptanoyl phosphatidylcholine, as regulated by the concentration of NaCl, can be kinetically equivalent to that for a monolayer of dioctanoyl phosphatidylcholine at the appropriate surface pressure. This indicates that the concept of interfacial quality is not restricted to monolayers but is applicable to micellar suspensions of phospholipids. Possibly this *quality* factor relates to the orientation of the acyl–ester bonds (Thuren *et al.*, 1984). However, as stated earlier, this kinetically derived concept is far from understood and will require more detailed knowledge of the substrate and the phospholipase under study.

These studies pointed out one of the limitations to the use of monolayers, namely, the low binding of enzyme to substrate. Likewise, it has been shown that binding of the pancreatic phospholipase to monolayers leads to inactivation (Rietsch *et al.*, 1977). Using [^{125}I] phospholipase A$_2$ from porcine pancreas, it was found that enzyme bound to the monolayer could be transferred between troughs and that the apparent inactivation process was more rapid after than before transfer. By transferring the enzyme adsorbed to the film to a compartment with no bulk-phase enzyme, adsorption of new enzyme to the film did not occur and the true decay rate of the enzyme could be determined (Rietsch *et al.*, 1977). The inactivation rate was inversely proportional to the surface pressure of the film. At low pressures, the adsorption to the lipid appears to be irreversible, unlike the situation when phospholipase A$_2$ binds to micelles which is a reversible process. In addition, over a period of time, enzyme binds to the Teflon trough which accounts for additional irreversible loss of enzyme. It was concluded in this study that the kinetics observed are dependent on the inactivation rate in addition to the absorptive rate and that the kinetics are linear when these two are at equilibrium. These results lead to a modification of the original scheme of Verger *et al.* (1973) by including a k_i term in which E*, the activated enzyme at the interface, is converted to E*$_i$, the inactive enzyme which remains adsorbed to the film.

Low surface pressures can lead to denaturation of some but not all phospholipases. Cohen *et al.* (1976) described nonlinear surface pressure dependency in the range of 3.5–8 dynes/cm for two phospholipases A, one from pancreas and one from bee venom. Another phospholipase from pancreas and the enzymes from *C. adamanteus* and *C. atrox* were linearly dependent on surface pressure up to the maximal pressure for penetration. It was concluded that the two pressure-dependent enzymes undergo an air–water interfacial denaturation in the spaces between phospholipid molecules. These results are important from the technical point of view but also raise the very interesting question as to how such subtle differences in the protein structure of the closely related phospholipases would produce different interfacial characteristics. In all probability, this will be true to a certain extent for all phospholipases since they will have a tendency to associate with the air–water interface.

Radiolabeled porcine pancreatic phospholipase A$_2$ was used effectively to determine the relation between enzyme binding to the monolayer, the induction period τ, and catalytic activity in a more quantitative manner (Pattus *et al.*, 1979*a*). Both [^{3}H] and [^{125}I] amidated phospholipases A$_2$ were used to minimize problems that might result from the labeling process. The [^{3}H] amide moieties were on the ε-amino groups of the lysines while the [^{125}I] was present on Tyr-69. As shown in Fig. 10-25, the two preparations behaved in a similar fashion except that the [^{125}I]-labeled enzyme appeared to penetrate the film more easily at the higher pressures (Pattus *et al.*, 1979*a*).

As a consequence of this greater binding capacity, the induction (lag phase, τ) was minimal and hydrolytic rates were higher at higher pressures for the iodinated enzyme. However, the minimal specific activities of the two preparations were identical. The specific activity increased with pressure up to a constant value which was achieved at 10 dynes/cm. These results demonstrated that there was a direct correlation between the capacity of the enzyme to adsorb to the interface and its capacity to degrade phospholipid.

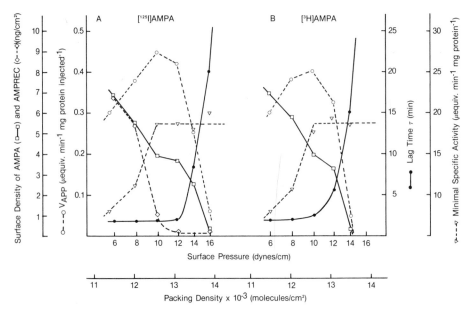

Figure 10-25. Influence of surface pressure, i.e., packing density, on the activity, lag time, and surface density of [^{125}I] ε-amidated phospholipase A$_2$ (A) and [^{3}H] ε-amidated phospholipase A$_2$ (B) acting on L-di-C10-PC(1,2-diacyl-*sn*-glycero-3-phosphatidylcholine with two identical acyl chains with 10 atoms of carbon) monolayer. Tris–acetate buffer (10 mM), pH 6.0; NaCl (0.1 M); CaCl$_2$ (0.02 M). Enzyme (15 μg) was injected. Volume of the reaction compartment was 210 ml; surface was 92 cm^2. The points on the specific activity curves were obtained by multiplying the ratio V_{app}/surface density by 1000/92. (From Pattus *et al.,* 1979a.)

Interestingly, the proenzyme penetrated at low pressures but had no observable catalytic activity. Apparently, the binding of the zymogen was nonspecific and the active site was not oriented toward the lipid film. Likewise, the zymogen did not bind to micelles composed of an equal mixture of didecanoyl phosphatidylcholine and myristoyl lysophosphatidylcholine, possibly because of the surface pressure characteristic of the micelle (Pieterson *et al.,* 1974).

The desorption process does not occur at an appreciable rate at low pressures (4 dynes/cm) as previously mentioned. However, if the enzyme is allowed to adsorb and then the pressure is increased to 18 dynes/cm, hydrolysis stops as the consequence of rapid enzyme desorption. This could be a general mechanism for the regulation of phospholipases.

It was pointed out in this study that it is more accurate to express hydrolysis as area per phospholipid molecule than in surface-pressure units. While this would be one and the same for a given phospholipid, it does not hold for phospholipids with different acyl moieties. As shown in Fig. 10-26, the induction time, τ, varied at a fixed pressure according to the acyl composition of the phospholipid (Pattus *et al.,* 1979a).

There is an inverse relation between τ and the acyl chain length; that is, the lag period increased more rapidly with pressure for the shorter acyl chain phosphatidylcholines. The lag period correlated directly with the amount of enzyme bound to the monolayer. Surface areas lower than 75 Å²/molecule prevented enzyme binding (1.3×10^{14} molecules/cm). This phenomenon was found to be dependent on surface charge as well; more negatively charged phospholipids are penetrated more rapidly by the pancreatic enzyme. The amount of enzyme adsorbed to the film increased with increased acyl chain length. These results demonstrated that hydrophobic interaction is important in the adsorptive process, in addition to charge effects. It was proposed by Pattus *et al.* (1979a) that a hydrophobic portion of the enzyme must cross the energy barrier formed by the polar head groups and then bind to the hydrophobic acyl chains. Since a conformational change has been proposed to occur in the adsorption process, the crossing of the polar energy barrier could be minimized by a favorable conformation of the enzyme that allows penetration through the polar region followed by a conformational change that permits strong hydrophobic interaction.

Monolayer studies were used by Gorter and Grendel (1925) to develop the postulate that the red cell membrane is comprised of a bilayer of lipid. This conclusion was based on the observation that the area of lipid covered by the monolayer was twice that of the surface area of the cells from which the lipid was extracted. In that study, however, the surface pressure was not determined or related to surface pressure of the cellular membrane. As a consequence, significant error in their comparison of film to cellular area could exist. At the time, no method existed to measure the surface pressure

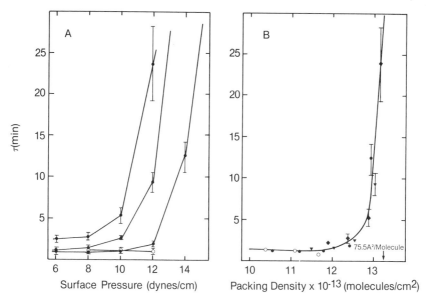

Figure 10-26. Influence of surface pressure (A) and lipid packing (B) of a lecithin film on the lag time. ○, Di-C8-PC; ●, di-C9-PC; ▼, di-C10-PC; ◆, di-C12-PC. Standard buffer, pH 8.0. [125I] ε-amidated phospholipase A₂ (8 μg) was injected. Abbreviations: di-C8-PC, di-C9-PC, di-C10-PC, di-C12-PC = 1,2-diacyl-*sn*-glycero-3-phosphatidylcholine with two identical acyl chains with 8, 9, 10, and 12 atoms of carbon, respectively. (From Pattus *et al.*, 1979a.)

within a cell membrane. Demel *et al.* (1975) investigated the problem of cell surface-pressure measurements using the comparative capacity of several phospholipases to attack red cell membranes and monolayers of various lipids at optimal surface pressures. Table 10-3 shows that phospholipases that cannot attack red cell membranes cannot penetrate and degrade monolayers of the appropriate substrate above a surface pressure of 31.0 dynes/cm. From this it was concluded that the pressure of the red cell membrane was somewhere between 31.0 and 34.8 dynes/cm. Allowing for certain assumptions, Demel and coworkers calculated that the ratio of monolayer area to cell surface area would be in the range of 1.3–1.7 and that the remainder of the membrane is composed of protein and glycolipid. To obtain a ratio of 2 : 1, the surface pressure of the film would have had to be 9 dynes/cm. Since this study was done, it has been appreciated that the presence of other components in the membrane will alter the maximal pressure permitting activity. Even so, the estimates of membrane surface pressure remain fairly accurate (Verger and Pieroni, 1986). Additional examples of this use of phospholipases are given in Chapter 11.

Water-insoluble amphipaths incorporate into monolayers and influence phospholipases in two ways. First, they can alter the surface-charge characteristics and, as a consequence, either enhance or inhibit hydrolysis as was found with aggregated phospholipids. Second, they can alter molecular packing and influence the penetration of the enzyme and the optimal surface pressure at which an enzyme can be active. The latter possibility cannot easily be studied with aggregates. The group of Hendrickson (Willman and Hendrickson, 1978; Hendrickson *et al.*, 1981) found that the inclusion of dicetyl phosphate in monolayers of didecanoyl phosphatidylcholine decreased the activity of the porcine pancreatic phospholipase A_2 at low and optimal pressures (10 dynes/cm) but allowed for activity at pressures as high as 20 dynes/cm (Hendrickson *et al.*, 1981) (Fig. 10-27).

On the other hand, the activity of the phospholipase A_2 from *C. ada-*

Table 10-3. Comparison of the Effects of Different Phospholipases on Erythrocyte Membranes and the Maximal Surface Pressure at Which These Phospholipases Can Hydrolyze Monomolecular Films[a]

Phospholipases	Erythrocytes	Monolayers (dynes/cm)
Phospholipase A_2 (pig pancreas)	−	16.5
Phospholipase D (cabbage)	−	20.5
Phospholipase A_2 (*C. adamanteus*)	−	23.0
Phospholipase C (*B. cereus*)	−	31.0
Phospholipase A_2 (*N. naja*)	+	34.8
Phospholipase A_2 (bee venom)	+	35.3
Sphingomyelinase (*S. aureus*)	+	>40.0
Phospholipase C (*C. welchii*)	+	>40.0
Phospholipase C (*B. cereus*) Sphingomyelinase (*S. aureus*)	+	>40.0

[a] From Demel *et al.* (1975).

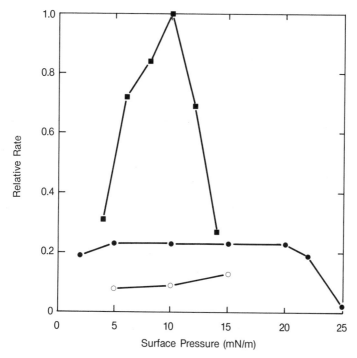

Figure 10-27. Relative rate (arbitrary units) of porcine pancreatic phospholipase A_2 versus surface pressure. Pure dipalmitoylphosphatidylcholine (■); 2 mole % dicetyl phosphate in dipalmitoyl-phosphatidylcholine (●); 5 mole % dicetyl phosphate in dipalmitoylphosphatidylcholine (○). (From Hendrickson *et al.*, 1981.)

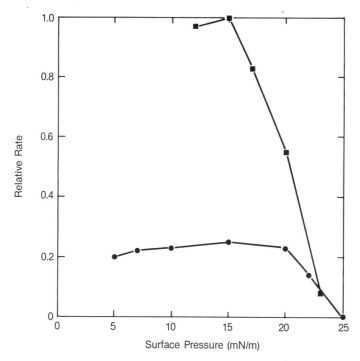

Figure 10-28. Relative rate (arbitrary units) of *Crotalus adamanteus* phospholipase A_2 versus surface pressure. Pure dipalmitoylphosphatidylcholine (■); 10 mole % dicetyl phosphate in dipalmitoyl-phosphatidylcholine (●). (From Hendrickson *et al.*, 1981.)

manteus did not demonstrate a higher cutoff pressure with dicetyl phosphate present in the monolayer (Fig. 10-28). Somewhat different results were obtained when the films were made negative by substituting acidic phospholipids for zwitterionic phosphatidylcholine. The pancreatic phospholipase A_2 was more active on the acidic phospholipids and had high catalytic rates at pressures up to 40 dynes/cm. The *C. adamanteus* enzyme was weakly active on the acidic phospholipids but could act on the acidic lipids at 33 dynes/cm. These results indicate that the binding sites of the two must be different. Whether or not this relates to the protein structure in the catalytic site, including residue 53, is unclear at present. The pancreatic enzyme bound to negatively charged amphipaths that were not necessarily substrates as shown by its preference for anionic substrates yet was inhibited by dicetyl phosphate.

The cause for the enhanced penetration of the pancreatic enzyme was more than a charge effect since didecanoyl glycerol also increased the cutoff pressure for both phospholipases. In part, the enhanced penetration could be the result of nonideal mixing of the monolayer that was shown to occur. If so, then nonideal mixing in some way would be inhibitory. It is not clear how this occurred but it would suggest that the effect would most likely be on k_{cat} although activation of the enzyme forming E* could be involved. Perhaps a better understanding of the events involved would be obtained if the results were expressed as area per molecule rather than in surface pressure, as previously described.

CONFORMATION:		
di C_{12}PG		
[NaCl] , $\underline{M}$	0.0	1.0
AREA/MOLEC, $\overset{\circ}{A}^2$	64.0	76.5
PLA ACTIVITY		
A_1	0.0	1.0
A_2	1.0	0.03
$(PB)_2$PG		
[NaCl] , M	0.0	0.5
I_E/I_M	2.9	3.3
PLA ACTIVITY		
A_1	0.46	1.0
A_2	1.0	0.1

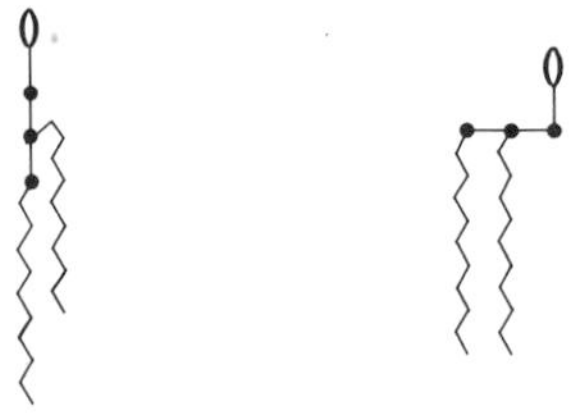

Figure 10-29. Schematic model relating the experimental results with the suggested changes in the phospholipid conformation controlling phospholipase A action. (From Thuren *et al.*, 1984.)

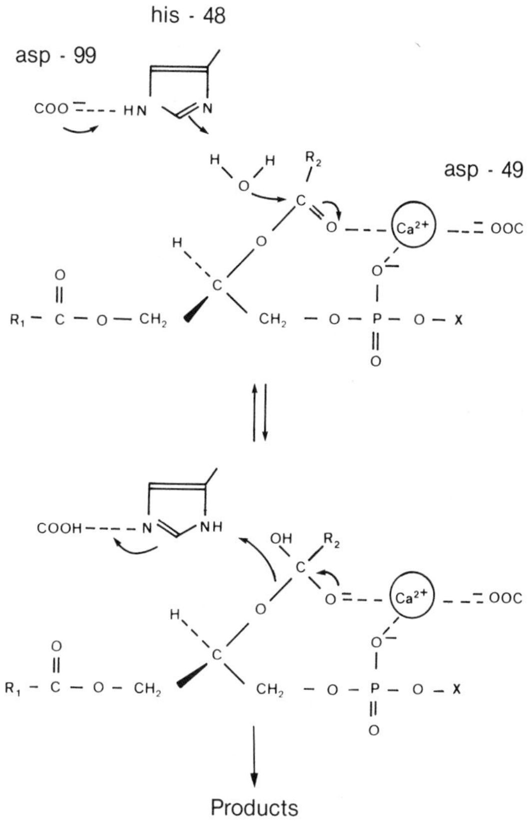

Figure 10-30. Proposed catalytic mechanism. (From Slotbloom *et al.*, 1982.)

Products

Monolayer studies were used in conjunction with fluorescently labeled substrate vesicles to determine the effect of acyl ester orientation on phospholipase A₂ activity (Thuren *et al.*, 1984). In this case, the surface pressure of a film of didodecyl phosphatidyglycerol was held constant at 15 dynes/cm and the area per molecule expanded from 63 to 75 Å² by bringing the subphase from 0 to 1 M NaCl. The increase in the cross-sectional area was thought to be a shift in the orientation of the glycerol moiety of the substrate from a parallel to perpendicular position, relative to the acyl chains (Fig. 10-29). This shift in the orientation of the glycerol moiety and consequently the acyl–ester bonds dramatically affected the attack of both acyl esters. The decreased activity of the phospholipase A₂ (porcine pancreatic) but increased activity of the phospholipase A₁ (lipoprotein lipase) are consistent with the ester at position 2 becoming more buried in the hydrophobic region while the ester at position 1 becomes more accessible to the interface. This interpretation was consistent with the finding that the I_e/I_m of 1,2-bis[(pyren-1-yl)butanoyl]phosphatidylglycerol vesicles increased upon the addition of NaCl, which suggested that an equal and parallel alignment of the fluorescent acyl chains occurred. Concomitant with this increase in fluorescent intensity was a shift in the phospholipase activities favoring attack at position 1 of the glycerol backbone.

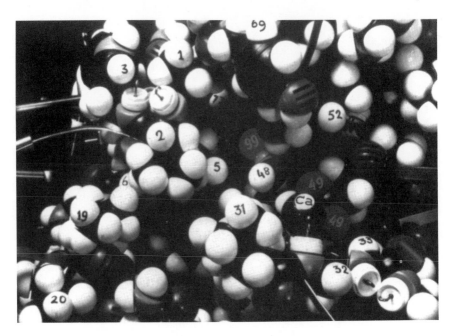

Figure 10-31. The space-filling model of bovine pancreatic phospholipase. (From Slotboom *et al.*, 1982.)

10.4. Mechanism of Phospholipase A₂ Hydrolysis

It has been proposed that the active site catalyzes a proton relay-type reaction common to serine esterases even though serine is not involved in catalysis. In this model, the immobilized water molecule serves the nucleophilic function normally fulfilled by serine. Although only demonstrated for the *C. adamanteus* enzyme, probably all act via an *O*-acyl cleavage mechanism (Wells, 1971) consistent with a proton relay system. A key feature to this proposal is that the proton relay system is buried in a hydrophobic *active-site wall* that is common to this entire class of phospholipases A₂. Interestingly, the surrounding wall does vary in composition and therefore can accommodate a number of lipids differing in their acyl composition.

The proposed catalytic mechanism is given in Fig. 10-30 taken from Slotboom *et al.* (1982).

In this proposal, Ca^{2+} binds to the phosphate and serves as a Lewis acid and polarizes the ester bond at the carbonyl oxygen. The water molecule, immobilized by the Asp-His pair, attacks the carbonyl of the substrate and donates a proton to His. The alkoxy oxygen of the glycerol backbone then retrieves the hydrogen from the His to complete the reaction. While not yet exhaustively studied, the Lys-49 phospholipase A₂ from *A. p. piscivorus* might operate by a similar mechanism (see Fig. 10-4).

While this model satisfies a number of experimental points, it is recognized that considerable data are yet required for it to be fully established. For example, data on the fitting of the substrate to the active site are lacking. However, a space-filling model of the system fits well with this proposal. In this model, the two acyl chains lie parallel and fit among the hydrophobic side chains of Leu-2, Leu-19, Leu-20, and Leu-31. One of the interesting questions yet to be resolved is how lipid molecules move between the organized interface into the active site. Molecular models of the enzyme have given some insight into this problem by their assignment of a lipophilic face of the enzyme in relation to the active site (Slotboom *et al.*, 1982) (Fig. 10-31).

Chapter 11

Function of Phospholipases

In the preceding chapters, descriptions of a wide variety of phospholipases have been covered. Emphasis has been placed on the characterization of these enzymes with emphasis on distribution, isolation, mechanism of action, and in a few instances, some reference was made to the function of the phospholipases in the metabolism of phospholipids. In general, those descriptions of function were limited to areas that have not been investigated extensively. The purpose of this chapter is to explore in greater depth some key areas of phospholipase function in which significant progress has been made or areas in which phospholipases have been used to study lipid and membrane composition.

11.1. General Considerations

As described in the Introduction, two general functions can be ascribed to the physiologic function of phospholipases. While somewhat simplistic, phospholipases serve in either a digestive or regulatory capacity. Quite a number of these have been described and can be either intercellular or extracellular. The first class to be discussed are those phospholipases that are considered regulatory in function. A number of these have been described, especially the phospholipases A_2 associated with the *arachidonate cascade* that provides substrate(s) for further metabolism to bioactive molecules. Likewise, the phospholipases C involved in the phosphatidylinositol cycle have been studied in a wide variety of cells and tissues. The localization of this type of regulatory phospholipase is not yet settled although it is an area of intense study. Another important group of regulatory phospholipases are involved in maintaining the acyl composition of membranous phospholipids in the deacylation–reacylation cycle. These are undoubtedly integral membrane proteins. A number of other phospholipases that might be assumed to have regulatory functions are not at all well understood. It is known that many membranes in prokaryotes and eukaryotes have phospholipases yet no function has been defined for them other than participating in the deacylation–reacylation cycle. The assignment of such a function, while possibly valid, is not based on experimental data but rather is the consequence of our limited understanding of these enzymes.

The second class of phospholipases to be covered is the digestive enzymes, in particular the pancreatic and venom phospholipases A_2. These enzymes are responsible for the hydrolysis of dietary phospholipid. The venom phospholipases are classified as digestive since, in addition to their toxic or lethal properties, digestion of the prey is initiated by their action. This is not to imply that these enzymes act alone in digestion and in fact may play a minor role in the overall digestive processes. However, their localized digestive action in capillary or site-specific locales is essential for the hunter. In addition, their action provides detergent products, lysophospholipid and fatty acid, that aid in the digestion of other lipids. Some of the venom phospholipases have taken on specialized function such as being site-directed toxins. There are intracellular phospholipases that serve a digestive role, also. Those that would be responsible for the digestion of phagocytized materials include cellular membranes. An example of intercellular digestive phospholipases are those found in lysosomes of many cells, especially phagocytic cells. In general, the digestive phospholipases are soluble enzymes as isolated even though they do have a high affinity for lipid.

A final section is then devoted to a brief description of the use of phospholipases as tools in chemical identification of lipids and as membrane probes. These studies aid not only in the probing of chemical and biologic structures, but also in helping to define further the characteristics of the enzymes under investigation.

11.2. The Phosphatidylinositol (PI) Cycle and the Arachidonate Cascade

Initial interest in phospholipase activity as a regulatory enzyme arose with the finding that acetylcholine stimulated pancreatic tissue to increase phosphatidylinositol turnover termed the PI cycle. The early work of Hokin and Hokin (1953) suggested that the initiation of the PI cycle was dependent on phospholipase C action. The rapid incorporation of inorganic ^{32}P into phosphatidic acid and phosphatidylinositol was not accompanied by a significant increase in mass of the phospholipid which suggested turnover rather than net synthesis of these compounds had occurred. Scheme 11-1 shows the proposed cycle which is initiated by phospholipase C cleavage of phosphatidylinositol. Since diacylglycerol is the immediate precursor of the phosphatidate being labeled, it was suggested that the phospholipase C specifically attacked phosphatidylinositol.

Table 11-1, taken in part from Michell *et al.* (1981), lists some but certainly not all tissues in which receptor-mediated hydrolysis of the phosphatidylinositols has been shown to occur. For a detailed review of mediator-induced PI cycle activity, the following reviews and book are suggested: Michell *et al.* (1981), Hawthorne (1982), Fain (1982), Snyderman *et al.* (1986), Berridge

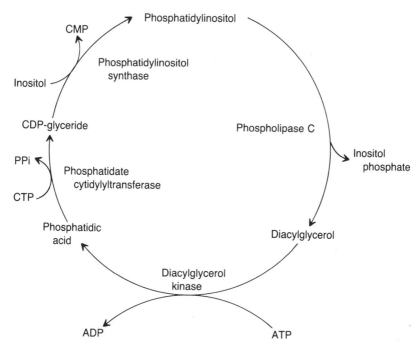

Scheme 11-1. Proposed cycle which is initiated by phospholipase C cleavage of phosphatidylinositol.

(1984), Abdel-Latif *et al.* (1985), and *Inositol and The Phosphoinositides* (Bleasdale *et al.,* 1985).

Several questions arise when Table 11-1 is studied.

1. What is the interrelationship between Ca^{2+} metabolism and phosphatidylinositol breakdown?
2. What substrates are involved (mono-, bis-, and trisphosphate inositol lipid) and what is the cellular localization of the phospholipase(s) C?
3. What is the coupling of the agonist–receptor complex to the phospholipase C and what is required for activation of the phospholipase C?
4. What are the consequences of phospholipase C action, as relates to the arachidonate cascade?

11.2.1. Calcium Regulation of Phospholipase C

The role of Ca^{2+} in the PI cycle historically was a subject of controversy. Recently, however, emphasis has focused on the primary importance of endogenous Ca^{2+}. Michell (1979) proposed that an enzyme (phospholipase C) coupled to an agonist receptor could release endogenous Ca^{2+} stores (Scheme 11-2).

Table 11-1. The Role of Ca^{2+} in Receptor-Stimulated Breakdown of Inositol Lipids

Tissue	Receptor-directed stimulus	Lipid broken down[a]	Effect of Ca^{2+} deprivation	Evoked by A23187 and Ca^{2+}?	References
Parotid gland	Acetylcholine	PtdIns	Unchanged	No	Jones and Michell (1975)
	Adrenalin	PtdIns	Unchanged	No	Jones and Michell (1978)
	Substance P	PtdIns	Unchanged	No	
Blowfly salivary gland	5-hydroxytryptamine	PtdIns	Unchanged	No	Fain and Berridge (1979), Berridge (1980, 1984)
Hepatocytes	Vasopressin	PtdIns	Unchanged	No	Billah and Michell (1979)
Lacrimal gland	Acetyl-β-methylcholine	PtdIns	Unchanged	?	Jones et al. (1979)
Synaptosomes	Electrical polarization	PtdIns	Unchanged	No	Pickard and Hawthorne (private communication) Griffin and Hawthorne (1978)
Synaptosomes	None known	PtdIns-4,5P₂ (and PtdIns-4P)	—	Yes	Griffin and Hawthorne (1978)
Platelets	Thrombin	PtdIns	Unchanged	Yes	Rittenhouse-Simmons (1981) Rittenhouse-Simmons and Deykin (1981) Lapetina et al. (1981a, 1981b)

Vas deferens	Acetylcholine	PtdIns	Unchanged	Yes	Egawa et al. (1981)
Lymphocytes	Phytohaemagglutinin	PtdIns	Decreased	?	Hui and Harmony (1980)
Pancreas	Acetylcholine	PtdIns	Abolished	Yes or no	Hokin-Neaverson (1977), Farese et al. (1980)
Polymorphonuclear leukocytes	f-Met-Leu-Phe	PtdIns	Abolished	Yes	Cockcroft et al. (1980)
	f-Met-Leu-Phe, GTP	$PtdIns-4,5P_2$	Unchanged	Yes	Snyderman et al. (1986)
Hepatocyte	Vasopressin	$PtdIns-4,5P_2$ (and PtdIns-4P)	Decreased	No	Kirk et al. (1981)
	Vasopressin + epinephrine + angiotensin II	PtdIns	Abolished	?	Prpic et al. (1982)
Paroid gland	Carbamylcholine	$PtdIns-4,5P_2$ (and PtdIns-4P)	Unchanged	No	Putney and Weiss (private communication)
Iris muscle	Acetylcholine or adrenalin	$PtdIns-4,5P_2$	Abolished	Yes	Abdel-Latif et al. (1985)
Erythrocyte	None known	$PtdIns-4,5P_2$ (and PtdIns-4P)	—	Yes	Lang et al. (1977)
Cultured epithelial cells	GTP	$PtdIns-4,5P_2$	—	—	Jackowski et al. (1986)

[a] Abbreviations: PtdIns, phosphatidylinositol; $PtdIns-4,5P_2$, phosphatidylinositol-4,5-bisphosphate; PtdIns-4P, phosphatidylinositol-4-phosphate; GTP, guanosine trisphosphate.

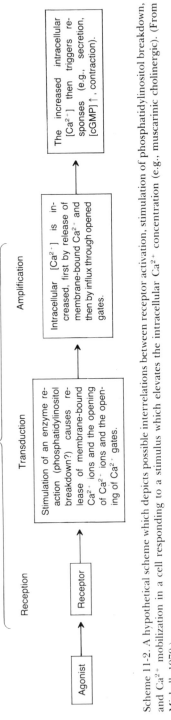

Scheme 11-2. A hypothetical scheme which depicts possible interrelations between receptor activation, stimulation of phosphatidylinositol breakdown, and Ca²⁺ mobilization in a cell responding to a stimulus which elevates the intracellular Ca²⁺ concentration (e.g., muscarinic cholinergic). (From Michell, 1979.)

The conclusion of Michell, which in general supports the views at the time of Hawthorne and others, seems appropriate with our current state of knowledge. "In some cells inositol lipid breakdown appears to be activated by receptors without the intervention of Ca^{2+}, in others a major component of inositol lipid breakdown may be a consequence of the mobilization of cytosolic Ca^{2+} by receptors, and in yet others both of these mechanisms may make large contributions. The realization that inositol lipid breakdown may be provoked by at least two distinct mechanisms introduces new difficulties into the design of definitive experimental tests of the hypothesis that receptors may employ inositol lipid breakdown as a reaction essential to coupling between receptor activation and Ca^{2+} mobilization" (Michell, 1979). We may therefore conclude that there is no clear or single answer to questions concerning the role of Ca^{2+} in the PI cycle.

More recent evidence has shown that hydrolysis of phosphatidylinositol bisphosphate requires either very low (0.1 μM) (Snyderman *et al.*, 1986) or no Ca^{2+} (Majerus *et al.*, 1985; Jackowski *et al.*, 1986). Since this substrate is generally thought to be key in the agonist response in a wide variety of cells, it can be thought that the Ca^{2+} response follows the initiation of the PI cycle rather than being causative. That is not to imply, however, that Ca^{2+} influx caused by ionophore or other agents cannot initiate the PI cycle. For example, addition of the Ca^{2+} ionophore A23187 directly activated phosphatidylinositol bisphosphate hydrolysis in the presence of pertussis toxin, an agent that blocks normal agonist-stimulated phospholipase C action in neutrophils (Verghese *et al.*, 1985) (to be covered in Section 11.2).

The question of Ca^{2+} regulation of the PI cycle was left somewhat open in studies on the muscarinic, cholinergic, and α_1 adrenergic receptor coupling to the PI cycle in iris smooth muscle (Abdel-Latif, 1983; Abdel-Latif *et al.*, 1985). These authors raised the question surrounding the Ca^{2+} requirement; "thus, while there could be some requirement for Ca^{2+} in the polyphosphoinositide effect in the iris, we have no experimental evidence to show that this phenomenon is Ca^{2+} regulated." As depicted in Scheme 11-3, activation of the phosphatidylinositol bisphosphate hydrolysis does lead to Ca^{2+} mobilization. It was suggested that removal of the polar head groups of the phosphatidylinositols would facilitate Na^+ and Ca^{2+} fluxes through channels which in turn would stimulate the Na^+, K^+ ATPase. Interestingly, Na^+ was required for the PI cycle to be completed following stimulation (Abdel-Latif and Akhtar, 1982; Abdel-Latif *et al.*, 1985; Abdel-Latif, 1986). While it is not clear from this study what effect trisphosphoinositol might have on Ca^{2+} mobilization, it is of interest to note that the PI cycle plays an essential role in neurotransmitter response. Such a role for the PI cycle in response to Ca^{2+} had previously been demonstrated in synaptosomes (Griffin *et al.*, 1979).

Considerable evidence now exists to show that phospholipase C regulates cellular Ca^{2+}. For example, Thomas *et al.* (1984) found that vasopressin stimulated the formation of trisphosphoinositol and bisphosphoinositol from the respective phospholipids in suspensions of hepatocytes. The cytosolic Ca^{2+} concentration from 160 to 400 nM arose primarily from a vesicular pool (mitochondria and endoplasmic reticulum). Addition of trisphosphoinositol

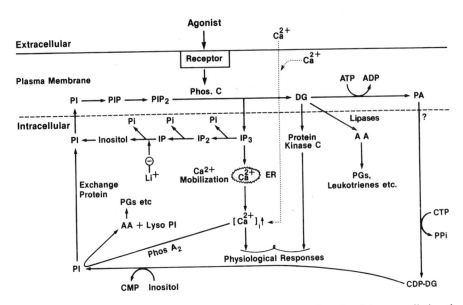

Scheme 11-3. Scheme showing phosphatidylinositol 4,5-bisphosphate breakdown mediating the step between activation of Ca^{2+}-mobilizing receptors in smooth muscle and other tissues and cellular responses. (From Abdel-Latif, 1986.)

to saponin-permeabilized cells caused similar effects. As a consequence, there was conversion of phosphorylase b to a and enhanced gluconeogenesis and glycogenolysis. This study and a variety of other studies indicate a special role for the degradation of the phosphatidylinositol bisphosphate that releases both diacylglycerol with its interesting biologic properties and inositol trisphosphate that mobilizes cellular Ca^{2+}. Similar events occurred upon thrombin stimulation of platelets (Siess and Binder, 1985). The rise in cytosolic Ca^{2+} in turn stimulated the activity of numerous Ca^{2+}-dependent reactions. In this way, the action of phospholipase C on phosphatidylinositol bisphosphate could be considered a part of a *second messenger* system. Rapid degradation of the trisphosphoinositol sequentially by separate phosphomonoesterases apparently regulated the duration of the Ca^{2+} mobilization (Majerus *et al.*, 1985). As will be pointed out later, this is a multifaceted stimulatory process since both diglyceride and arachidonate derived from diglyceride can lead to other amplification reactions. Likewise it has been reported (Kito *et al.*, 1986) that collagen and thrombin mobilization of Ca^{2+} could be separated from the activation of phospholipase C by the use of sodium fluoride, which inhibited the phospholipase C. However, sodium fluoride has many effects including the inhibition of phosphatases, so additional studies will be needed to explore further this interesting observation. This general area has been reviewed by Berridge (1984), one of the original proponents for the role of inositol trisphosphate in Ca^{2+} mobilization.

11.2.2. Substrate for the PI Cycle

Another intriguing question that remains to be resolved concerns the apparent substrate specificity of the phospholipase C. It is known that poly-phosphoinositides are minor components of the inositol lipids, as depicted in Scheme 11-4 (Majerus *et al.*, 1985). In this scheme, the relative proportions of the three phosphatidylinositol phosphates in platelets are indicated. It is proposed that once the phosphatidylinositol bisphosphate is cleaved, the response is magnified by a secondary hydrolysis of phosphatidylinositol. This secondary response is Ca^{2+}-dependent and would provide the bulk of the diacylglycerol for protein kinase C stimulation (Takai *et al.*, 1979) and arachidonate release. Recent evidence has brought into question the assumption that phosphatidylinositol is the sole donor of diacylglycerol in response to agonists. Daniel *et al.* (1986*b*) found that phorbol myristate acetate stimulated hydrolysis of phosphatidylcholine in Madin–Darby canine kidney cells. It is clear that the initial attack on phosphatidylinositol requires conditions leading to a high degree of specificity. This specificity could be met by a specific enzyme, by Ca^{2+} levels, or by arrangement of substrate within the membrane.

Separate phosphatases have been shown to degrade the inositol 1,4,5-trisphosphate either by the initial removal of the phosphate at position 5 or position 1 of the inositol (Majerus *et al.*, 1985) (Scheme 11-4). Lithium can block this cycle by inhibiting the inositol phosphate phosphatase. While this scheme satisfies a number of experimental results, more recent work shows that inositol tetraphosphates, including cyclic phosphates, are formed (Hansen *et al.*, 1986; Ishii *et al.*, 1986; Ross and Majerus, 1986). The interconversion of the inositol phosphates opens novel pathways of physiologic regulation.

Most studies on the substrate specificity of the cytosolic phospholipases C have indicated that these enzymes do not differentiate between the three phosphatidylinositols. It must be pointed out, however, that the optimal conditions for the hydrolysis of the different phosphatidylinositols vary considerably which could be a mechanism for specifying which of the substrates will be degraded (Graff *et al.*, 1984). It is of interest to note that some acyl specificity was found in the degradation of membranous phosphatidylinositol(s) of plate-

Scheme 11-4. The proposed sequence of reactions in phosphoinositide signal transduction in throm-bin-stimulated platelets. The sizes of the phos-phoinositide letters reflect their molecular masses. Therefore, most of the diglyceride is derived from phosphatidylinositol (PI). The question marks indicate that neither the magnitude of phospholipase C (PLC) hydrolysis of phosphatidylinositol 4-phosphate (PI 4-P) nor the route of metabolism of inositol 1,4-bisphosphate (I 1,4-P$_2$) is known.

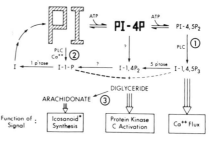

*Icosanoids include prostaglandins, thrombox-anes, leukotrienes, and other oxygenated derivatives of arachidonate and related polyunsaturated fatty acids. (From Majerus *et al.*, 1985.)

Table 11-2. Acyl Specificity of Phospholipase C

Class	Relative label in PI	Decrease in PI[a] (nmole/10^9 cells)	Relative degradation (thrombin)
Saturated	0.03	<0.05	0.41
Monoene	0.12	0.98	0.57
Diene	0.26	0.14	0.75
Trienen	0.20	0.09	0.79
Tetraenen	1.00	10.23	1.00
>Tetraenen[b]	1.30	1.19	1.05

[a] Based on mass analysis of molecular species in human platelets.
[b] Probably oleoyl-arachidonoyl phosphatidylinositol.

lets (Mahadevappa and Holub, 1983) (Table 11-2). In this study, the 1-stea-royl-(oleoyl)-2-arachidonoyl phosphatidylinositol(s) was degraded most rapidly. It will be of interest to see if the same species is degraded in the phosphatidylinositol bisphosphate pool.

Recent work by Jackowski *et al.* (1986) has added considerably to our understanding of the apparent specificity of phosphatidylinositol bisphosphate hydrolysis. Using a mink epithelial cell line, they demonstrated that the membrane fraction contained a phospholipase C that specifically degraded phosphatidylinositol bisphosphate. This enzyme, while not activated by Ca^{2+}, could be inhibited by 100 μM EDTA. Magnesium, on the other hand, stimulated a phosphatidylinositol bisphosphate phosphatase and caused an apparent inhibition of the phospholipase C by competing for the substrate. Interestingly, the activity of this phospholipase C was elevated in cells that had been transformed by either the v-*fms* or v-*fes* oncogenes that encode for tyrosine-specific protein kinases. It was suggested that this aberrantly high activity of the PI cycle could result in some of the transformed properties associated with these oncogenes. Related to this, Akhtar and Abdel-Latif (1978, 1980) reported that iris muscle phospholipase C that degrades polyphosphoinositides is membrane associated and requires Ca^{2+} at low concentrations (20 μM). Cockcroft *et al.* (1984) likewise demonstrated that human and rabbit neutrophil membranes degraded polyphosphoinositides in the presence of Ca^{2+} and the diacylglycerol could be converted to phosphatidic acid. Just how general the existence of this membrane-associated phospholipase C will be remains to be determined. Also, it remains to be determined what, if any, relation the membrane-associated enzymes have with the Ca^{2+}-requiring phospholipase C found in the soluble fraction of many cells.

11.2.3. Coupling of Agonist Receptor to Phospholipase C

Evidence is accumulating that the PI cycle and in particular the hydrolysis of phosphatidylinositol bisphosphate is regulated by a mechanism analogous to the regulation of the adenylate cyclase. The recent report of Jackowski *et al.* (1986) shows that the membrane-associated phospholipase in mink epithelial cells is dependent on GTP or its nonmetabolized analog Gpp(NH)p that has nitrogen rather than oxygen between the β and γ phosphorus atoms. As shown in Fig. 11-1, hydrolysis of phosphatidylinositol bisphosphate is also dependent on the presence of cholate.

The K_a for Gpp(NH)p was 3 μM, lower than that for GTP itself, 25 μM. This value agrees well with studies done in whole cells in which nonhydrolyzable GTP analogs stimulated inositol trisphosphate release in Sendai virustreated neutrophils (Barrowman *et al.*, 1986). Pertussis toxin, responsible for ribosylation and inactivation of the regulatory N proteins, was found to inhibit the hydrolysis of phosphatidylinositol bisphosphate stimulated by the chemotactic peptide f-Met-Leu-Phe (C. D. Smith *et al.*, 1985) (see Scheme 11-5). The toxin did not directly inhibit the phospholipase C as shown by the failure of the toxin to block the hydrolysis stimulated in a nonreceptor mechanism by the Ca^{2+} ionophore A23187 or concanavalin A. Similar observations were made using isolated membranes. Pertussis toxin treatment of the cells prior to isolation of the membranes blocked phospholipase C action at low Ca^{2+} concentrations. However, this inhibition could be overcome by higher Ca^{2+} concentrations. One possible interpretation of these results would be that the interaction of the phospholipase C with the N protein alters the affinity of the phospholipase C for Ca^{2+}. In this system, a low but physiologic level of Ca^{2+}, 0.1 μM, was required by the phospholipase C even in the presence of GTP and f-Met-Leu-Phe. A similar low level of Ca^{2+} was required by liver plasma membrane phospholipase C which was stimulated by vasopressin and

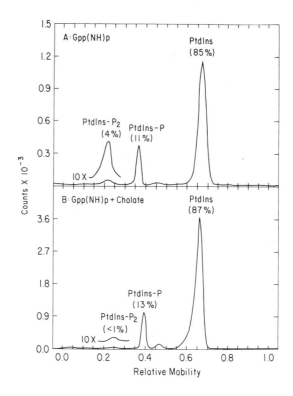

Figure 11-1. Detergent requirement for guanine nucleotide-dependent activation of phosphatidylinositol-4,5-bisphosphate hydrolysis in prelabeled membranes. The G2M (*v-fms*) cell line was labeled for 18 hr with 15 μCi/ml [³H]myo-inositol and a membrane fraction prepared. Degradation of phosphatidylinositol species was assessed in assays containing 65,000 cpm of labeled membranes and either 100 μM guanyl-5'-yl imidodiphosphate (panel A) or 100 μM guanyl-5'-yl imidodiphosphate plus 12 mM sodium cholate (panel B). The assays were extracted, the organic phases chromatographed on Silica Gel H thin-layer plates, and the distribution of radioactivity among the tritiated phosphatidylinositol species quantitated with the Bioscan imaging detector. Data accumulation time was 10 min in panel A and 20 min in panel B. (From Jackowski *et al.*, 1986.)

GTP (Uhing *et al.,* 1986). This is in conflict with a report by Wallace and Fain (1985) who indicated that the GTPγS (γ phosphorus is replaced by sulfur) stimulation of phospholipase C was Ca^{2+} independent. These results further substantiate the questioning about the requirement for Ca^{2+} in the PI cycle, a question that is somewhat difficult to resolve when one considers the problems of the low Ca^{2+} concentration required and the presence of endogenous Ca^{2+}. It was proposed that phosphorylation of proteins by protein kinases might serve as a feedback type of regulation, as depicted in Scheme 11-5. No strong evidence is available yet as to whether or not a regulatory protein similar to the purported phospholipase A_2 inhibitor, lipocortin, might be involved in regulation of this scheme (Blackwell *et al.,* 1980; Hirata *et al.,* 1980). The problem of regulation remains a fascinating area of research.

Additional suggestions that N proteins might regulate the phospholipase C were provided in studies on smooth muscle cells (Brock *et al.,* 1985). Both PMA and 1-oleoyl-2-acetylglycerol, potent stimulators of protein kinase C, blocked angiotension II stimulation of phospholipase C hydrolysis of phosphatidylinositol bisphosphate. It was presumed that these agents caused a phosphorylation of N protein(s) which dissociated the regulatory protein from the phospholipase C. If this were the case, the N protein(s) involved would have been different from those regulating the adenylate cyclase, based on differential effects of angiotensin II on cyclic AMP levels. A similar system may operate in human platelets based on the observation that PMA stimulated the thrombin-induced phospholipase C activity (Rittenhouse and Sasson, 1985).

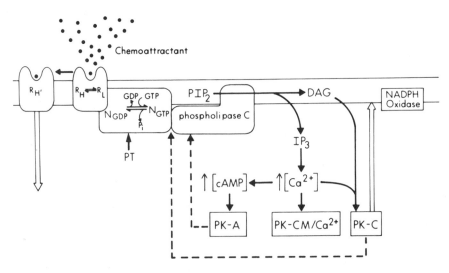

Scheme 11-5. A model for the regulation of phagocyte responses to chemoattractants. R_H, R_L, and $R_{H'}$ represent the reversible high, low, and irreversible high affinity states of the receptor, respectively. Other abbreviations: N, N-protein; PT, pertussis toxin; PIP_2, phosphatidylinositol 4,5-bisphosphate; DAG, diacylglycerol; IP_3, inositol phosphate; PK-C, Ca^{2+}-dependent protein kinase C; PK-CM/Ca^{2+}, membrane PK-C/Ca^{2+}; PK-A represents cAMP-dependent protein kinase and NADPH oxidase is the respiratory burst enzyme. → represents positive signals; – – → represents inhibitory signals; ⇒ represents subcellular redistribution. (From Snyderman *et al.,* 1986.)

In this case, the phospholipase C was measured by gas chromatographic analysis of inositol trisphosphate.

It also has been proposed that proteolysis of the platelet phospholipase C might lead to its activation (Ruggiero and Lapetina, 1985). A similar observation was made in studies on rat brain (Hirasawa *et al.*, 1982a) although subsequent studies suggested that proteolysis might be removing an inhibitor rather than causing direct activation of the enzyme (Irvine and Dawson, 1983). No matter which answer is correct, this mechanism of activation deserves further study.

11.2.4. The Arachidonate Cascade

11.2.4.1. Release of Arachidonate by Phospholipases C and A_2

From the foregoing discussion, it is clear that stimulation of the PI cycle, possibly initiated by phosphatidylinositol bisphosphate hydrolysis, results in a multitude of cellular responses. Thus far, only the roles of diacylglycerol and inositol trisphosphate as stimulators have been discussed. Another regulatory pathway is dependent on arachidonate released from phosphatidylinositol (phosphates) by the action of a phospholipase C and a coupled lipase (Bell *et al.*, 1979; Bell and Majerus, 1980). However, arachidonate also can be released from phospholipids by the direct action of a phospholipase A_2 (Walsh *et al.*, 1981). Considerable literature exists on the *arachidonate cascade* that is responsible for the production of the bioactive metabolites collectively known as the *eicosanoids* (Scheme 11-6). No attempt will be made here to cover this area and only those aspects that relate to the phospholipases will be included. Recent reviews and books on the metabolism and function of the

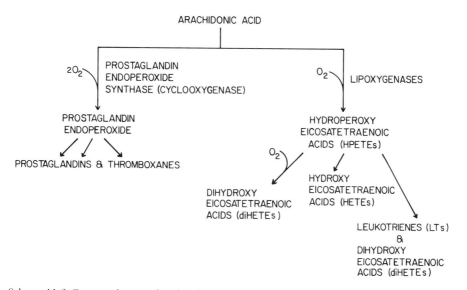

Scheme 11-6. Oxygenation reactions involving arachidonic acid. (From Smith and Borgeat, 1985.)

eicosanoids are available (Pace-Asciak and Granstrom, 1983; Bailey, 1985; Smith and Borgeat, 1985).

Questions concerning the relative contributions of the phospholipase C–lipase pathway versus phospholipase A_2 to the arachidonate cascade have been investigated vigorously over the past decade and recent studies have begun to unravel the relative interplay between the phospholipases A_2 and C. The work of Rittenhouse (1979, 1981, 1984) has helped to clarify the interplay between phospholipases A_2 and C in platelets. She showed by the use of cyclooxygenase inhibitors that in the presence of ionomycin, phospholipase A_2 releases arachidonate that is metabolized by the PGH_2 synthase that in turn activates the phospholipase C (Scheme 11-7). Such results appear to be consistent with the results of Siess and coworkers (1983) who showed that free arachidonate stimulated the PI cycle and that this stimulation could be blocked by aspirin, a cyclooxygenase inhibitor. On the other hand, Siess *et al.* (1984) found that the early events of platelet activation such as shape change induced by low concentrations of thrombin or platelet-activating factor induce the PI cycle but not the arachidonate cascade. Subsequent to the stimulation of phospholipase C action, phospholipase A_2 liberated arachidonate. It would therefore appear that when coupled to external agonists such as thrombin or platelet-activating factor, mobilization of Ca^{2+} by inositol trisphosphate can trigger phospholipase A_2 to release arachidonate and the latter stimulation can be duplicated by Ca^{2+} ionophores. The study of Okajima and Ui (1984) and Bokoch and Gilman (1984) demonstrated that pertussis toxin blocked f-Met-Leu-Phe-stimulated arachidonate release that could be overcome by ionophore. These studies, plus those described earlier, provide strong evidence

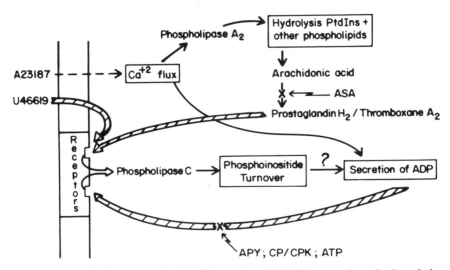

Scheme 11-7. Proposed route by which Ca^{2+} ionophore (A23187) causes the activation of phospholipase C and secretion. Aspirin inhibits cyclooxygenase. Apyrase and phosphocreatine/creatine kinase (PC/CK) convert ADP to nonstimulatory products. ATP competes with ADP for binding sites. Specific subcellular locations are not defined here. (From Rittenhouse, 1984.)

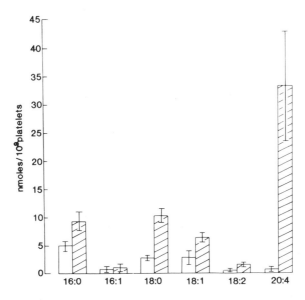

Figure 11-2. Average increases in free fatty acids induced by thrombin using the platelets from eight different donors. Unlabeled, gel-filtered platelet suspensions were preincubated with 100 μM 3-amino-1-[*m*-(trifluoromethyl)phenyl]-2-pyrazoline for 1 min at 37°C and then incubated with 5 units/ml thrombin for a further 5 min at 37°C. Free fatty acids were extracted and analyzed. Results shown are the mean ± 1 SD. (From J.B. Smith *et al.*, 1985.)

that there is a very close coupling of phospholipases C and A₂ activation. Further studies on how the events initiated by these enzymes relate to the events surrounding the activities of nucleotide cyclases and protein kinases will be most rewarding.

Smith and coworkers did a rather detailed analytical study to quantitate the amount of arachidonate arising from the action of phospholipases A_2 and C in platelets. According to their calculation, the bulk of arachidonate arises from phosphatidylcholine through the action of a phospholipase A_2. To quantitate the amount of arachidonate and other fatty acids released upon stimulation by thrombin, they used inhibitors of both the cyclooxygenase and lipoxygenase pathways (J. B. Smith *et al.*, 1985). Quantitative gas–liquid chromatographic analysis showed that there was an increase of nearly 50 nmoles of free fatty acid per 10^9 platelets upon stimulation, 70% of which was arachidonate (Fig. 11-2). By comparison, they estimated that only 4–5 nmoles of arachidonate arose by hydrolysis of diacylglycerol (Mauco *et al.*, 1984). The bulk of the diacylglycerol, however, was rapidly converted to phosphatidic acid, a commonly recognized phenomenon. They proposed that the phosphatidic acid formed in the PI cycle plus the fatty acid released via the action of the diacylglycerol lipase could cause a perturbation of the membrane, thereby activating the phospholipase A_2 (Smith and Kozlowski, 1986). They postulated that the action of the phospholipase C, acting on phosphatidylinositol bisphosphate, triggers the phospholipase A_2 to hydrolyze phosphatidylcholine yielding arachidonate for eicosanoid synthesis. Presumably, the Ca^{2+}

mobilized by the released inositol trisphosphate would stimulate the phospholipase A_2. Based on the analysis of fatty acids released, the phospholipase A_2 involved appears to be rather specific for arachidonate. Other studies have suggested the quantitative conversion of diglyceride to phosphatidic acid permits a phospholipase A_2 to release significant quantities of arachidonate from the phosphatidic acid (Lapetina *et al.,* 1978, 1981*a*). These two pathways in platelets could account for the release of some but not necessarily all the arachidonate from phosphatidylinositol. In that regard, Rittenhouse demonstrated that a significant quantity of phosphatidylinositol was degraded by platelet phospholipase A_2 in response to A23187. An inherent difficulty in accessing these various pathways is being able to quantitate the mass of the phosphatidic acid formed and turned over, studies not yet adequately done. In any event, the maintenance of phosphatidylinositol levels is thought to occur via recycling of diacylglycerol in the PI cycle based on analysis of the acyl composition of phosphatidylinositol following platelet stimulation by thrombin (Broekman *et al.,* 1981) (Table 11-3).

11.2.4.2. Reincorporation of Arachidonate into Phospholipids

Generally, only small quantities of lysophospholipids accumulate and these rapidly disappear, depending on the cell studied. The introduction of arachidonate or other fatty acids into lysophospholipid is an interesting facet of this problem. The reesterification with long-chain acids appears to be rather specific for arachidonate in neutrophils (Chilton *et al.,* 1983). Two mechanisms for reacylation have been described, one that is CoA dependent but ATP

Table 11-3. Changes in Fatty Acid Content of Phosphatidylinositol and Phosphatidic Acid in Thrombin-Stimulated Human Platelets[a]

	0 min	1 min	5 min	30 min
Phosphatidylinositol				
Palmitate	1.6	2.2	1.2	1.9
Stearate	55.0	36.1	31.1	41.5
Oleate	7.5	4.8	4.4	6.8
Linoleate	0.6	0.4	0.5	0.7
Arachidonate	59.5	35.3	33.9	43.8
Total fatty acids/2	62.1	39.4	35.6	47.4
Organic phosphate	61.8	35.1	41.2	47.5
Phosphatidic acid				
Palmitate	0.4	0.9	1.4	1.2
Stearate	0.7	9.5	7.0	1.6
Oleate	0.4	1.5	1.6	0.7
Linoleate	0.1	0.2	0.4	0.2
Arachidonate	0.7	9.7	6.8	1.3
Total fatty acids/2	1.2	10.9	8.6	2.5
Organic phosphate	1.7	11.8	8.4	3.0

[a] The fatty acid composition of phosphatidylinositol and phosphatidic acid was determined in control (0 min) platelets and 1, 5, and 30 min following addition of thrombin (0.3 unit/10^8 platelets).

Scheme 11-8. Possible pathways of biosynthesis of arachidonoyl phosphatides. 1-Acyllysophosphatides (LPL) are generated by phospholipase A_2-mediated deacylation of phospholipids (PL) synthetized *de novo* [(1)]. Arachidonoyl phosphatides (AA-PL) are subsequently formed by reacylation of lysophosphatides with free arachidonic acid (after conversion to arachidonoyl-CoA thioester) [(2), (3)] and/or direct transacylation of esterified arachidonate to lysophosphatides [(4)]. AA, arachidonic acid; FA, fatty acid. (From Kramer *et al.*, 1984.)

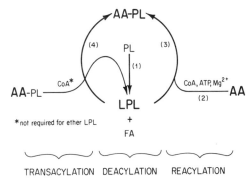

independent and one that is CoA independent, both involving an acyl transferase reaction. The arachidonate that is transferred does not involve a free acyl intermediate. In platelet, the CoA-independent reaction produced only phosphatidylethanolamine, primarily the plasmalogen form (Kramer *et al.*, 1984). The CoA-dependent acylation was shown to be different from the classical deacylation–reacylation pathway (Scheme 11-8).

The preference for acyl acceptors was lysophosphatidylserine > lysophosphatidylethanolamine > lysophosphatidylinositol; in all cases, phosphatidylcholine was the acyl donor. It was proposed in this paper that lysophosphatidylcholine, generated through the action of a phospholipase A_2, is preferentially reacylated by arachidonoyl CoA that is synthesized by the arachidonate-specific acyl CoA synthetase (Wilson *et al.*, 1982). The arachidonate is then transferred to the other phospholipids that are poor substrates for direct reacylation. Unlike the acyltransferase reaction in liver (Irvine and Dawson, 1979), this sequence does not account for the reacylation of lysophosphatidylinositol since it was shown that the acyl composition of phosphatidylinositol changes upon stimulation.

11.2.4.3. Arachidonate and Platelet-Activating Factor

Lysophosphatidylcholine, including the ether analog, can be reesterified by acetyl CoA. Since much of the phosphatidylcholine in circulating white blood cells contains an ether linkage in position 1 (Sugiura *et al.*, 1982; Mueller *et al.*, 1985), the acetylation process gives rise to the platelet-activating factor (Wykle *et al.*, 1986). Platelet-activating factor, discovered in the late 1970s (Benveniste *et al.*, 1979; Blank *et al.*, 1979; Demopoulos *et al.*, 1979), has a number of biologic activities that were the subject of a recent meeting (Benveniste and Arnoux, 1983). Wykle's group (Chilton *et al.*, 1983, 1984) has proposed a cycle involving the metabolism of both platelet-activating factor and arachidonate in neutrophils in which a phospholipase A_2 is responsible for the release of arachidonate from phosphatidylcholine within alkyl group at position 1 (Scheme 11-9).

This is an attractive hypothesis since it is known that there is interplay between platelet-activating factor and the lipoxygenase product 5-hydroxy-

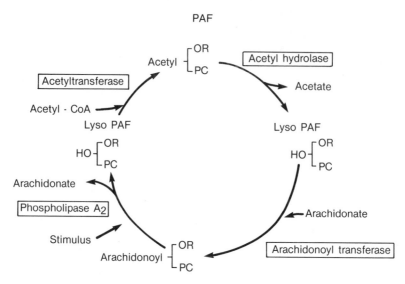

Scheme 11-9. Proposed platelet-activating factor cycle in human neutrophils. (From Chilton *et al.*, 1984.)

eicosatetraenoate (5-HETE). For example, 5-HETE reduces the concentration of platelet-activating factor causing neutrophil aggregation some 100-fold (O'Flaherty *et al.*, 1983; O'Flaherty, 1985). Theoretically then, very low levels of activity of this cycle could produce sufficient products to elicit biologic responses. The universality of this is yet to be determined.

11.2.4.4. Arachidonate from Bis(Lyso)Phosphatidic Acid

While many cells respond to stimulation by deacylating phosphatidyl-choline and phosphatidylinositol and in some cases phosphatidylethanol-amine, rabbit alveolar macrophages have a high level of a novel phospholipid, lyso(bis)phosphatidic acid (Mason *et al.*, 1972) which is deacylated in response to PMA (Cochran *et al.*, 1986).

$$
\begin{array}{ccc}
\text{O} & & \text{O} \\
\parallel & & \parallel \\
\text{RCOC} & & \text{COCR} \\
| & & | \\
\text{HOC} \quad \text{O} & & \text{C—OH} \\
| \quad \parallel & & | \\
\text{C—O P O—OC} \\
| \\
\text{O}^-
\end{array}
$$

This lipid comprises about 15% of the total phospholipid of alveolar macrophages and has a high content of arachidonate (Cochran *et al.*, 1986). Interestingly, macrophages from other sources have a much lower content of lyso(bis)phosphatidate. When rabbit alveolar macrophages are stimulated with the Ca^{2+} ionophore, arachidonate is released from phosphatidylcholine and phosphatidylethanolamine. The arachidonate released in response to A23187 is not converted to eicosanoids, however, by the rabbit cells. If the cells are stimulated with PMA or zymosan, lyso(bis)phosphatidate and phosphatidylinositol are also deacylated and lipoxygenase products are formed. It would appear that either the lyso(bis)phosphatidate- and phosphatidylinositol-derived arachidonate are a specific pool for eicosanoid synthesis or a phagocytosis is required to activate both the lipoxygenase and cyclooxygenase. It is of interest to note that lyso(bis)phosphatidate is synthesized via a transacylation of lysophosphatidyl glycerol and that the donor is phosphatidylinositol, at least in rat liver lysosomes (Matsuzawa *et al.*, 1978). These results also raise the possibility that a phospholipase A_1, rather than a phospholipase A_2, could be involved in deacylation of lyso(bis)phosphatidate in rabbit alveolar macrophages.

11.2.5. Regulation of Phospholipase A_2

The control of phospholipases in the arachidonate cascade has presented a challenge that is not yet met despite the considerable efforts of numerous laboratories. It is generally considered the regulation of the arachidonate cascade is at the initiation stage, the release of the arachidonate from phospholipid. Some evidence has been presented that factors such as fatty acids or their oxidized products could serve as inhibitors (Ballou and Cheung, 1983). Just how such inhibitors would function physiologically is unclear at present but certainly this is an attractive hypothesis. This would therefore provide a coupling between the enzyme system producing the eicosanoid and the initiating phospholipase. Such a regulation of the deacylation by the cyclooxygenase of synovial fluid cells was postulated by Robinson *et al.* (1981). We reached a similar conclusion in our studies using rat embryo fibroblasts (REF) and a herpes simplex virus type 2 transformed REF, rat fibrosarcoma (RFS) (C. R. Krebs, L. S. Kaceia, and M. Waite, unpublished observations). Both cell lines responded to PMA stimulation by deacylating arachidonate but only the RFS had cyclooxygenase activity. If, however, indomethacin was added with PMA to block cyclooxygenase activity, the RFS cells had no deacylation activity as well. While it is possible that indomethacin inhibited the phospholipase, as was found with the phospholipase isolated from neutrophils (Franson *et al.*, 1980), this appears not to be the case in REF cells since indomethacin did not influence deacylation activity. The details of such a regulation remain to be worked out but a mechanism by which the deacylation was regulated by the prostanoid product is certainly not unique and is appealing.

Not all systems have a coupling between the deacylation system and the cyclooxygenase. Madin–Darby canine kidney cells, for example, rapidly deacylate phospholipids in response to PMA even in the presence of cyclooxy-genase inhibitors (Beaudry *et al.*, 1983, 1985). Likewise, stimulation by the Ca^{2+} ionophore, A23187, causes deacylation without significant cyclooxygen-ase activity. Neutrophils respond to challenge with opsonized zymosan by deacylating phospholipids yet do not produce HETEs or leukotrienes (Waite *et al.*, 1979). Clearly, these coupling systems (or the lack thereof) are dependent on the cell type as well as the stimulant used.

Over the past several years, work on a group of proteins that inhibit phospholipase A_2 has been ongoing in three laboratories (Cloix *et al.*, 1983; Flower *et al.*, 1984; Hirata, 1984). Initially, these investigators termed the proteins lipomodulin, macrocortin, and renocortin, respectively. The reviews by Hirata and Flower summarize the historical development of this research and the adoption of the term lipocortin for these inhibitory proteins and, for the reader interested in inflammation, these reviews are recommended.

Basically, these investigations began as a search for the mechanism by which steroids block inflammation. Independently, Hirata and Flower re-ported that proteins were synthesized that suppressed phospholipase A_2 ac-tivity. Using rat peritoneal macrophages there appeared to be a two-step response to steroids; there was the initial release of lipocortin from the cells followed by the resynthesis of the lipocortin upon depletion of cellular stores (Flower *et al.*, 1984). The whole depletion–repletion cycle takes about 5 hr *in vitro*.

The mechanism by which lipocortin acts has been suggested by Hirata *et al.* (1984). They proposed that a specific tyrosine kinase phosphorylates li-pocortin which, when phosphorylated, can no longer interact with and inhibit phospholipase activity. In turn, the tyrosine kinase, suggested to be identical to pp60[src] or MT (middle-sized tumor) antigen, is regulated by protein kinases. Protein kinase C was thought to stimulate the tyrosine kinase, whereas protein kinase A would be inhibitory. Scheme 11-10 summarizes this proposal.

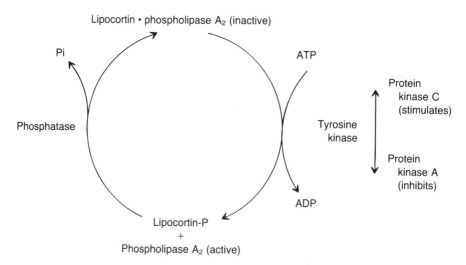

Scheme 11-10. Proposed interaction of lipocortin, phospholipase, and protein kinases.

The acceptance of such a model is far from certain since a great deal of experimental information is lacking. The fact that PMA stimulates phospholipase A$_2$ action (measured as arachidonate release) is consistent with the action of protein kinase C leading to phosphorylation and concomitant upregulation of deacylation (Waite *et al.*, 1979; Beaudry *et al.*, 1985; Cochran *et al.*, 1985). Studies from our laboratory have shown that inhibitors of protein kinase C block phospholipase A$_2$ stimulation by PMA in MDCK cells without blocking the PMA stimulation of cyclooxygenase synthesis (Parker *et al.*, 1987). On the other hand, it has been shown that elevated cAMP levels, presumably activating protein kinase A, also stimulate deacylation (Chiappe de Cingolani *et al.*, 1972; Gullis and Rowe, 1975; Lindgren *et al.*, 1978). This conclusion is not universal, however, since Lapetina *et al.* (1977, 1978) found that deacylation was blocked by cAMP.

Another proposed role for lipocortin is as a stimulator of suppressor T lymphocytes (Hirata and Iwata, 1983). It was thought that lipocortin induced maturation of precursor T cells into suppressor T cells, whereas there was no interaction with the helper T cells. Further investigations of the genetics as well as the functionality of lipocortin will be useful in appreciating all the facets of this protein(s).

The question of the size and structure of lipocortin was elegantly approached by Pepinsky *et al.* (1986) and Wallner *et al.* (1986). They were able to clone the protein from a human lymphoma cell line and rat peritoneal

Table 11-4. Amino Acid Differences between Rat and Human Sequences of Lipocortin

Amino acid	Rat	Human
12	Cysteine	Tryptophan
13	Leucine	Phenylalanine
16	Lysine	Asparagine
17	Glutamine	Glutamic acid
24	Alanine	Threonine
28	Tyrosine	Serine
41	Serine	Threonine
112	Methionine	Leucine
124	Leucine	Arginine
211	Glutamine	Glutamic acid
232	Histine	Glutamine
235	Lysine	Arginine
239	Asparagine	Lysine
241	Arginine	Threonine
245	Glutamine	Lysine
251	Alanine	Valine
283	Tyrosine	Histidine
284	Glutamic acid	Glutamine
289	Alanine	Valine
293	Arginine	Histidine
295	Threonine	Alanine
310	Glutamic acid	Aspartic acid
313	Valine	Alanine

exudates and found that antibodies to these preparations cross reacted. The cloned product had a molecular weight of 37,000 daltons, close to the primary size of protein isolated by Flower (40,000–45,000 daltons) and Hirata (45,000 daltons). Furthermore, antibodies to a phospholipase A_2 inhibitor (8000-dalton protein) cross reacted with the cloned lipocortin, which suggested to the authors that these inhibitory proteins are highly conserved. Comparison of the amino acid sequences of rat and human lipoprotein is given in Table 11-4.

A major question concerning the synthesis of lipocortin arose in this study. Analysis of the level of mRNA for lipocortin upon treatment of rats with steroids showed a severalfold increase, as expected. However,the quantity of lipocortin present in the peritoneal cells exhibiting the high level of mRNA did not increase in a 2-hr period of steroid treatment. In this case, the amount of lipocortin was quantitated by immunoprecipitation. Possible explanations for the lack of cellular lipocortin could be (1) increased synthesis is equaled by increased secretion, (2) the initiation of protein synthesis had not yet begun, and (3) the increase in activity noted by Flower and Hirata was activation of lipocortin already present in the cells at this particular time point. None of the answers is totally satisfying, and therefore considerably more research is required in this area before we can have a real appreciation as to how these purported proteins are regulating phospholipase A_2 in the arachidonate cascade.

11.3. Phospholipases in Digestion of Dietary Fat

The digestion and absorption of dietary fat are a complex series of events requiring various tissues and organs, enzymes, and cofactors. Furthermore, there is interplay between the digestion of the two major dietary lipids, triacylglycerol and phospholipid which is primarily phosphatidylcholine. Humans consume in the range of 2–4 g of phospholipid per day that emulsifies the more abundant triacylglycerol estimated to be in 40-fold excess (Borgstrom, 1980). The ratio of triacylglycerol to phospholipid is decreased from 40 : 1 to 10 : 1 by the phospholipid in bile. Three enzymes are involved in the digestive process beginning with the lingual lipase that degrades 10–20% of the triacylglycerol in the stomach. The activity in the intestine is initiated by the introduction of the pancreatic and bile juices that contain the essential activated lipase, phospholipase, colipase (that facilitates binding of lipase to lipid), and the bile salt–phospholipid emulsion. The interplay between these agents was outlined by Borgstrom (1980) *in vitro* using purified enzymes and coenzymes and the commercial triacylglycerol emulsion Intralipid as substrate. Basically, he and his colleagues demonstrated that phospholipid blocked the action of lipase on triacylglycerol although hydrolysis could occur after a substantial lag period if both colipase and bile salts were present. The presence of the phospholipase markedly stimulated the hydrolysis of triacylglycerol by hydrolyzing a portion of phospholipid (Fig. 11-3).

The lag period seen prior to the initiation of triacylglycerol hydrolysis

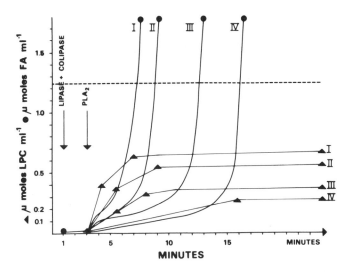

Figure 11-3. Effect of phospholipase A₂ on the hydrolysis of (▲) Intralipid phosphatidylcholine (formation of lysophosphatidylcholine = LPC) and on the lipase-catalyzed hydrolysis of the Intralipid triacylglycerol (●). One milliliter 20% Intralipid was diluted to 10 ml; final concentration of buffer: 150 mM NaCl, 1 mM Ca²⁺, 2 mM Tris-HCl, pH 8.0, and 4 mM taurodeoxycholate. Incubated at 40°C and tritiated. Lipase and colipase were added at zero time to a final concentration of 10⁻⁷ M and 2 × 10⁻⁷ M. The final concentrations of phospholipase A₂ were: I, 10 μg/ml; II, 5 μg/ml; III, 2.5 μg/ml; and IV, 1 μg/ml. Phosphatidylcholine = 1.24 μmoles/ml. (From Borgstrom, 1980.)

was dependent on the amount of lysophosphatidylcholine formed by the action of the phospholipase A₂. The reduction in the lag phase could be reproduced by the addition of fatty acid to the emulsion which suggests that the role phospholipase A₂ plays in facilitating the action of the lipase is to provide fatty acid. Indeed, some binding of the lipase (but not colipase) occurred without fatty acid and the addition of high concentrations of bile salt reduced this lag period by emulsifying the phospholipid. The effect of phospholipid hydrolysis was to enhance lipase and colipase binding as was shown directly by measuring the amount of these proteins in the soluble fraction under various conditions (Fig. 11-4).

Phospholipase A₂ treatment markedly stimulated binding of lipase and, to a lesser extent, colipase to the oil phase. The slight release of lipase and colipase into the aqueous phase was the result of a high percentage of monoacylglycerol and diacylglycerol that did not bind these proteins as well as triacylglycerol.

While these experiments give considerable insight into the interplay between phospholipid and triacylglycerol digestion, it should be recalled that the ratio of bile salt to phospholipid is crucial in regulating phospholipid hydrolysis (Chapter 8). It would therefore appear that total hydrolysis of phospholipid in the oil emulsion is dependent on the ratio of dietary and bile fat. This in turn regulates the hydrolysis of dietary triacylglycerol.

Is the action of the phospholipase A₂ all that is required for the uptake of dietary and bile phospholipid or is the removal of both fatty acids required

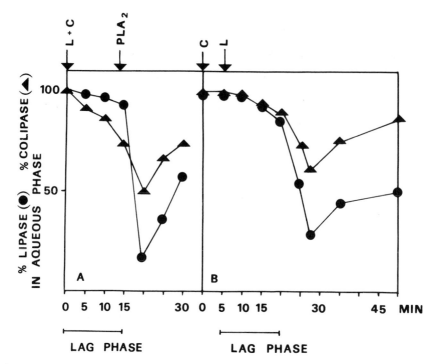

Figure 11-4. Distribution of lipase (●) and colipase (▲) to the aqueous phase when fatty acids were generated by: (A) phospholipase A_2 in the presence of lipase and colipase; and (B) lipase and colipase. Conditions of the experiments: (A) 4 mM taurodeoxycholate, 1 mM Ca^{2+}, pH 8.0. Final concentration of lipase = 2×10^{-8} M, colipase = 3×10^{-8} M, and phospholipase A_2 = 1×10^{-7} M. Lipase + colipase was added from the start. Phospholipase A_2 was added 14 min later. (B) 4 mM taurodeoxycholate, 1 mM Ca^{2+}, pH 7.5. Final concentration of lipase = 2×10^{-8} M, colipase = 3×10^{-8} M. Colipase was added at zero time followed 5 min later by lipase. Sampling was done at 5-min intervals. The lag time was 26 min. Samples of 0.7 ml were taken as indicated and immediately centrifuged to give an oil and an aqueous phase. The latter was sampled and used for determination of lipase and colipase. (From Borgstrom, 1980.)

for the transport of phospholipid across the mucosal wall of the intestine? It was shown in the late 1960s that lysophosphatidylcholine, derived by the action of phospholipase A_2 on dietary phosphatidylcholine, was utilized for the synthesis of the chylomicra phosphatidylcholine (Scow *et al.*, 1967). In an elegant experiment 1-[^{3}H]palmitoyl-2-[^{14}C]linoleoyl-[^{32}P]phosphatidylcholine was fed to rats with corn oil. The chylomicra obtained from the intestinal lymph contained phosphatidylcholine with a [^{3}H] : [^{32}P] ratio identical to that in the diet, whereas the [^{14}C] : [^{32}P] ratio was only 20% of the original ratio. These results indicate that the acyl group in position 1 but not in position 2 of the glycerol was retained during the conversion of dietary to lymph phosphatidylcholine.

While this mechanism for the absorption of dietary phosphatidylcholine is widespread, it does not appear to be universal. For example, the guinea pig pancreas does not contain the zymogen phospholipase A_2 found in most other species. Instead, it contains a fully activated phospholipase A_1 and lysophospholipase that is thought to be responsible for phospholipid degradation in the intestine (Fauvel *et al.*, 1981*b*). Likewise, the guinea pig pancreas lacks the classical triacylglycerol lipase but does not possess two cationic lipases with high phospholipase activity. Why such a dramatic evolutionary divergence from other rodent digestive mechanisms occurred remains intriguing.

Is the pancreas the only organ producing phospholipase involved in intestinal phospholipid degradation? Mansbach *et al.* (1982) and Verger *et al.* (1982) demonstrated that the intestinal cryptal cells contained a very potent phospholipase A_2 active toward phosphatidylglycerol. The enzyme was shown to be particulate primarily and was not due to bacterial contamination or derived from the pancreas. This enzyme had a high degree of homology with the pancreatic phospholipase A_2 with one major distinction; at position 4 it contained asparagine rather than glutamine found in most pancreatic and venom phospholipases A_2. It was thought perhaps that this, plus its basic character, might account for its unusual substrate specificity.

The function of the intestinal phospholipase A_2 in phospholipid digestion was shown by comparing phosphatidylcholine uptake in control rats and rats with pancreatic fistulas that diverted flow of pancreatic juice. Both groups of rats absorbed 85% of the phosphatidylcholine infused into the bile duct. These authors pointed out that the intestinal enzyme could function at times when pancreatic juice is not secreted, to recover phospholipid from sloughed intestinal cells, and to digest the phosphatidylglycerol from plant and intestinal bacterial sources.

Some insight into how this intercellular phospholipase A_2 can function in the digestion of extracellular phospholipid was given by Mansbach (1984) who described its possible activity in Paneth cells. These cells are thought to be phagocytic and therefore could be involved in the uptake of bacteria which contain a high content of phosphatidylglycerol. Likewise, these cells are secretory and could secrete enzyme into the lumen where it could function. A number of points concerning this system are yet to be established including its substrate specificity relative to the composition of the "normal" dietary phospholipid. Nonetheless, this enzyme could be a major factor in phospholipid digestion.

11.4. Phospholipases in Lipoprotein Metabolism

Degradation of phospholipids in lipoproteins is now thought to be important in the uptake of circulating lipids by cells, in particular liver cells. While no attempt will be made here to describe lipoprotein metabolism in any detail, it is worthwhile to point out that both the extrahepatic lipoprotein lipase and the hepatic lipase have phospholipase A_1 activity. Some recent studies have indicated that the hepatic lipase may very well function as a phospholipase involved in high-density lipoprotein (HDL) metabolism (Kuusi *et al.*, 1980; van Tol *et al.*, 1980). In this capacity, the hepatic lipase could serve to shuttle the cholesterol ester contained in HDL_2 to the liver and other tissues rich in the hepatic lipases where the enzymes degrade a portion of the phospholipid reducing the size of the particle and, as a consequence, forcing cholesterol ester out of the particle to be absorbed by the cell. Likewise, the fatty acid and lysophosphatidylcholine formed can either be absorbed by the cell or by blood components such as albumin and circulating cells.

The hepatic lipase is thought to be synthesized and secreted from the parenchymal cells of the liver (Schoonderwoerd *et al.*, 1983; Leitersdorf *et al.*, 1984). Once secreted, the enzyme is selectively absorbed by the endothelial cells of the liver (Jansen *et al.*, 1978*b*; Kuusi *et al.*, 1979*d*) or by other target tissues (Jansen and Hulsmann, 1980). In the latter case, the tissues absorbing the hepatic lipase are actively involved in cholesterol metabolism (i.e., adrenal glands that synthesize steroids). Scheme 11-11 shows the proposed role for lipoprotein phospholipid metabolism by the hepatic lipase.

In this proposal *heavy HDL* is HDL_3, whereas *PL-rich HDL* is HDL_2. Also shown is a possible regulatory role for the apoprotein C. Other apoproteins that have been thought to regulate the hepatic lipase include AI and AII (Kubo *et al.*, 1982; Jahn *et al.*, 1983), apo CI, CII, CIII (Kinnunen and Ehn-holm, 1976), and apo E (Windler *et al.*, 1980; Lippiello *et al.*, 1981, 1985). The role of the apoproteins in regulation of hepatic lipase awaits more detailed study. Likewise, the substrate specificity relative to its physiologic function remains open. If indeed its function is to degrade phospholipids associated with HDL, then its name would be more appropriately *lipoprotein phospholipase*. In this capacity, it would serve the role of channeling phospholipids to certain target tissues just as lipoprotein lipase channels triglyceride to other tissues (Jansen and Hulsmann, 1980). This work was thoroughly covered recently in the volume *Lipases* (Borgstrom and Brockman, 1984).

Together these studies on the metabolism of dietary and lipoprotein phospholipid demonstrate the role phospholipases play not only in the transport of phospholipid but also in the metabolism and transport of other lipids. Alterations in their activities can therefore markedly affect many other metabolic functions.

Another phospholipase is thought to be involved in lipoprotein metabolism. In this case, a phospholipase A_2 activity was shown to degrade the

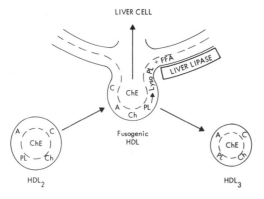

Scheme 11-11. Proposed role for lipoprotein phospholipid metabolism by the hepatic lipase. (From van't Hooft *et al.*, 1981.)

phosphatidylcholine of LDL (Steinbrecher *et al.*, 1984; Parthasarathy *et al.*, 1985). They first noted that endothelial cells modified LDL that caused a much faster rate of LDL uptake by macrophages. However, similar modifications of phospholipid degradation and lipid peroxidation occurred when LDL was incubated under oxidative conditions (i.e., with Cu^{2+}). Indirect evidence gave some insight into this process; bromphenacyl bromide blocked hydrolysis and oxidation, suggesting that the degradation was catalyzed by a phospholipase A_2. They proposed that phospholipase and oxidation were "reciprocally reinforcing," raising the possibility that oxidative changes could initiate phospholipase action. Possibly this effect could be manifested by a perturbation in lipid packing that favors phospholipase action. Indeed, this could be a more general phenomenon than is currently appreciated.

11.5. Action of Snake-Venom Enzymes

A major function of snake-venom phospholipases in digestion is somewhat analogous to the function of pancreatic phospholipases. By injecting digestive enzymes into its prey, the process of digestion begins even as the animal is being engulfed. Venoms, of course, have the function of also killing or incapacitating the prey, thereby providing a means other than speed and strength (used by racers and constrictors) for managing more mobile and dexterous animals. It was suggested in a review by Yang (1974) that the evolution of more toxic secretion in the "saliva" (venom) would be a strong evolutionary force. It appears that in some cases minor changes in protein structure determine if the phospholipase A_2 is a toxin (as compared with nontoxic enzyme), whereas in other cases major changes in phospholipase A_2 structure lead to complex formation with other proteins. In the latter case, the auxiliary protein can target the complex to specific tissue sites in the body. Related to these changes in structures are the mechanisms by which the toxins function.

Toxic phospholipases that are isolated in a simple, unconjugated form most often if not always function as anticoagulants. Phospholipids, primarily phosphatidylserine and phosphatidylcholine, are required as an interface for the interaction of the clotting proteins and clot formation (Verheij *et al.*, 1980*b*). Degradation of the diacyl phospholipid prevents clotting from occurring and hemorrhaging occurs. It was shown by these workers that it is the removal of the diacyl phospholipid rather than the production of lysophospholipid and fatty acid that is responsible for the anticoagulant properties of the phospholipases. It also was shown that the catalytic activity of the enzyme is essential for its effect on coagulation; inactivation of the active site rendered the *V. berus* enzyme incapable of blocking coagulation even though significant binding of the enzyme to phospholipid occurred.

Considerable insight into the characteristics of the phospholipase A_2 that are required for their anticoagulation properties was obtained in a study of 26 purified phospholipids from the Viperidae, Crotalidae, and Elapidae families (Verheij *et al.*, 1980*b*). Two general characteristics were shown to be important in clotting: (1) a high isoelectric point and (2) the capacity to penetrate films of phospholipases at high surface pressures. These phospholipases were then divided into three groups based on the concentration of enzyme required to alter the clotting time as shown in Fig. 11-5.

The range of concentration of enzyme used was from less than 1 µg/ml of basic *N. nigricollis* with essentially no effect to 500 µg/ml of the pancreatic enzyme. While a high isoelectric point was found to be one factor regulating clotting, it certainly is not sufficient to account for all observations.

A number of phospholipases A$_2$ with isoelectric points of nine or greater fall into group III with little, if any, effect on clotting. Also, phospholipases from the same species, *N. nigricollis,* that vary in the substitution of asparagine (basic) for aspartate (acidic), had markedly different effects on clotting and phospholipase activity when measured on *E. coli* membranes (Evans *et al.,* 1980). The other criterion that must be met is the enzyme's capacity to act at high surface pressures.

A limited number of venom phospholipases have neurotoxic rather than anticoagulant activity. These enzymes achieve their specificity and targeting through conjugation to another protein(s) in some but not all cases. It is not yet clear as to how the targeting is effected although it would appear that specific receptors must be involved. Once bound to the target tissue, the action of the phospholipase A$_2$ may accompany the neurotoxic effect in some but not all cases. All seem to act primarily through a common mechanism of presynaptic inhibition of cholinergic transmission. Myotoxin, on the other

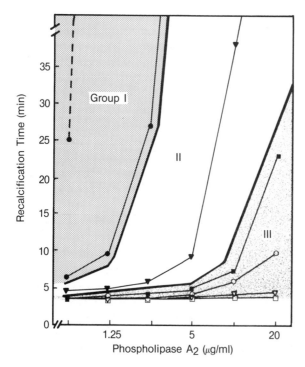

Figure 11-5. Reduction between the concentration of phospholipases and recalcification time. Clotting system: 0.2 ml plasma, 0.1 ml phospholipase A$_2$ solution in buffer. After 60-sec incubation at 37°C, 0.2 ml 25 mM CaCl$_2$ was added. Clotting time with buffer added instead of enzyme was 3.75 ± 0.15 min. (●– – –●) Basic *N. nigricollis;* (●———●) *V. berus;* (▼———▼) neutral *Agkistrodon halys blomhoffii;* (■———■) *Hemachatus haemachatus;* (○———○) *Notechis scutatus,* fraction 11-5; (▽———▽) pig pancreas; (□———□) β-bungarotoxin. The heavy solid lines indicate the division of the phospholipases into three classes of anticoagulant activity. (From Verheij *et al.,* 1980*b.*)

Table 11-5. Some Toxins, Their Origins, and the Number of Components

Name of toxin	Species	Number of components
β-Bungarotoxin	*Bungarus multicinctus*	2
Notexin	*Notechis scutatus scutatus*	1
Notechis II-5	*Notechis scutatus scutatus*	1
Myotoxin	*Enhydrina schistosa*	1
Taipoxin	*Oxyuranus scutellatus scutellatus*	3
Crotoxin	*Crotalus terrificus terrificus*	2

hand, has greater myotoxic activity than neurotoxicity. The neurotoxic effect leads to leakage of acetylcholine from motor nerve terminals and to death as the result of failure of respiratory muscles (Fohlman *et al.*, 1976). Table 11-5 contains a list of some toxins, their origin, and the number of components (Dowdall *et al.*, 1979).

It was found in a model study employing vesicular bodies obtained from the electric organ of *Torpedo* that these toxins vary considerably in their capacity to block choline uptake ranging from an IC_{50} of 3.7×10^{-11} M for notexin to 2.5×10^{-7} M for β-bungarotoxin. The latter value was comparable to that obtained with nontoxic phospholipases. The action of the phospholipase is rather subtle and does not cause a leakage of choline from the cells by a lytic action. Furthermore, there appears to be a lack of correlation between the phospholipase A_2 activity and inhibition of choline uptake. From this it appears that phospholipase A_2 action functions by modifying membrane structure that in turn inactivates the choline transport system. Evidence also exists that phospholipase action might not be required for neurotoxicity. Binding of a bromphenacyl bromide-treated taipoxin without phospholipase activity leads to altered motion of nitroxide spin-label stearic acid *Torpedo* nerve membranes (Dowdall *et al.*, 1979). Since the modified taipoxin retained capacity to inhibit choline uptake, it was suggested that the perturbation caused by taipoxin binding could be sufficient to inhibit choline uptake.

It is of interest to note that the three components of taipoxin are all homologous to phospholipases. The α and β subunits are homologous to venom and pancreatic phospholipases A_2, respectively, while the γ subunit is homologous to the prophospholipase A_2 of pancreas (Fohlman *et al.*, 1977). Of the three, only a basic α subunit has toxicity and the capacity to perturb membranes.

β-Bungarotoxin, on the other hand, a dimer composed of two unequal peptides that are linked via a disulfide bond, appears to have a biphasic effect. The first is independent of phospholipase activity whereas the second involves the degradation of the membrane (Caratsch *et al.*, 1981). These two effects of the toxin could be differentiated on the basis of dose-response curves, sensitivity to temperature, and dependency on Ca^{2+}. These workers found that the end plate potential of frog muscle was reduced some 70% in 30 min under conditions thought not to involve phospholipase action. It would therefore appear that β-bungarotoxin is similar to taipoxin in that the toxic response can be separated from the hydrolytic activity, at least in part.

It is presumed that the larger of the two subunits of the β-bungarotoxin with a molecular weight of 13,500 daltons is the phospholipase A_2, rather than the 7000-dalton molecular weight peptide. Thus far, it has not been possible to separate the two in an active form to test this possibility. Monoclonal antibodies to the dimer have been useful in mapping the site(s) of binding to membranes (Strong *et al.*, 1984). It appears that the antibodies isolated by these workers were against the A (13,500-daltons molecular weight) chain although the possibility that the epitope spanned both peptides could not be ruled out. The antibodies bound to the region Gly-53 to Lys-58, based on calculated antigenicity, and somewhat reduced phospholipase activity against liposomes. This region is adjacent to a catalytic center and could account for some inhibition of activity. The four different isotoxins could be classified based on their interactions with the antibodies; these differed only in the region of residues 55–60 which would be consistent with that region being epitopic. A heterogeneous antibody preparation blocked both phospholipase activity (90%) and the inhibition of the end plate potential in frog muscle. If, however, the binding of β-bungarotoxin is essential for its activity, any agent that blocks binding should block both activities, consistent with the studies on taipoxin. Monoclonal antibodies to the β subunit would be useful in probing its function in binding.

It is of interest from the evolutionary point of view how these various toxins have evolved. Indeed, these toxins apparently have assumed a function different from the more primitive simple digestive event and now are highly specialized. If such is an evolutionary advantage, it is curious relatively few snakes developed this type of toxin, although many other toxins are present in venoms (Yang, 1974). Also, it is of interest that many of the venom phospholipases have a rather weak capacity to attack membranous structures, at least when studied in pure form. This raises the question as to how they can function digestively *in situ*. Are there other components of venom (or other components of the snake digestive tract) that aid in their attack? On one hand, essential features of the digestive phospholipases A_2 have been rigorously conserved yet minor genetic changes have lead products with markedly different characteristics. This remains a fascinating area of chemical, biologic, and evolutionary research.

11.6. Phospholipases as Structural Probes

For many decades, phospholipases have been useful tools in probing both the structure of membranes as well as the structure of phospholipid molecules. Only brief examples of these studies will be given here along with references for reviews on the subject of phospholipases as probes.

11.6.1. Phospholipases as Probes of Membranes

Phospholipases of all types have been used to probe the structure of membranes. The membranes studied range from artificial bilayers as model membranes to the membranes of intact cells. In addition, they have been used to probe the lipid requirement of membrane-associated protein such as enzymes, receptors, and transport systems. The phospholipases used have most often been the phospholipases A$_2$ from pancreas or venoms while the phospholipases C have been from bacteria and the phospholipases D from plants or bacteria. The use of phospholipases in such studies must be judicious; it is essential that the purity of the phospholipase employed be known (especially protease activity) as well as its enzymologic characteristics. The latter requires knowledge of the enzyme's substrate specificity, capacity to penetrate lipid in organized membranes, and its recognition of charge of the surface. Some studies have not recognized these points and, as a consequence, interpretation of their results is clouded. Details on the purification and substrate specificity of phospholipases were given in the earlier chapters so their specificity will be mentioned only to indicate which phospholipid was undergoing hydrolysis. A number of excellent reviews on this subject have appeared over the past couple of decades and only a few examples of the use of phospholipases as membrane probes are given here (Bernheimer *et al.*, 1974; Avigad, 1976; op den Kamp, 1981).

11.6.1.1. *Phospholipases as Probes of Membrane Lipid Organization*

The erythrocyte has been the most widely studied cell using phospholipases. This choice is logical since it contains a single membrane, is readily available, and has a wide range of lipid composition, depending on the species of origin. Furthermore, the erythrocyte can be manipulated under the appropriate conditions to remain sealed with only one side available to attack, leaky with both sides accessible, or inverted with only the inner leaflet of the bilayer primarily accessible to hydrolysis. By showing that each side yields 50% hydrolysis and that "leaky ghosts" have essentially 100% hydrolysis, it is assumed that there is site-specific hydrolysis. Phospholipases have been used to study the molecular distribution of lipid between the two halves of the monolayer, surface pressure, and area of erythrocytes and requirements for integrity. In a classic study, it was shown that most erythrocytes have the choline phospholipids, sphingomyelin, and phosphatidylcholine on the exterior side, while phosphatidylethanolamine and phosphatidylserine are in the interior half of the membrane (Zwaal *et al.*, 1973; Zwaal, 1978) (Fig. 11-6).

The phospholipases used in this and studies on platelets were the phospholipase A$_2$ from *N. naja* venom, bee venom, and porcine pancreas; phospholipase C from *B. cereus* and *C. perfringens;* and sphingomyelinase (phospholipase C) from *S. aureus.*

Shortly after this report, a study was done to probe and expand these results (Martin *et al.*, 1975). This paper actually set the stage for more detailed studies on the dependency of hydrolysis of erythrocyte on the donor species, the ATP content of the cell, the conditions employed in hydrolysis, and the type of phospholipase employed. The use of different phospholipases, even from a single species, has been important in describing domains of phospholipid within the membrane.

A simple but illustrative experiment that demonstrates the importance

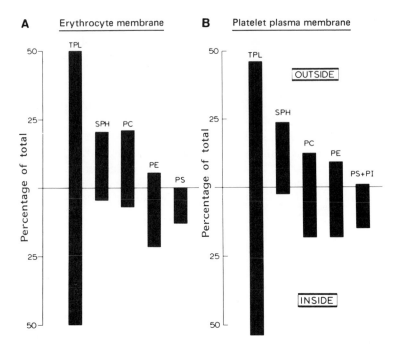

Figure 11-6. Asymmetric distribution of phospholipids between inner and outer layer of human red cell membranes (A) and pig platelet surface membranes (B). TPL, total phospholipid; SPH, sphingomyelin; PC, phosphatidylcholine; PE, phosphatidylethanolamine; PS, phosphatidylserine; PI, phosphatidylinositol. (From Zwaal, 1978.)

of erythrocyte ATP content is shown in Fig. 11-7. In the absence of glucose the ATP decreased 8- to 10-fold, while in the presence of glucose ATP was maintained at near-normal levels. Similar to the findings of Zwaal *et al.* (1973), phosphatidylcholine was initially attacked and hydrolysis ceased if glucose was present. However, if glucose were absent, phosphatidylethanolamine and phosphatidylserine on the inner leaflet of the membrane were degraded with the initiation of hemolysis.

Similar studies were done by Chap *et al.* (1977, 1982) on platelet plasma membranes. Similar to the erythrocyte, the outer monolayer of the platelet plasma membrane was rich in the choline phospholipids, whereas the inner monolayer was rich in the acidic phospholipids including phosphatidylinositol. The latter observation is of significance when one considers the functioning of the PI cycle (see Section 11.2). Other studies by this group demonstrated that free arachidonate is incorporated into the phospholipids (phosphatidylinositol) of the inner monolayer (Chap *et al.*, 1981). These results are consistent with the earlier report that about 60% of the cellular arachidonate is in the plasma membrane and 90% of the arachidonate in the plasma membrane is in the inner monolayer (Perret *et al.*, 1979).

Advantage was taken of the knowledge of the capacity of various highly purified phospholipases to penetrate monolayers of phospholipid to estimate the surface pressure of the erythrocyte (Demel *et al.*, 1975; also see Chapter

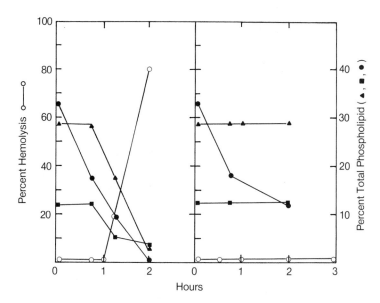

Figure 11-7. Effect of glucose on the hydrolysis of human erythrocyte membrane phospholipids by the basic phospholipase A_2 from *A. halys blomhoffii*. A 5% suspension of cells was incubated at 37°C. Left-hand panel, no added glucose; right-hand panel, 5 mM glucose. Phosphatidylcholine (PC) (●———●); phosphatidylethanolamine (PE) (▲———▲); phosphatidylserine (PS) (■———■); and extent of hemolysis (○———○). (From Martin *et al.*, 1975.)

10). Comparison of the enzymatic activity on erythrocytes and monolayers at fixed surface pressures of phospholipid are given in Table 10-3.

The four enzymes that cannot attack erythrocyte membranes have activity on monolayers held at varying pressures below 31 dynes/cm. On the other hand, four of the phospholipases could degrade erythrocyte membranes and could penetrate and hydrolyze the monolayer of the appropriate phospholipid. From this they concluded that the lateral surface pressure was 33–35 dynes/cm. While there are a number of uncertainties in extrapolating from the attack of a phospholipase on a pure lipid film to the attack of a complex biologic membrane, the results appear to be rather accurate in their predictions. Chap *et al.* (1977) use the same approach to show that the lateral pressure of the platelet membrane was about 34 dynes/cm.

Based on a surface pressure of 33–35 dynes/cm, they could calculate the average area per molecule of lipid in the erythrocyte, the area of the erythrocyte, and hence the area of the erythrocyte membrane covered by lipid. The phospholipid bilayer area, according to their calculations, would cover about 75% of the erythrocyte, leaving about 25% for the associated proteins. This is in good agreement with estimates of the protein content of erythrocyte membranes. These studies do not allow an assignment of the relative percentages of total lipid to a given half of the bilayer, however.

Shukla and Hanahan (1982a) extended these observations to show that the acidic and basic phospholipases A_2 from *A. halys blomhoffii* could recognize different domains of phosphatidylcholine in human erythrocyte membranes.

Based on the patterns of fatty acids released by the two enzymes, it was concluded that the domains were comprised of phosphatidylcholine species that differed in their acyl composition. Apparently, the two phospholipases bound to "different affinity sites on the membrane." While the concept of domains had been described previously in other systems, this paper very nicely points out the advantage of studying the comparative activity of different phospholipases. These workers also were able to show an age-dependent difference in erythrocyte membrane structure based on their susceptibility to phospholipase treatment (Shukla and Hanahan, 1982*b*).

As described earlier, treatment of ATP-rich erythrocytes with phospholipase A_2 seldom causes direct hemolysis unless the cell is mechanically perturbed (usually a temperature shift) or another component such as albumin or detergent is added that, by itself, does not disrupt the cell (Condrea *et al.*, 1964; Murofushi *et al.*, 1973; Zwaal *et al.*, 1973; Gul and Smith, 1974). For a number of years, this was a puzzle since it was known that the hydrolytic products when added to erythrocytes caused lysis. These results are now more clearly understood based on the finding of Jain *et al.* (1980) that an equimolar mixture of lysophosphatidylcholine and fatty acid will form a stable bilayer.

Phospholipase C treatment often produces the so-called *hot–cold* hemolysis of erythrocytes, in particular when ruminant erythrocytes are treated with *S. aureus* phospholipase C specific for sphingomyelin (Colley *et al.*, 1973; Bernheimer *et al.*, 1974). Following hydrolysis at 37°C, little hemolysis occurred unless the cells were cooled to 4°C. This is thought to be the result of a phase separation of more solid particles in the membranes, which in *Tetrahymena* is about 17°C (Wunderlich *et al.*, 1975). The conversion of phosphoglyceride to diacylglycerol causes a number of structural changes in the membrane including the formation of oil droplets, viewed as black dots in the membrane using phase microscopy (Kemp and Howatson, 1966; Coleman *et al.*, 1970; Bowman *et al.*, 1971; Colley *et al.*, 1973). This is accompanied by a loss in surface area. As pointed out by Avigad (1976), a crucial factor in the release of hemoglobin from the cell depends to a large extent on the state of organization of the inner bilayer. Also like the results with phospholipase A_2, the energy state of the cell can regulate its susceptibility to phospholipase C attack. For example, energy-depleted erythrocytes treated with phospholipase C from *C. perfringens* were degraded more rapidly than control cells (Frish *et al.*, 1973) and restoration of the ATP levels increased resistance to hemolysis (Gazitt *et al.*, 1975). Whether this has to do with some energy-dependent interaction between phospholipid and protein (perhaps via the cytoskeleton) remains to be seen.

Many proteins associated with membranes are known to require phospholipid for their function as shown by loss of function upon extraction with organic solvents or by treatment with phospholipases and the restoration of activity upon the reintroduction of the appropriate phospholipid(s). Likewise, it is important to demonstrate that products of hydrolysis such as fatty acids, lysophospholipid, or diglyceride do not remain and alter the function of the protein upon reconstitution of the protein with phospholipid. Table 11-6 lists

Table 11-6. Some Sites of Phospholipase A₂ Action

1. Receptors—hormone, immunoglobulins
2. Enzymes—mitochondrial, plasma membrane, microsomal, nuclear, leukocyte, bacteria, serum
3. Lipoproteins
4. Cell fusion
5. Cell transport and secretion
6. Whole organelles—mitochondria, lysosomes, microsomes
7. Muscle function
8. Excitable membranes
9. Microtubule–phospholipid interaction
10. Cell locomotion
11. Blood clotting
12. Rhodopsin
13. Viral infectivity

some types of membrane and cell functions that have been altered by phospholipase treatment.

The review of Avigad (1976) from which this list was taken is replete with a much broader description of the enzymes used for these studies. Only two types of examples will be given here, one on the effects of phospholipase action on intact erythrocytes and the other on the use of phospholipases in probing the phospholipid requirement of assorted enzymes and receptors. It does appear that phospholipase C treatment of erythrocytes leads to changes in lipid–protein interaction, although the exact nature of the changes are not settled (Lenard and Singer, 1968; Wallach, 1969; Zwaal *et al.*, 1973).

11.6.1.2. Regulation of Membrane Proteins by Phospholipases

Some membrane-associated enzymes whose dependence on phospholipid can be demonstrated by regulation with phospholipase treatment are given in Table 11-7. The study of Roelofsen and Schatzmann (1977) was of particular interest since it not only showed the requirement of the ATPase for phospholipid but also suggested that it is present on the internal side of the membrane. Both phospholipase A₂ and phospholipase C inactivated the enzyme of erythrocyte ghosts, whereas the ATPase remained fully active in the intact cells where treated in a comparable manner. Some residual activity remained when the ghosts were treated with phospholipase A₂ unless albumin was used to remove fatty acids and lysophospholipids. A number of phospholipids including lysophosphatidylcholine restored activity of the ATPase. This restoration of activity was an essential feature in experiments showing the requirement of phospholipid for enzymatic activity. Interestingly, sphingomyelinase did not inactivate the ATPase nor could sphingomyelin restore activity to the enzyme inactivated by phospholipase C treatment.

As can be appreciated at this point, the use of phospholipases to probe intact cellular membranes and their associated proteins is complex and has

Table 11-7. Membrane-Associated Enzymes Whose Dependence on Phospholipid Can Be Demonstrated by Regulation with Phospholipase Treatment

Membrane enzyme	Location	Phospholipase[a]	Lipid required	Reference
$(Ca^{2+} + Mg^{2+})$ ATPase	Red cell (inner leaflet)	PLA_2 (pancreas) PLA_2 (N. naja) Sphingomyelinase PLC (B. cereus)	Any glycerolipid	Roelofsen and Schatzmann (1977)
3-Oxosteroid Δ^4–Δ^5 isomerase	—	PLA_2 (N. naja)	Total phospholipid	Geynet et al. (1976)
Gonadotropin receptor	Corpus luteum plasma membrane	PLA_2 (V. russelli) PLA_2 (C. terrificus) PLC (C. perfringens) PLC (B. cereus) PLD (cabbage)	Inhibition by lysophospholipid and fatty acid	Azhar and Menon (1976) Azhar et al. (1976)
CTP:phosphocholine cytidylyltransferase	Cytosol	PLA_2 (C. adamanteus) PLC (B. cereus)	Stimulation by lysophospholipid	Choy and Vance (1978)
$(Na^+ + K^+)$ ATPase kidney	Purified from membranes	PLC (S. aureus) PS decarboxylase	Requires negative lipid	dePont et al. (1978)
Fc receptors	Macrophage	PLC (B. cereus)	Phosphatidylcholine	Itonaga et al. (1984)

[a] Abbreviations: PLA, phospholipase A; PLC, phospholipase C; PLD, phospholipase D; PS, phosphatidylserine.

far-reaching effects on the cell. It is not possible to draw simple conclusions from such studies. However, by the combined approaches of biophysics, lipid chemistry and enzymology, and cell biology, such studies can be quite enlightening.

11.6.2. Analysis of Phospholipid Structure Using Phospholipases

Numerous studies have been made using phospholipases to determine structural characteristics of phospholipids, some of which are covered in Chapters 6 and 7 of *Lipolytic Enzymes* (Brockerhoff and Jensen, 1974). Perhaps the simplest and most common use has been phospholipase A_2 treatment of phospholipids to determine the acyl composition at positions 1 and 2 of the glycerol. Phospholipases can be useful in identifying reaction products employing photoactivatable cross-linked phospholipids (Gupta *et al.*, 1979). The book on *Lipases* by Brockerhoff and Jensen is recommended for more details on the use of phospholipases in structural analysis. Here an example of such an analysis is given, taken from the work of Bonsen *et al.* (1965, 1967), who elucidated the structure of the bacterial phosphatidylglycerol that contains an amino acid. To this end, they synthesized 1-oleoyl-2-palmitoyl-3-phosphoryl-1'-(3'-O-L-lysyl)glycerol and compared its structure with the natural product isolated from *S. aureus* (Scheme 11-12).

Three phospholipases were used to compare the structures of the natural and synthesized lipid : phospholipase A_2 from *C. adamanteus*, phospholipase C from *B. cereus*, and phospholipase D from savoy cabbage or brussels sprout. While these were not highly purified preparations of the enzymes, they were adequate for this study. In this case, phospholipase A_2 allowed the identification of the acyl composition in positions 1 and 2 of the glycerol. Phospholipase C was used to remove the phosphoglycerolysine. Subsequent studies showed that the lysine was at the free primary hydroxyl of phosphatidylglycerol. Phospholipase C was also used to remove the glycerol phosphate. Since it was not a substrate for glycerol-3-phosphate dehydrogenase, it was concluded that the product was glycerol-1-phosphate. Phospholipase D yielded phosphatidic acid plus glycerol-3-lysine. These results demonstrated that the two lipids were identical structurally as well as stereochemically, with the exception of the acyl composition.

Phospholipases D are used often in the preparation of phospholipid since it is a transphosphatidylase, a reaction that efficiently leads to the substitution of one base group for another. While phosphatidic acid is always a product of the reaction, base substitution can be the favored reaction because of the high affinity of the enzyme for alcohols, relative to water. The use of phospholipase D for substrate preparation is conveniently detailed by Eibl and Kovatchev (1981).

This very powerful approach to phospholipid analysis has been used by a number of investigators to provide us with the detailed knowledge of phospholipid structure. This knowledge in turn has been crucial in our understanding of how phospholipids form membranes and other aggregated structures. An essential component toward this end was the understanding of

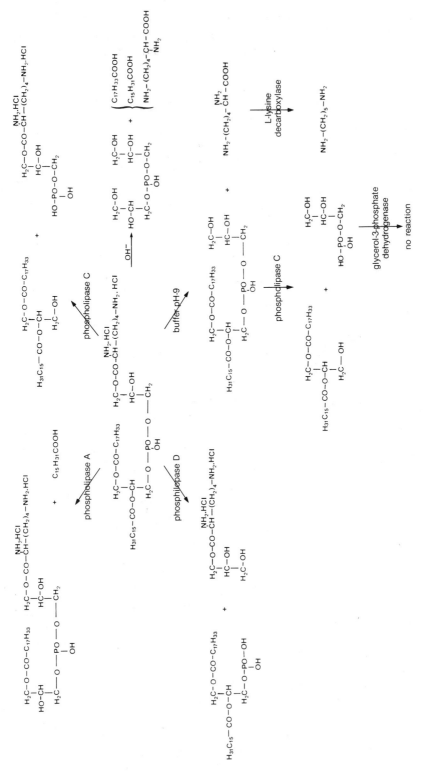

Scheme 11-12. Chemical and enzymic degradation of 1-oleoyl-2-palmitoylglycerol-3-phosphoryl-1'-(3'-O-L-lysyl)glycerol. (From Bonsen *et al.*, 1967.)

phospholipases, studies elegantly carried out in the laboratories of Brocker-hoff, van Deenen, Hanahan, Renkonen, Wells, Dennis, Little, Dawson, and others.

11.7. Summary

This chapter has covered some functions and regulations of phospholipases and attempted to put into a broader context the significance of this ubiquitous group of enzymes. Indeed, this group of enzymes is so diverse in structure, mechanism of action, and function that the only property in common for these enzymes is their phospholipid substrate. However, the approaches to their study hold a number of common threads, especially the need for an understanding of the physical properties of the substrate aggregate. For a number of years, these enzymes have been used experimentally to probe phospholipid and membrane structure and to understand better membrane–protein interactions. With newer emphasis on cellular phospholipases and their physiologic roles in cells, the field should continue to grow and contribute to the knowledge of a number of disciplines in the biologic and physical sciences.

References

Aakre, S.-E., and Little, C., 1982, Inhibition of *Bacillus cereus* phospholipase C by univalent anions, *Biochem. J.* **203:**799.

Aalmo, K., Hansen, L., Hough, E., Jynge, K., Krane, J., Little, C., and Storm, C. B., 1984, An anion binding site in the active centre of phospholipase C from *Bacillus cereus, Biochem. Int.* **8:**27.

Aarsman, A. J., and van den Bosch, H., 1979, A comparison of acyl-oxyester and acyl-thioester substrates for some lipolytic enzymes, *Biochim. Biophys. Acta* **572:**519.

Aarsman, A. J., van Deenen, L. L. M., and van den Bosch, H., 1976, Studies on lysophospholipids. VII. Synthesis of acyl-thioester analogs of lysolecithin and their use in a continuous spectrophotometric assay for lysophospholipases, a method with potential applicability to other lipolytic enzymes, *Bioorgan. Chem.* **5:**241.

Aarsman, A. J., Hille, J. D. R., and van den Bosch, H., 1977, Purification of lysophospholipases: Application of a continuous spectrophotometric assay using thioester substrate analogs, *Biochim. Biophys. Acta* **489:**242.

Aarsman, A. J., Neys, F., and van den Bosch, H., 1984, A simple and versatile affinity column for phospholipase A_2, *Biochim. Biophys. Acta* **792:**363.

Abdel-Latif, A. A., 1983, Metabolism of phosphoinositides, in *Handbook of Neurochemistry* (A. Lajtha, ed.), vol. 3, pp. 91–131, Plenum Press, New York.

Abdel-Latif, A. A., 1986, Calcium-mobilizing receptors, polyphosphoinositides, and the generation of second messengers, *Pharmacol. Rev.* **38:**227.

Abdel-Latif, A., and Akhtar, R. A., 1982, Cations and the acetylcholine-stimulated ^{32}P-labelling of phosphoinositides in the rabbit iris, in *Phospholipids in the Nervous System* (L. A. Horrocks, G. B. Ansell, and G. Porcellatti, eds.), vol. 1: *Metabolism,* pp. 251–264, Raven Press, New York.

Abdel-Latif, A. A., Akhtar, R. A., and Hawthorne, J. N., 1977, Acetylcholine increases the breakdown of triphosphoinositide of rabbit iris muscle prelabelled with [^{32}P]phosphate, *Biochem. J.* **162:**61.

Abdel-Latif, A. A., Smith, J. P., and Akhtar, R. A., 1985, Polyphosphoinositides and muscarinic cholinergic and α_1-adrenergic receptors in the iris smooth muscle, in *Inositol and Phosphoinositides, Metabolism and Regulation* (J. B. Bleasdale, J. Eichberg, and G. Hauser, eds.), p. 275, Humana Press, Clifton, NJ.

Abe, M., Okamoto, N., Doi, O., and Nojima, S., 1974, Genetic mapping of the locus for detergent-resistant phospholipase A (p1dA) in *Escherichia coli* K-12, *J. Bacteriol.* **119:**543.

Abe, T., Alema, S., and Miledi, R., 1977, Isolation and characterization of presynaptically acting neurotoxins from the venom of *Bungarus* snakes, *Eur. J. Biochem.* **80:**1.

Abita, J.-P., Lazdunski, M., Bonsen, P. P. M., Pieterson, W. A., and deHaas, G. H., 1972, Zymogen–enzyme transformations—on the mechanism of activation of phospholipase A, *Eur. J. Biochem.* **30:**37.

Ackman, R. G., 1981, Flame ionization detection applied to thin layer chromatography on coated quartz rods, *Methods Enzymol.* **72:**205.

Akamatsu, Y., Ono, Y., and Nojima, S., 1967, Studies on the phospholipid metabolism in *Mycobacterium phlei, J. Biochem.* **61**:96.

Akhtar, R. A., and Abdel-Latif, A. A., 1978, Studies on the properties of the triphosphoinositide phosphomonoesterase and phosphodiesterase of rabbit iris smooth muscle, *Biochim. Biophys. Acta* **527**:159.

Akhtar, R. A., and Abdel-Latif, A. A., 1980, Requirement for calcium ions in acetylcholine-stimulated phosphodiesteratic cleavage of phosphatidyl-myo-inositol 4,5-bisphosphate in rabbit iris smooth muscle, *Biochem. J.* **192**:783.

Akhtar, R. A., and Abdel-Latif, A. A., 1982, Effects of Na^+, Ca^{2+} and acetylcholine on the phosphoinositide and ATP-phosphate turnover in ^{32}P-labeled rabbit iris smooth muscle, *J. Neurochem.* **39**:1374.

Albright, F. R., White, D. A., and Lennarz, W. J., 1973, Studies on the enzymes involved in the catabolism of phospholipids in *Escherichia coli, J. Biol. Chem.* **248**:3968.

Allan, D., Thomas, P., and Limbrick, A. R., 1982, Microvesiculation and sphingomyelinase action in chicken erythrocytes treated with ionophore A23187 and Ca^{2+}, *Biochim. Biophys. Acta* **693**:53.

Allgyer, T. T., and Wells, M. A., 1978, Kinetic anomalies associated with phospholipase A_2 hydrolysis of micellar substrates, *Adv. Exp. Med. Biol.* **101**:153.

Allgyer, T. T., and Wells, M., 1979, Phospholipase D from savoy cabbage: Purification and preliminary kinetic characterization, *Biochemistry* **18**:5348.

Andersson, T., Drakenberg, T., Forsen, S., Wieloch, T., and Lindstrom, M., 1981, Calcium binding to porcine pancreatic prophospholipase A_2 studied by ^{43}Ca NMR, *FEBS Lett.* **123**:115.

Angus, W. W., and Lester, R. L., 1972, Turnover of inositol and phosphorus-containing lipids in *Saccharomyces cerevisiae:* Extracellular accumulation of glycerophosphorylinositol derived from phosphatidylinositol, *Arch. Biochem. Biophys.* **151**:483.

Angus, W. W., and Lester, R. L., 1975, The regulated catabolism of endogenous and exogenous phosphatidylinositol by *Saccharomyces cerevisiae* leading to extracellular glycerophosphorylinositol and inositol, *J. Biol. Chem.* **250**:22.

Apitz-Castro, R. J., Mas, M. A., Cruz, M. R., and Jain, M. K., 1979, Isolation of homogeneous phospholipase A_2 from human platelets, *Biochem. Biophys. Res. Commun.* **91**:63.

Apitz-Castro, R. J., Cruz, M. R., Mas, M. A., and Jain, M. K., 1981, Further studies on a phospholipase A_2 isolated from human platelet plasma membranes, *Thromb. Res.* **23**:347.

Apitz-Castro, R., Jain, M. K., and deHaas, G. H., 1982, Origin of the latency phase during the action of phospholipase A_2 on unmodified phosphatidylcholine vesicles, *Biochim. Biophys. Acta* **688**:349.

Arai, H., Inoue, K., Natori, Y., Banno, Y., Nozawa, Y., and Nojima, S., 1985, Intracellular phospholipase activities of *Tetrahymena pyriformis, J. Biochem.* **97**:1525.

Arai, M., Matsunaga, K., and Murao, S., 1974, Purification and some properties of phospholipase C of *Pseudomonas schuylkilliensis, J. Agric. Chem. Soc. (Japan)* **7**:409.

deAraujo, P. S., Rosseneu, M. Y., Kremer, J. M. H., van Zoelen, E. J. J., and deHaas, G. H., 1979, Structure and thermodynamic properties of the complexes between phospholipase A_2 and lipid micelles, *Biochemistry* **18**:580.

Argiolas, A., and Pisano, J. J., 1983, Facilitation of phospholipase A_2 activity by mastoparans, a new class of mast cell degranulating peptides from wasp venom, *J. Biol. Chem.* **258**:13697.

Astrachan, L., 1973, The bond hydrolyzed by cardiolipin-specific phospholipase D, *Biochim. Biophys. Acta* **296**:79.

Atwal, A. S., Eskin, N. A. M., and Henderson, H. M., 1979, Isolation and characterization of phospholipase D from fababeans, *Lipids* **14**:913.

Audet, A., Nantel, G., and Proulx, P., 1974, Phospholipase A activity in growing *Escherichia coli* cells, *Biochim. Biophys. Acta* **348**:334.

Audet, A., Cole, R., and Proulx, P., 1975, Polyglycerophosphatidate metabolism in *Escherichia coli, Biochim. Biophys. Acta* **380**:414.

Aurebekk, B., and Little, C., 1977a, Phospholipase C from *Bacillus cereus:* Evidence for essential lysine residues, *Biochem. J.* **161**:159.

Aurebekk, B., and Little, C., 1977b, Functional arginine in phospholipase C of *Bacillus cereus, Int. J. Biochem.* **8**:757.

Avigad, G., 1976, Microbial phospholipases, in *Mechanisms in Bacterial Toxinocology* (A. W. Bernheimer, ed.), pp. 99–167, John Wiley & Sons, New York.

Azhar, S., and Menon, K. M. J., 1976, Gonadotropin receptors in plasma membranes of bovine *Corpus luteum*. I. Effect of phospholipases on the binding of ^{125}I-choriogonadotropin by membrane-associated and solubilized receptors, *J. Biol. Chem.* **251**:7398.

Azhar, S., Hajra, A. K., and Menon, K. M. J., 1976, Gonadotropin receptors in plasma membranes of bovine *Corpus luteum*. II. Role of membrane phospholipids, *J. Biol. Chem.* **251**:7405.

Baer, E., and Buchnea, D., 1959, Synthesis of saturated and unsaturated L-α-lecithins—acylation of the cadmium chloride compound of L-α-glycerol phosphorylcholine, *Can. J. Biochem. Physiol.* **37**:953.

Bailey, J. M., 1985, *Prostaglandins, Leukotrienes, and Lipoxins: Biochemistry, Mechanism of Action, and Clinical Applications*, Plenum Press, New York.

Baker, R. R., and Thompson, W., 1973, Selective acylation of 1-acylglycerophosphorylinositol by rat brain microsomes: Comparison with 1-acylglycerophosphorylcholine, *J. Biol. Chem.* **248**:7060.

Ballou, L. R., and Cheung, W. Y., 1983, Marked increase of human platelet phospholipase A_2 activity *in vitro* and demonstration of an endogenous inhibitor, *Proc. Natl. Acad. Sci. USA* **80**:5203.

Bamberger, M., Lund-Katz, S., Phillips, M. C., and Rothblat, G. H., 1985, Mechanism of the hepatic lipase induced accumulation of high-density lipoprotein cholesterol by cells in culture, *Biochemistry* **24**:3693.

Bangham, A. D., 1968, Membrane models with phospholipids, in *Progress in Biophysics and Molecular Biology* (J. A. V. Butler and D. Noble, eds.), vol. 18, ch. 2, pp. 31–95, Pergamon Press, New York.

Bangham, A. D., and Dawson, R. M. C., 1959, The relation between the activity of a lecithinase and the electrophoretic charge of the substrate, *Biochem. J.* **72**:486.

Bangham, A. D., and Dawson, R. M. C., 1962, Electrokinetic requirements for the reaction between *Cl. perfringens* α toxin (phospholipase C) and phospholipid substrates, *Biochim. Biophys. Acta* **59**:103.

Bangham, A. D., Standish, M. M., and Watkins, J. C., 1965, Diffusion of univalent ions across lamellae of swollen phospholipids, *J. Mol. Biol.* **13**:238.

Barenholz, Y., Roitman, A., and Gatt, S., 1966, Enzymatic hydrolysis of sphingolipids. II. Hydrolysis of sphingomyelin by an enzyme from rat brain, *J. Biol. Chem.* **241**:3731.

Barrett-Bee, K., Hayes, Y., Wilson, R. G., and Ryley, J. F., 1985, A comparison of phospholipase activity, cellular adherence and pathogenicity of yeasts, *J. Gen. Microbiol.* **131**:1217.

Barrowman, M. M., Cockcroft, S., and Gomperts, B. D., 1986, Two roles for guanine nucleotides in the stimulus-secretion sequence of neutrophils, *Nature* **319**:504.

Bartels, C. T., and van Deenen, L. L. M., 1966, The conversion of lysophosphoglycerides by homogenates of spinach leaves, *Biochim. Biophys. Acta* **125**:395.

Beare, J. L., and Kates, M., 1967, Properties of the phospholipase B from *Penicillium notatum*, *Can. J. Biochem.* **45**:101.

Beaudry, G. A., Daniel, L. W., King, L., and Waite, M., 1983, Stimulation of deacylation in Madin–Darby canine kidney cells: 12-O-Tetradecanoyl-phorbol-13-acetate stimulates rapid phospholipid deacylation, *Biochim. Biophys. Acta* **750**:274.

Beaudry, G. A., Waite, M., and Daniel, L. W., 1985, Regulation of arachidonic acid metabolism in Madin–Darby canine kidney cells: Stimulation of synthesis of the cyclooxygenase system by 12-O-tetradecanoyl-phorbol-13-acetate, *Arch. Biochem. Biophys.* **239**:242.

Belcher, J. D., Sisson, P. J., and Waite, M., 1985, Degradation of mono-oleoylglycerol, trioleoylglycerol and phosphatidylcholine in emulsions and lipoproteins by rat hepatic acylglycerol lipase, *Biochem. J.* **229**:343.

Belfanti, S., and Arnaudi, C., 1932, Sur une lecithase du pancreas produlsant la lysocithine, *Bull. Soc. Int. Microbiol.* **4**:399.

Bell, R. L., and Majerus, P. W., 1980, Thrombin-induced hydrolysis of phosphatidylinositol in human platelets, *J. Biol. Chem.* **255**:1790.

Bell, R. L., Kennerly, D. A., Stanford, N., and Majerus, P. W., 1979, Diglyceride lipase: A pathway for arachidonate release from human platelets, *Proc. Natl. Acad. Sci. USA* **76**:3238.

Bell, R. M., Mavis, R. D., Osborn, M. J., and Vagelos, R. P., 1971, Enzymes of phospholipid metabolism: Localization in the cytoplasmic and outer membrane of the cell envelope of *Escherichia coli* and *Salmonella typhimurium, Biochim. Biophys. Acta* **249:**628.

Benveniste, J., and Arnoux, B., 1983, *Platelet-Activating Factor and Structurally Related Ether-Lipids,* Elsevier Science Publishers, Amsterdam.

Benveniste, J., Tence, M., Varenne, P., Bidault, J., Boullet, C., and Polonsky, J., 1979, Semi-synthese et structure proposee du facteur activant les plaquettes (PAF): PAF-acether, un alkyl ether analogue de la lysophosphatidylcholine, *C. R. Acad. Sci. (Paris)* **289D:**1017.

Bereziat, G., Wolf, C., Colard, O., and Polonovski, J., 1978, Phospholipases of plasma membranes of adipose tissue: Possible intermediaries for insulin action, in *Enzymes of Lipid Metabolism* (S. Gatt, L. Freysz, and P. Mandel, eds.), pp. 191–197, Plenum Press, New York.

Berka, R. M., and Vasil, M. L., 1982, Phospholipase C (heat-labile hemolysin) of *Pseudomonas aeruginosa:* Purification and preliminary characterization, *J. Bacteriol.* **152:**239.

Berka, R. M., Gray, G.L., and Vasil, M. L., 1981, Studies of phospholipase C (heat-labile hemolysin) in *Pseudomonas aeruginosa, Infect. Immunol.* **34:**1071.

Bernheimer, A. W., Avigad, L. S., and Kim, K. S., 1974, Staphylococcal sphingomyelinase (β-hemolysin), *Ann. NY Acad. Sci.* **236:**292.

Bernheimer, A. W., Linder, R., and Avigad, L. S., 1980, Stepwise degradation of membrane sphingomyelin by *Corynebacterial* phospholipases, *Infect. Immun.* **29:**123.

Berridge, M. J., 1980, Receptors and calcium signalling, *Trends Pharmacol. Sci.* **1:**419.

Berridge, M. J., 1983, Rapid accumulation of inositol trisphosphate reveals that agonists hydrolyse polyphosphoinositides instead of phosphatidylinositol, *Biochem. J.* **212:**849.

Berridge, M. J., 1984, Inositol trisphosphate and diacylglycerol as second messengers, *Biochem. J.* **220:**345.

Berridge, M. J., Dawson, R. M. C., Downes, C. P., Heslop, J. P., and Irvine, R. F., 1983, Changes in the levels of inositol phosphates after agonist-dependent hydrolysis of membrane phosphoinositides, *Biochem. J.* **212:**473.

Bertsch, L. L., Bonsen, P. P. M., and Kornberg, A., 1969, Biochemical studies of bacterial sporulation and germination. XIV. Phospholipids in *Bacillus megaterium, J. Bacteriol.* **98:**75.

Bevers, E. M., Singal, S. A., op den Kamp, J. A. F., and van Deenen, L. L. M., 1977, Recognition of different pools of phosphatidylglycerol in intact cells and isolated membranes of *Acholeplasma laidlawii* by phospholipase A₂, *Biochemistry* **16:**1290.

Billah, M. M., and Michell, R. H., 1979, Phosphatidylinositol metabolism in rat hepatocytes stimulated by glycogenolytic hormones: Effects of angiotensin, vasopressin, adrenaline, ionophore A23187 and calcium-ion deprivation, *Biochem. J.* **182:**661.

Billah, M. M., Lapetina, E. G., and Cuatrecasas, P., 1981, Phospholipase A₂ activity specific for phosphatidic acid: A possible mechanism for the production of arachidonic acid in platelets, *J. Biol. Chem.* **256:**5399.

Bird, R. A., Low, M. G., and Stephen, J., 1974, Immunopurification of phospholipase C (α toxin) from *Clostridium perfringens, FEBS Lett.* **44:**279.

Bjørklid, E., and Little, C., 1980, The isoelectric point of phospholipase C from *Bacillus cereus, FEBS Lett.* **113:**161.

Bjørnstad, P., 1965, Phospholipase activity in rat liver microsomes studied by the use of endogenous substrates, *Biochim. Biophys. Acta* **116:**500.

Blackwell, G. J., Carnuccio, R., DiRosa, M., Flower, R. J., Parente, L., and Persico, P., 1980, Macrocortin: A polypeptide causing the antiphospholipase effect of glucocorticoids, *Nature* **287:**147.

Blank, M. L., Snyder, F., Byers, L. W., Brooks, B., and Muirhead, E. E., 1979, Antihypertensive activity of an alkyl ether analog of phosphatidylcholine, *Biochem. Biophys. Res. Commun.* **90:**1194.

Blaschko, H., Jerome, D. W., Robb-Smith, A. H. T., Smith, A. D., and Winkler, H., 1968, Biochemical and morphological studies on catecholamine storage in human phaeochromocytoma, *Clin. Sci.* **34:**453.

Bleasdale, J. E., Eichberg, J., and Hauser, G., 1985, *Inositol and Phosphoinositides, Metabolism and Regulation,* Humana Press, Clifton, NJ.

Bligh, E. G., and Dyer, W. J., 1959, A rapid method for total lipid extraction and purification, *Can. J. Biochem. Physiol.* **37:**911.

Boffa, G. A., Boffa, M.-C., and Winchenne, J.-J., 1976, A phospholipase A₂ with anticoagulant activity. I. Isolation from *Vipera berus* venom and properties, *Biochim. Biophys. Acta* **429**:828.

Bokay, A., 1877–1878, Ueber die verdaulichkeit des nucleins und lecithins, *Hoppe Seyler's Z. Physiol. Chem.* **1**:157.

Bokoch, G. M., and Gilman, A. G., 1984, Inhibition of receptor-mediated release of arachidonic acid by pertussis toxin, *Cell* **39**:301.

Bonney, R. J., Wightman, P. D., Davies, P., Sadowski, S. J., Kuehl, F. A., Jr., and Humes, J. L., 1978, Regulation of prostaglandin synthesis and of the selective release of lysosomal hydrolases by mouse peritoneal macrophages, *Biochem. J.* **176**:433.

Bonsen, P. P. M., deHaas, G. H., and van Deenen, L. L. M., 1965, Synthesis and enzymatic hydrolysis of an *O*-alanyl ester of phosphatidylglycerol, *Biochim. Biophys. Acta* **106**:93.

Bonsen, P. P. M., deHaas, G. H., and van Deenen, L. L. M., 1967, Synthetic and structural investigations on 3-phosphatidyl-1'-(3'-*O*-L-lysyl)glycerol, *Biochemistry* **6**:1114.

Bonventre, P. F., and Johnson, C. E., 1971, *Bacillus cereus* toxins, in *Microbial Toxins* (T. C. Montie, S. Kadis, and S. J. Ajl, eds.), vol. III, pp. 415–435, Academic Press, New York.

Borgstrom, B., 1980, Importance of phospholipids, pancreatic phospholipase A₂, and fatty acid for the digestion of dietary fat: *In vitro* experiments with the porcine enzymes, *Gastroenterology* **78**:954.

Borgstrom, B., and Brockman, H. L., 1984, *Lipases*, Elsevier, Amsterdam.

Borochov, H., Zahler, P., Wilbrandt, W., and Shinitzky, M., 1977, The effect of phosphatidylcholine to sphingomyelin mole ratio on the dynamic properties of sheep erythrocyte membrane, *Biochim. Biophys. Acta* **470**:382.

van den Bosch, H., 1980, Intracellular phospholipases A, *Biochim. Biophys. Acta* **604**:191.

van den Bosch, H., 1982, Phospholipases, in *Phospholipids* (J. N. Hawthorne and G. B. Ansell, eds.), vol. 4, pp. 313–357, Elsevier Biomedical, Amsterdam.

van den Bosch, H., and Aarsman, A. J., 1979, A review on methods of phospholipase A determination, *Agents Actions* **9/4**:382.

van den Bosch, H., and deJong, J. G. N., 1975, Studies on lysophospholipases. IV. The subcellular distribution of two lysolecithin hydrolyzing enzymes from beef liber, *Biochim. Biophys. Acta* **398**:244.

van den Bosch, H., Bonte, H. A., and van Deenen, L. L. M., 1965, On the anabolism of lysolecithin, *Biochim. Biophys. Acta* **98**:648.

van den Bosch, H., van der Elzen, H. M., and van Deenen, L. L. M., 1967, On the phospholipases of yeast, *Lipids* **2**:279.

van den Bosch, H., Aarsman, A. J., Slotboom, A. J., and van Deenen, L. L. M., 1968, On the specificity of rat liver lysophospholipase, *Biochim. Biophys. Acta* **164**:215.

van den Bosch, H., Aarsman, A. J., and van Deenen, L. L. M., 1974, Isolation and properties of a phospholipase A₁ activity from beef pancreas, *Biochim. Biophys. Acta* **348**:197.

Botes, D. P., and Viljoen, C. C., 1974, Purification of phospholipase A from *Bitis gabonica*, *Toxicon* **12**:619.

Bowman, M. H., Ottolenghi, A. C., and Mengel, C. E., 1971, Effects of phospholipase C on human erythrocytes, *J. Membr. Biol.* **4**:156.

Brady, R. O., Kanfer, J. N., Mock, M. B., and Frederickson, D. S., 1966, The metabolism of sphingomyelin. II. Evidence of an enzymatic deficiency in Niemann–Pick disease, *Proc. Natl. Acad. Sci. USA* **55**:366.

Braganca, B. M., Sambray, Y. M., and Ghadially, R. C., 1969, Simple method for publication of phospholipase A from cobra venom, *Toxicon* **7**:151.

Brittain, H. G., Richardson, F. S., and Martin, R. B., 1976, Terbium(III) emission as a probe of calcium(II) binding sites in proteins, *J. Am. Chem. Soc.* **98**:8255.

Brock, T. A., Rittenhouse, S. E., Powers, C. W., Ekstein, L. S., Gimbrone, M. A., Jr., and Alexander, R. W., 1985, Phorbol ester and 1-oleoyl-2-acetylglycerol inhibit angiotensin activation of phospholipase C in cultured vascular smooth muscle cells, *J. Biol. Chem.* **260**:14158.

Brockerhoff, H., and Jensen, R. G., 1974, *Lipolytic Enzymes*, Academic Press, New York.

Broekman, M. J., Ward, J. W., and Marcus, A. J., 1980, Phospholipid metabolism in stimulated human platelets: Changes in phosphatidylinositol, phosphatidic acid, and lysophospholipids, *J. Clin. Invest.* **66**:275.

Broekman, M. J., Ward, J. W., and Marcus, A. J., 1981, Fatty acid composition of phosphati-dylinositol and phosphatidic acid in stimulated platelets: Persistence of arachidonyl-stearyl structure, *J. Biol. Chem.* **256**:8271.

Brumley, G., and van den Bosch, H., 1977, Lysophospholipase-transacylase from rat lung: Iso-lation and partial purification, *J. Lipid Res.* **18**:523.

Brunie, S., Bolin, J., Gewirth, D., and Sigler, P. B., 1985, The refined crystal structure of dimeric phospholipase A₂ at 2.5 Å: Access to a shielded catalytic center, *J. Biol. Chem.* **260**:9742.

Bruzik, K., and Tsai, M.-D., 1984, Phospholipids chiral at phosphorus: Synthesis of chiral phos-phatidylcholine and stereochemistry of phospholipase D, *Biochemistry* **23**:1656.

Bugaut, M., Kuksis, A., and Myher, J. J., 1985, Loss of stereospecificity of phospholipases C and D upon introduction of a 2-alkyl group into rac-1,2-diacylglycero-3-phosphocholine, *Biochim. Biophys. Acta* **835**:304.

Burns, R. A., El-Sayed, M. Y., and Roberts, M. F., 1982, Kinetic model for surface-active enzymes based on the Langmuir adsorption isotherm: Phospholipase C (*Bacillus cereus*) activity toward dimyristoyl phosphatidylcholine/detergent micelles, *Proc. Natl. Acad. Sci. USA* **79**:4902.

Caratsch, C. G., Maranda, B., Miledi, R., and Strong, P. N., 1981, A further study of the phos-pholipase-independent action of β-bungarotoxin at frog end-plates, *J. Physiol.* **319**:179.

Cate, R. L., and Bieber, A. L., 1978, Purification and characterization of mojave (*Crotalus scutulatus scutulatus*) toxin and its subunits, *Arch. Biochem. Biophys.* **189**:397.

Chalifour, R., and Kanfer, J. N., 1980, Microsomal phospholipase D of rat brain and lung tissues, *Biochem. Biophys. Res. Commun.* **96**:742.

Chalifour, R., and Kanfer, J. N., 1982, Fatty acid activation and temperature perturbation of rat brain microsomal phospholipase D, *J. Neurochem.* **39**:299.

Chalifour, R. J., Taki, T., and Kanfer, J. N., 1980, Phosphatidylglycerol formation via trans-phosphatidylation by rat brain extracts, *Can. J. Biochem.* **58**:1189.

Chang, S. L., 1979, Pathogenesis of pathogenic *Naegleria omeba, Folia Parasitol. (Praha)* **26**:195.

Chap, H. J., Zwaal, R. F. A., and van Deenen, L. L. M., 1977, Action of highly purified phos-pholipases on blood platelets: Evidence for an asymmetric distribution of phospholipids in the surface membrane, *Biochim. Biophys. Acta* **467**:146.

Chap, H., Mauco, G., Perret, B., Plantavid, M., Laffont, F., Simon, M. F., and Douste-Blazy, L., 1981, Studies on topological distribution of arachidonic acid replacement in platelet phos-pholipids and on enzymes involved in the phospholipid effect accompanying platelet acti-vation, *Agents Actions* **11**:538.

Chap, H., Perret, B., Mauco, G., Plantavid, M., Laffont, F., Simon, M. F., and Douste-Blazy, L., 1982, Organization and role of platelet membrane phospholipids as studied with phospho-lipases A₂ from various venoms and phospholipases C from bacterial origin, *Toxicon* **20**:291.

Chargaff, E., and Cohen, S. S., 1939, On lysophosphatides, *J. Biol. Chem.* **129**:619.

Chen, J.-S., and Barton, P. G., 1971, Studies of dialkyl ether phospholipids. II. Requirement for a liquid-crystalline substrate for hydrolysis by cabbage leaf phospholipase D, *Can. J. Biochem.* **49**:1362.

Chetal, S., Wagle, D. S., and Nainawatee, H. S., 1982, Phospholipase D activity in leaves of water-stressed wheat and barley, *Biochem. Physiol. Pflanzen* **177**:92.

Chiappe de Cingolani, G. E., van den Bosch, H., and van Deenen, L. L. M., 1972, Phospholipase A and lysophospholipase activities in isolated fat cells: Effect of cyclic 3'-5'-AMP, *Biochim. Biophys. Acta* **260**:387.

Chilton, F. H., O'Flaherty, J. T., Ellis, J. M., Swendsen, C. L., and Wykle, R. L., 1983, Metabolic fate of platelet activating factor in neutrophils, *J. Biol. Chem.* **258**:6357.

Chilton, F. H., Ellis, J. M., Olson, S. C., and Wykle, R. L., 1984, 1-O-Alkyl-2-arachidonoyl-sn-glycero-3-phosphocholine: A common source of platelet-activating factor and arachidonate in human polymorphonuclear leukocytes, *J. Biol. Chem.* **259**:12014.

Choy, P. C., and Vance, D E., 1978, Lipid requirements for activation of CTP:phosphocholine cytidylyltransferase from rat liver, *J. Biol. Chem.* **253**:5163.

Chu, H. P., 1949, The lecithinase of *Bacillus cereus* and its comparison with *Clostridium welchii* α-toxin, *J. Gen. Microbiol.* **3**:255.

Clarke, N. G., Irvine, R. F., and Dawson. R. M. C., 1981, Formation of bis(phosphatidyl)inositol and phosphatidic acid by phospholipase D action on phosphatidylinositol, *Biochem. J.* **195**:521.

Cloix, J. F., Colard, O., Rothhut, B., and Russo-Marie, F., 1983, Characterization and partial purification of "renocortins:" Two polypeptides formed in renal cells causing the anti-phospholipase-like action of glucocorticoids, *Br. J. Pharmacol.* **79**:313.

Cochran, F. R., Connor, J. R., Roddick, V. L., and Waite, M., 1985, Lyso(bis)phosphatidic acid: A novel source of arachidonic acid for oxidative metabolism by rabbit alveolar macrophages, *Biochem. Biophys. Res. Commun.* **130**:800.

Cochran, F. R., Roddick, V. L., Connor, J. R., Thornburg, J. T., and Waite, M., 1987, Regulation of arachidonic acid metabolism in resident and BCG-activated macrophages: Role of lyso(bis)phosphatidic acid, *J. Immunol.* **138**:1877.

Cockcroft, S., Bennett, J. P., and Gomperts, B. D., 1980, f-MetLeuPhe-induced phosphatidylinositol turnover in rabbit neutrophils is dependent on extracellular calcium, *FEBS Lett.* **110**:115.

Cockcroft, S., Baldwin, J. M., and Allan, D., 1984, The Ca^{2+}-activated polyphosphoinositide phosphodiesterase of human and rabbit neutrophil membranes, *Biochem. J.* **221**:477.

Cohen, H., Shen, B. W., Snyder, W. R., Law, J. H., and Kezdy, F. J., 1976, The surface pressure dependency of the enzymatic hydrolysis of lipid monolayers, *J. Colloid Interface Sci.* **56**:240.

Colard-Torquebiau, O., Bereziat, G., and Polonovski, J., 1975, Stude des phospholipases A des membranes plasmiques de foie de rat, *Biochimie* **57**:1221.

Cole, R., and Proulx, P., 1975, Phospholipase D activity of gram-negative bacteria, *J. Bacteriol.* **124**:1148.

Cole, R., and Proulx, P., 1977, Further studies on the cardiolipin phosphodiesterase of *Escherichia coli*, *Can. J. Biochem.* **55**:1228.

Cole, R., Benns, G., and Proulx, P., 1974, Cardiolipin specific phospholipase D activity in *Escherichia coli* extracts, *Biochim. Biophys. Acta* **337**:325.

Coleman, K., Dougan, G., and Arbuthnott, J. P., 1983, Cloning, and expression in *Escherichia coli* K-12, of the chromosomal hemolysin (phospholipase C) determinant of *Pseudomonas aeruginosa*, *J. Bacteriol.* **153**:909.

Coleman, R., Finean, J. B., Knutton, S., and Limbrick, A. R., 1970, A structural study of the modification of erythrocyte ghosts by phospholipase C, *Biochim. Biophys. Acta* **219**:81.

Colley, C. M., Zwaal, R. F. A., Roelofsen, B., and van Deenen, L. L. M., 1973, Lytic and non-lytic degradation of phospholipids in mammalian erythrocytes by pure phospholipases, *Biochim. Biophys. Acta* **307**:74.

Comfurius, P., and Zwaal, R. F. A., 1977, The enzymatic synthesis of phosphatidylserine and purification by CM–cellulose column chromatography, *Biochim. Biophys. Acta* **488**:36.

Condrea, E., deVries, A., and Mager, J., 1964, Hemolysis and splitting of human erythrocyte phospholipids by snake venoms, *Biochim. Biophys. Acta* **84**:60.

Connor, A. M., Brimble, P. D., and Choy, P. C., 1981, A simple method for the preparation of phosphatidylcholine labelled at 2-acyl position, *Prep. Biochem.* **11**:91.

Contardi, A., and Ercoli, A., 1932, Uber die enzymatische spaltung der lecithine und lysocithine, *Biochem. Z.* **261**:275.

Cooper, M. F., and Webster, G. R., 1970, The differentiation of phospholipase A_1 and A_2 in rat and human nervous tissue, *J. Neurochem.* **17**:1543.

Cooper, P. H., and Hawthorne, J. N., 1975, Phosphomonoesterase hydrolysis of polyphosphoinositides in rat kidney: Properties and subcellular localization of the enzyme system, *Biochem. J.* **150**:537.

Cullis, P. R., and Hope, M. J., 1985, Physical properties and functional roles of lipids, in *Biochemistry of Lipids and Membranes* (D. E. Vance and J. E. Vance, eds.), Benjamin-Cummings, Menlo Park, CA.

Cullis, P. R., and deKruijff, B., 1978, The polymorphic phase behavior of phosphatidylethanolamines of natural and synthetic origin—a ^{31}P-NMR study, *Biochim. Biophys. Acta* **513**:31.

Cursons, R. T. M., Brown, T. J., and Keys, E. A., 1978, Virulence of pathogenic free-living amebae, *J. Parasitol.* **64**(4):744.

van Dam-Mieras, M. C. E., Slotboom, A. J., Pieterson, W. A., and deHaas, G. H., 1975, The interaction of phospholipase A_2 with micellar interfaces: The role of the N-terminal region, *Biochemistry* **14**:5387.

Daniel, L. W., King, L., and Kennedy, M., 1987, Phospholipase activity of bacterial toxins, in

Microbial Toxins: Tools in Enzymology (S. Harshman, ed.), a volume of *Methods in Enzymology* (S. P. Colowick and N. O. Kaplan, eds. in chief), Academic Press, Orlando, FL.

Daniel, L. W., Waite, M., and Wykle, R. L., 1986, A novel mechanism of diglyceride formation: 12-*O*-Tetradecanoyl-phorbol-13-acetate stimulates the cyclic breakdown and resynthesis of phosphatidylcholine, *J. Biol. Chem.* **261:**9128.

Davidson, F. M., and Long, C., 1958, The structure of naturally-occurring phosphoglycerides, *Biochem. J.* **69:**458.

Davies, J. T., and Rideal, E. K., 1963, *Interfacial Phenomena*, Academic Press, London.

Dawson, R. M. C., 1956, The phospholipase B of liver, *Biochem. J.* **64:**192.

Dawson, R. M. C., 1966, The metabolism of animal phospholipids and their turnover in cell membranes, in *Essays in Biochemistry* (P. N. Campbell and G. D. Greville, eds.), vol. 2, pp. 69–115, Academic Press, New York.

Dawson, R. M. C., 1967, The formation of phosphatidylglycerol and other phospholipids by the transferase activity of phospholipase D, *Biochem. J.* **102:**205.

Dawson, R. M. C., 1973, Specificity of enzymes involved in the metabolism of phospholipids, in *Form and Function of Phospholipids* (G. B. Ansell, R. M. C. Dawson, and J. N. Hawthorne, eds.), p. 97, Elsevier, Amsterdam.

Dawson, R. M. C., and Hemington, N. L., 1967, Some properties of purified phospholipase D and especially the effect of amphipathic substances, *Biochem. J.* **102:**76.

Dawson, R. M. C., and Hemington, N. L., 1974, An inhibitor of phospholipase D in saliva, *Biochem. J.* **143:**427.

Dawson, R. M. C., and Thompson, W., 1964, The triphosphoinositide phosphomonoesterase of brain tissue, *Biochem. J.* **91:**244.

Dawson, R. M. C., Freinkel, N., Jungalwala, F. B., and Clarke, N., 1971, The enzymic formation of myoinositol 1:2-cyclic phosphate from phosphatidylinositol, *Biochem. J.* **122:**605.

Dawson, R. M. C., Hemington, N. L., Miller, N. G. A., and Bangham, A. D., 1976, On the question of an electrokinetic requirement for phospholipase C action, *J. Membr. Biol.* **29:**179.

Dawson, R. M. C., Hemington, N., and Irvine, R. F., 1980, The inhibition and activation of Ca^{2+}-dependent phosphatidylinositol phosphodiesterase by phospholipids and blood plasma, *Eur. J. Biochem.* **112:**33.

Dawson, R. M. C., Irvine, R. F., Hirasawa, K., and Hemington, N. L., 1982, Hydrolysis of phosphatidylinositol by pancreas and pancreatic secretion, *Biochim. Biophys. Acta* **710:**212.

Dawson, R. M. C., Hemington, N. L., and Irvine, R. F., 1983a, Diacylglycerol potentiates phospholipase attack upon phospholipid bilayers: Possible connection with cell stimulation, *Biochem. Biophys. Res. Commun.* **117:**196.

Dawson, R. M. C., Irvine, R. F., Hemington, N. L., and Hirasawa, K., 1983b, The alkaline phospholipase A_1 of rat liver cytosol, *Biochem. J.* **209:**865.

Dawson, R. M. C., Irvine, R. F., Bray, J., and Quinn, P. J., 1984, Long-chain unsaturated diacylglycerols cause a perturbation in the structure of phospholipid bilayers rendering them susceptible to phospholipase attack, *Biochem. Biophys. Res. Commun.* **125:**836.

DeBony, J., and Dennis, E. A., 1981, Magnetic nonequivalence of the two fatty acid chains in phospholipids of small unilamellar vesicles and mixed micelles, *Biochemistry* **20:**5256.

Deems, R. A., and Dennis, E. A., 1975, Characterization and physical properties of the major form of phospholipase A_2 from cobra venom (*Naja naja naja*) that has a molecular weight of 11,000, *J. Biol. Chem.* **250:**9008.

Deems, R. A., Eaton, B. R., and Dennis, E. A., 1975, Kinetic analysis of phospholipase A_2 activity toward mixed micelles and its implications for the study of lipolytic enzymes, *J. Biol. Chem.* **250:**9013.

van Deenen, L. L. M., and deHaas, G. H., 1963, The substrate specificity of phospholipase A, *Biochim. Biophys. Acta* **70:**538.

Delezenne, C., and Fourneau, E., 1914, Constitution du phosphotase hemolysant (lysocithine) proveant de l'action du venin de cobra sur le vitellufs de l'oeut de poule, *Bull. Soc. Chim.* **15:**421.

Demel, R. A., Geurts van Kessel, W. S. M., Zwaal, R. F. A., Roelofsen, B., and van Deenen, L. L. M., 1975, Relation between various phospholipase actions on human red cell membranes and the interfacial phospholipid pressure in monolayers, *Biochim. Biophys. Acta* **406:**97.

Demopoulos, C. A., Pinckard, R. N., and Hanahan, D. J., 1979, Platelet-activating factor: Evidence for 1-*O*-alkyl-2-acetyl-*sn*-glyceryl-3-phosphorylcholine as the active component (a new class of lipid chemical mediators), *J. Biol. Chem.* **254**:9355.

Dennis, E. A., 1973*a*, Phospholipase A₂ activity towards phosphatidylcholine in mixed micelles: Surface dilution kinetics and the effect of thermotropic phase transitions, *Arch. Biochem. Biophys.* **158**:485.

Dennis, E. A., 1973*b*, Kinetic dependence of phospholipase A₂ activity on the detergent Triton X-100, *J. Lipid Res.* **14**:152.

Dennis, E. A., 1983, Phospholipases, in *The Enzymes* (P. Boyer, ed.), vol. XVI, ch. 9, pp. 307–353, Academic Press, New York.

Dennis, E. A., and Pluckthun, A., 1985, Mechanism of interaction of phospholipase A₂ with phospholipid substrates and activators, in *Enzymes of Lipid Metabolism*, Plenum Press, New York.

Der, O. M., and Sun, G. Y., 1981, Degradation of arachidonoyl-labeled phosphatidylinositols by brain synaptosomes, *J. Neurochem.* **36**:355.

Dickens, B. F., and Thompson, G. A., Jr., 1982, Phospholipid molecular species alterations in microsomal membranes as an initial key step during cellular acclimation to low temperature, *Biochemistry* **21**:3604.

Dijkstra, B. W., Drenth, J., Kalk, K. H., and Vandermaelen, P. J., 1978, Three-dimensional structure and disulfide bond connections in bovine pancreatic phospholipase A₂, *J. Mol. Biol.* **124**:53.

Dijkstra, B. W., Kalk, K. H., Hol, W. G. J., and Drenth, J., 1981, Structure of bovine pancreatic phospholipase A₁ at 1.7 Å resolution, *J. Mol. Biol.* **147**:97.

Dijkstra, B. W., Kalk, K. H., Drenth, J., deHaas, G. H., Egmond, M. R., and Slotboom, A., 1984, Role of the N-terminus in the interaction of pancreatic phospholipase A₂ with aggregated substrates: Properties and crystal structure of transaminated phospholipase A₂, *Biochemistry* **23**:2759.

Dils, R. R., and Hubscher, G., 1961, Metabolism of phospholipids, *Biochim. Biophys. Acta* **46**:505.

Dinur, D., Kantrowitz, E. R., and Hajdu, J., 1981, Reaction of Woodward's reagent K with pancreatic porcine phospholipase A₂: Modification of an essential carboxylate residue, *Biochem. Biophys. Res. Commun.* **100**:785.

Doery, H. M., Magnusson, B. J., Cheyne, I. M., and Gulasekharam, J., 1963, A phospholipase in staphylococcal toxin which hydrolyses sphingomyelin, *Nature* **198**:1091.

Doery, H. M., Magnusson, B. J., Gulasekharam, J., and Pearson, J. E., 1965, The properties of phospholipase enzymes in staphylococcal toxins, *J. Gen. Microbiol.* **40**:283.

Doherty, F. J., and Rowe, C. E., 1980, The intracellular localization of a Ca²⁺-stimulated phospholipase A₁ in rat brain, *Brain Res.* **197**:113.

Doi, O., and Nojima, S., 1971, Phospholipase C from *Pseudomonas fluorescens*, *Biochim. Biophys. Acta* **248**:234.

Doi, O., and Nojima, S., 1975, Lysophospholipase of *Escherichia coli*, *J. Biol. Chem.* **250**:5208.

Doi, O., and Nojima, S., 1976, Nature of *Escherichia coli* mutants deficient in detergent-sensitive and/or detergent-resistant phospholipase A, *J. Biochem.* **80**:1247.

Doi, O., Ohki, M., and Nojima, S., 1972, Two kinds of phospholipase A and lysophospholipase in *Escherichia coli*, *Biochim. Biophys. Acta* **260**:244.

Donne-Op den Kelder, G. M., deHaas, G. H., and Egmond, M. R., 1983, Localization of the second calcium ion binding site in porcine and equine phospholipase A₂, *Biochemistry* **24**:2470.

Dowdall, M. J., Fohlman, J. P., and Watts, A., 1979, Presynaptic action of snake venom neurotoxins and cholinergic systems, *Adv. Cytopharmacol.* **3**:63.

Drakenberg, T., Andersson, T., Forsen, S., and Wieloch, T., 1984, Calcium ion binding to pancreatic phospholipase A₂ and its zymogen: A ⁴³Ca NMR study, *Biochemistry* **23**:2387.

Dufton, M. J., and Hider, R. C., 1983, Classification of phospholipases A₂ according to sequence: Evolutionary and pharmacological implications, *Eur. J. Biochem.* **137**:545.

Dufton, M. J., Eaker, D., and Hider, R. C., 1983, Conformational properties of phospholipases A₂: Secondary-structure prediction, circular dichroism and relative interface hydrophobicity, *Eur. J. Biochem.* **137**:537.

Duncombe, W. G., 1963, The colorimetric c micro-determination of long-chain fatty acids, *Biochem. J.* **88**:7.

Durkin, J. P., Pickwell, G. V., Trotter, J. T., and Shier, W. T., 1981, Phospholipase A₂ electrophoretic variants in reptile venoms, *Toxicon* **19**:535.

Dutilh, C., van Doren, P. J., Verheul, F. E. A. M., and deHaas, G. H., 1975, Isolation and properties of prophospholipase A₂ from ox and sheep pancreas, *Eur. J. Biochem.* **53**:91.

Egawa, K., Sacktor, B., and Takenawa, T., 1981, Ca²⁺-dependent and Ca²⁺-independent degradation of phosphatidylinositol in rabbit vas deferens, *Biochem. J.* **194**:129.

Ehnholm, C., Shaw, W., Greten, H., and Brown, W. V., 1975, Purification from human plasma of a heparin-released lipase with activity against triglyceride and phospholipids, *J. Biol. Chem.* **250**:6756.

Eibl, H., and Kovatchev, S., 1981, Preparation of phospholipids and their analogs by phospholipase D, *Methods Enzymol.* **72**:632.

van Eijk, J. H., Verheij, H., Dijkman, R., and deHaas, G. H., 1983, Interaction of phospholipase A₂ from *Naja melanoleuca* snake venom with monomeric substrate analogs: Activation of the enzyme by protein–protein or lipid–protein interactions, *Eur. J. Biochem.* **132**:183.

van Eijk, J. H., Verheij, H. M., and deHaas, G. H., 1984a, Characterization of phospholipase A₂ of *Naja melanoleuca* snake venom modified at the N-terminal region, *Eur. J. Biochem.* **139**:51.

van Eijk, J. H., Verheij, H. M., and deHaas, G. H., 1984b, Interaction of native and modified *Naja melanoleuca* phospholipases A₂ with the fluorescent probe 8-anilinonaphthalene-1-sulfonate, *Eur. J. Biochem.* **140**:407.

Einset, E., and Clark, W. L., 1958, The enzymatically catalyzed release of choline from lecithin, *J. Biol. Chem.* **231**:703.

El-Maghrabi, M. R., Waite, M., Rudel, L. L., and King, V., 1977, Metabolism of lipoprotein acylglycerols by liver parenchymal cells, *Biochim. Biophys. Acta* **572**:52.

El-Sayed, M. Y., DeBose, C. D., Coury, L. A., and Roberts, M. F., 1985, Sensitivity of phospholipase C (*Bacillus cereus*) activity to phosphatidylcholine structural modifications, *Biochim. Biophys. Acta* **837**:325.

Elsbach, P., 1980, Degradation of microorganisms by phagocytic cells, *Rev. Infect. Dis.* **2**:106.

Elsbach, P., and Rizack, M. A., 1963, Acid lipase and phospholipase activity in homogenates of rabbit polymorphonuclear leukocytes, *Am. J. Physiol.* **205**:1154.

Elsbach, P., Goldman, J., and Patriarca, P., 1972, Phospholipid metabolism by phagocytic cells: Observations on the fate of phospholipids of granulocytes and ingested *Escherichia coli* during phagocytosis, *Biochim. Biophys. Acta* **280**:33.

Elsbach, P., Weiss, J., Franson, R. C., Beckerdite-Quagliata, S., Schneider, A., and Harris, L., 1979, Separation and purification of a potent bactericidal/permeability increasing protein and a closely associated phospholipase A₂ from rabbit polymorphonuclear leukocytes, *J. Biol. Chem.* **254**:11000.

Elsbach, P., Weiss, J., and Kao, L., 1985, The role of intramembrane Ca²⁺ in the hydrolysis of the phospholipids of *Escherichia coli* by Ca²⁺-dependent phospholipases, *J. Biol. Chem.* **260**:1618.

Emilsson, A., and Sundler, R., 1984, Differential activation of phosphatidylinositol deacylation and a pathway via diphosphoinositide in macrophages responding to zymosan and ionophore A23187, *J. Biol. Chem.* **259**:3111.

Epstein, B., and Shapiro, B., 1958, Lecithinase and lysolecithinase of intestinal mucosa, *Biochem. J.* **71**:615.

Erbland, J. F., and Marinetti, G. V., 1965a, The enzymatic acylation and hydrolysis of lysolecithin, *Biochim. Biophys. Acta* **106**:128.

Erbland, J. F., and Marinetti, G. V., 1965b, The metabolism of lysolecithin in rat liver particulate systems, *Biochim. Biophys. Acta* **106**:139.

Esselman, M. T., and Liu, P. V., 1961, Lecithinase production by gram-negative bacteria, *J. Bacteriol.* **81**:939.

Etienne, J., Gruber, A., and Polonovski, J., 1980, Activite phospholipasique A₂ du serum de rat: Association de deux proteines, *Biochim. Biophys. Acta* **619**:693.

Etienne, J., Gruber, A., and Polonovski, J., 1982, Phospholipases A₂ activables des plaquettes sanguines de rat, *Biochimie* **220**:82.

Evans, H. J., Franson, R., Qureshi, G. D., and Moo-Penn, W. F., 1980, Isolation of anticoagulant proteins from cobra venom (*Naja nigricollis*): Identity with phospholipases A$_2$, *J. Biol. Chem.* **255**:3793.

Evenberg, A., Meyer, H., Gaastra, W., Verheij, H. M., and deHaas, G. H., 1977a, Amino acid sequence of phospholipase A$_2$ from horse pancreas, *J. Biol. Chem.* **252**:1189.

Evenberg, A., Myer, H., Verheij, H. M., and deHaas, G. H., 1977b, Isolation and properties of prophospholipase A$_2$ and phospholipase A$_2$ from horse pancreas and horse pancreatic juice, *Biochim. Biophys. Acta* **491**:265.

Fain, J. N., 1982, Involvement of phosphatidylinositol breakdown in elevation of cytosol Ca^{2+} by hormones and relationship to prostaglandin formation, in *Hormone Receptors: Horizons in Biochemistry and Biophysics* (L. D. Kohn, ed.), vol. 6, ch. 11, pp. 237–276, John Wiley & Sons, New York.

Fain, J. N., and Berridge, M. J., 1979, Relationship between hormonal activation of phosphatidylinositol hydrolysis, fluid secretion and calcium flux in the blowfly salivary gland, *Biochem. J.* **178**:45.

Fairbairn, D., 1948, The preparation and properties of a lysophospholipase from *Penicillium notatum*, *J. Biol. Chem.* **173**:705.

Farese, R V., Larson, R. E., and Sabir, M. A., 1980, Effects of Ca^{2+} ionophore A23187 and Ca^{2+} deficiency on pancreatic phospholipids and amylase release *in vitro*, *Biochim. Biophys. Acta* **633**:479.

Farooqui, A. A., Taylor, W. A., Pendley, C. E., II, Cox, J. W., and Horrocks, L. A., 1984, Spectrophotometric determination of lipases, lysophospholipases, and phospholipases, *J. Lipid Res.* **25**:1555.

Fauvel, J., Bonnefis, M.-J., Sarda, L., Chap, H., Touvenot, J.-P., and Douste-Blazy, L.,1981a, Purification of two lipases with high phospholipase A$_1$ activity from guinea pig pancreas, *Biochim. Biophys. Acta* **663**:446.

Fauvel, J., Bonnefis, M.-J., Chap, H., Thouvenot, J.-P., and Douste-Blazy, L., 1981b, Evidence for the lack of classical secretory phospholipase A$_2$ in guinea-pig pancreas, *Biochim. Biophys. Acta* **666**:72.

Fleer, E. A. M., Verheij, H. M., and deHaas, G. H., 1978, The primary structure of bovine pancreatic phospholipase A$_2$, *Eur. J. Biochem.* **82**:261.

Fleer, E. A. M., Verheij, H. M., and deHaas, G. H., 1981, Modification of carboxylate groups in bovine pancreatic phospholipase A$_2$, *Eur. J. Biochem.* **113**:283.

Flower, R. J., Wood, J. N., and Parente, L., 1984, Macrocortin and the mechanism of action of the glucocorticoids, in *Advances in Inflammation Research* (I. Otterness, R. Capetola, and S. Wong, eds.), vol. 7, Raven Press, New York.

Fohlman, J., Eaker, D., Karlsson, E., and Thesleff, S., 1976, Taipoxin, an extremely potent presynaptic neurotoxin from the venom of the Australian snake Taipan (*Oxyuranus s. scutellatus*): Isolation, characterization, quaternary structure and pharmacological properties, *Eur. J. Biochem.* **68**:457.

Fohlman, J., Lind, P., and Eaker, D., 1977, Taipoxin, an extremely potent presynaptic snake venom neurotoxin: Elucidation of the primary structure of the acidic carbohydrate-containing taipoxin-subunit, a prophospholipase homolog, *FEBS Lett.* **84**:367.

Forst, S., Weiss, J., and Elsbach, P., 1982, The role of phospholipase A$_2$ lysines in phospholipolysis of *Escherichia coli* killed by a membrane-active neutrophil protein, *J. Biol. Chem.* **257**:14055.

Forst, S., Weiss, J., Blackburn, P., Frangione, B., Goni, F., and Elsbach, P., 1986, Amino acid sequence of a basic *Agkistrodon halys blomhoffii* phospholipase A$_2$: Possible role of NH$_2$-terminal lysines in action on phospholipids of *Escherichia coli*, *Biochemistry* **25**:4309.

Fossum, K., and Hoyem, T., 1963, Phospholipase D activity in a non-haemolytic coryneform bacteria, *Acta Pathol. Microbiol. Scand.* **57**:295.

Franson, R. C., and van den Bosch, H., 1982, Lysophospholipase activity of bovine adrenal medulla—a reevaluation, *Biochim. Biophys. Acta* **711**:75.

Franson, R., and Waite, M., 1973, Lysosomal phospholipases A$_1$ and A$_2$ of normal and *Bacillus Calmetta-Guerin*-induced alveolar macrophages, *J. Cell Biol.* **56**:621.

Franson, R., and Waite, M., 1978, The relation between calcium requirement, substrate charge and rabbit polymorphonuclear leukocyte phospholipase A$_2$ activity, *Biochemistry* **17**:4029.

Franson, R., and Weir, D. L., 1982, Isolation and characterization of a membrane-associated, calcium-dependent phospholipase A_2 from rabbit lung, *Lung* **160**:275.

Franson, R., Waite, M., and LaVia, M., 1971, Identification of phospholipase A_1 and A_2 in the soluble fraction of rat liver lysosomes, *Biochemistry* **10**:1942.

Franson, R., Waite, M., and Weglicki, W. B., 1972, Phospholipase A activity of lysosomes of rat myocardial tissue, *Biochemistry* **11**:472.

Franson, R., Beckerdite, S., Wang, P., Waite, M., and Elsbach, P., 1973, Some properties of phospholipases of alveolar macrophages, *Biochim. Biophys. Acta* **296**:365.

Franson, R., Patriarca, P., and Elsbach, P., 1974, Phospholipid metabolism by phagocytic cells: Phospholipases A_2 associated with rabbit polymorphonuclear leukocyte granules, *J. Lipid Res.* **15**:380.

Franson, R. C., Pang, D. C., Towle, D. W., and Weglicki, W. B., 1978a, Phospholipase A activity of highly enriched preparations of cardiac sarcolemma from hamster and dog, *J. Mol. Cell. Cardiol.* **10**:921.

Franson, R. C., Dobrow, R., Weiss, J., Elsbach, P., and Weglicki, W. B., 1978b, Isolation and characterization of a phospholipase A_2 from an inflammatory exudate, *J. Lipid Res.* **19**:18.

Franson, R. C., Pang, D. C., and Weglicki, W. B., 1979, Modulation of lipolytic activity in isolated canine cardiac sarcolemma by isoproterenol and propranolol, *Biochem. Biophys. Res. Commun.* **90**:956.

Franson, R. C., Eisen, D., Jesse, R., and Lanni, C., 1980, Inhibition of highly-purified mammalian phospholipases A_2 by non-steroidal antiinflammatory agents, *Biochem. J.* **186**:633.

Frei, E., and Zahler, P., 1979, Phospholipase A_2 from sheep erythrocyte membranes: Ca^{2+} dependence and localization, *Biochim. Biophys. Acta* **550**:450.

Frish, A., Gazitt, Y., and Loyter, A., 1973, Metabolically controlled hemolysis of chicken erythrocytes, *Biochim. Biophys. Acta* **291**:690.

Fujino, Y., 1952, Studies on the conjugated lipids. IV. On the enzymatic hydrolysis of sphingomyelin, *J. Biochem.* **39**:55.

Fung, C. K., and Proulx, P. R., 1969, Metabolism of phosphoglycerides in *E. coli.* III. The presence of phospholipase A, *Can. J. Biochem.* **47**:371.

Gaines, G. L., Jr., 1966, *Insoluble Monolayers at Gas–Liquid Interfaces,* p. 59, Interscience Publishers, New York.

Garcia, A., Newkirk, J., and Marvis, R., 1975, Lung surfactant synthesis: A Ca^{2+}-dependent microsomal phospholipase A_2 in lung, *Biochem. Biophys. Res. Commun.* **64**:128.

Gatt, S., 1968, Purification and properties of phospholipase A_1 from rat and calf brain, *Biochim. Biophys. Acta* **159**:304.

Gatt, S., 1976, Magnesium-dependent sphingomyelinase, *Biochem. Biophys. Res. Commun.* **68**:235.

Gatt, S., and Gottesdiner, T., 1976, Solubilization of sphingomyelinase by isotonic extraction of rat brain lysosomes, *J. Neurochem.* **26**:421.

Gatt, S., Herzl, A., and Barenholz, Y., 1973, Hydrolysis of sphingomyelin liposomes by sphingomyelinase, *FEBS Lett.* **30**:281.

Gatt, S., Barenholz, Y., Goldberg, R., Dinur, T., Beasley, G., Leibovitz-Ben Gershon, Z., Rosenthal, J., Desnick, R. J., Devine, E. A., Shafit-Zagardo, B., and Tsuruki, F., 1981, Assay of enzymes of lipid metabolism with colored and fluorescent derivatives of natural lipids, *Methods Enzymol.* **72**:351.

Gazitt, Y., Ohad, I., and Loyter, A., 1975, Changes in phospholipid susceptibility toward phospholipases induced by ATP depletion in avian and amphibian erythrocyte membranes, *Biochim. Biophys. Acta* **382**:65.

deGeus, P., van Die, I., Bergmans, H., Tommassen, J., and deHaas, G., 1983, Molecular cloning of pldA, the structural gene for outer membrane phospholipase of *E. coli.* K-12, *Mol. Gen. Genet.* **190**:150.

deGeus, P., van Die, I., Bergmans, H., Tommassen, J., and deHaas, G., 1984, The pro- and mature forms of the *E. coli* K-12 outer membrane phospholipase A are identical, *Eur. Mol. Biol. Org. J.* **3**:1799.

Geynet, P., DePaillerets, C., and Alfsen, A., 1976, Lipid requirement of membrane-bound 3-oxosteroid Δ^4-Δ^5-isomerase: Studies on beef adrenocortical microsomes, *Eur. J. Biochem.* **71**:607.

Glenn, J. L., and Straight, R. C., 1982, The rattlesnakes and their venom yield and lethal toxicity, in *Rattlesnake Venoms: Their Actions and Treatment* (A. T. Tu, ed.), ch. 1, pp. 3–119, Marcel Dekker, New York.

Glenny, A. T., and Stevens, N. F., 1935, Staphylococcus toxins and antitoxins, *J. Pathol. Bacteriol.* **40**:201.

van Golde, L. M. G., Fleischer, B., and Fleischer, A., 1971, Some studies on the metabolism of phospholipids in golgi complex from bovine and rat liver in comparison to other subcellular fractions, *Biochim. Biophys. Acta* **249**:318.

Gorter, E., and Grendel, F., 1925, Bimolecular layers of lipoids on chromocytes of blood, *J. Exp. Med.* **41**:439.

Gow, J. A., and Robinson, J., 1969, Properties of purified staphylococcal β-hemolysin, *J. Bacteriol.* **97**:1026.

Graff, G., Nahas, N., Nikolopoulou, M., Natarajan, V., and Schmid, H. H. O., 1984, Possible regulation of phospholipase C activity in human platelets by phosphatidylinositol 4′,5′-bis-phosphate, *Arch. Biochem. Biophys.* **228**:229.

Gray, N. C. C., and Strickland, K. P., 1982a, The specificity of a phospholipase A_2 purified from the 106,000 g pellet of bovine brain, *Lipids* **17**:91.

Gray, N. C. C., and Strickland, K. P., 1982b, The purification and characterization of a phospholipase A_2 activity from the 106,000 g pellet (microsomal fraction) of bovine brain acting on phosphatidylinositol, *Can. J. Biochem.* **60**:108.

Griffin, H. D., and Hawthorne, J. N., 1978, Calcium-activated hydrolysis of phosphatidyl-myo-inositol 4-phosphate and phosphatidyl-myo-inositol 4,5-bisphosphate in guinea-pig synaptosomes, *Biochem. J.* **176**:541.

Griffin, H. D., Hawthorne, J. N., and Sykes, M., 1979, A calcium requirement for the phosphatidylinositol response following activation of presynaptic muscarinic receptors, *Biochem. Pharmacol.* **28**:1143.

Gronchi, V., 1936, Ricerche sulla lecitinasi "A" del pancreas; estrazione dell'enzima e sua azione in vitro, *Sperimentale* **90**:223.

Grossman, S., Cobley, J., Hogue, P. K., Kearney, E. B., and Singer, T. P., 1973, Relation of phospholipase D activity to the decay of succinate dehydrogenase and of covalenty-bound flavin in yeast cells undergoing glucose repression, *Arch. Biochem. Biophys.* **158**:744.

Grossman, S., Oestreicher, G., Houge, P. K., Cobley, J. G., and Singer, T. P., 1974, Microanalytical determination of the activities of phospholipases A, C, and D and of their mixtures, *Anal. Biochem.* **58**:301.

Gul, S., and Smith, A. D., 1974, Haemolysis of intact human erythrocytes by purified cobra venom phospholipase A_2 in the presence of albumin and Ca^{2+}, *Biochim. Biophys. Acta* **367**:271.

Gullis, R. J., and Rowe, C. E., 1975, The stimulation of the phospholipase A_2-acylation system of synaptic membranes of brain by cyclic nucleotides, *Biochem. J.* **148**:567.

Gupta, C. M., Radhakrishnan, R., Gerber, G. E., Olsen, W. L., Quay, S. C., and Khorana, H. G., 1979, Intermolecular crosslinking of fatty acyl chains in phospholipids: Use of photoactivable carbene precursors, *Proc. Natl. Acad. Sci. USA* **76**:2595.

deHaas, G. H., Heemskerk, C. H. T., van Deenen, L. L. M., Baker, R. W. R., Galli-Hatchard, J., Magee, W. L., and Thompson, R. H. S., 1963, The mode of action of human pancreatic phospholipase A, in *Biochemical Problems of Lipids* (A. C. Frazer, ed.), Biochim. Biophys. Library, vol. 1, pp. 244–250, Elsevier, Amsterdam.

deHaas, G. H., Sarda, L., and Roger, J., 1965, Positional specific hydrolysis of phospholipids by pancreatic lipase, *Biochim. Biophys. Acta* **106**:638.

deHaas, G. H., Postema, N. M., Nieuwenhuizen, W., and van Deenen, L. L. M., 1968, Purification and properties of phospholipase A from porcine pancreas, *Biochim. Biophys. Acta* **159**:103.

deHaas, G. H., Bonsen, P. P. M., Pieterson, W. A., and van Deenen, L. L. M., 1971, Studies on phospholipase A and its zymogen from porcine pancreas. III. Action of the enzyme on short-chain lecithins, *Biochim. Biophys. Acta* **239**:252.

Habermann, E., and Breithaupt, H., 1978, Mini-review: The crotoxin complex—an example of biochemical and pharmacological protein complementation, *Toxicon* **16**:19.

Hachimori, Y., Wells, M. A., and Hanahan, D. J., 1977, Observations on the phospholipase A_2 of *Crotalus atrox:* Molecular weight and other properties, *Biochemistry* **10**:4084.

Specificity of PLA2 established.

Hakata, H., Kambayashi, J., and Kosaki, G., 1982, Purification and characterization of phosphatidylinositol-specific phospholipase C from bovine platelets, *J. Biochem.* **92**:929.

Halpert, J., Eaker, D., and Karlsson, E., 1976, The role of phospholipase activity in the action of a presynaptic neurotoxin from the venom of *Notechis scutatus scutatus* (Australian tiger snake), *FEBS Lett.* **61**:72.

Hanahan, D. J., 1952, The enzymatic degradation of phosphatidylcholine in diethyl ether, *J. Biol. Chem.* **195**:199.

Hanahan, D. J., and Chaikoff, I. L., 1947, A new phospholipid-splitting enzyme specific for the ester linkage between the nitrogenous base and the phosphoric acid grouping, *J. Biol. Chem.* **169**:699.

Hanahan, D. J., and Chaikoff, I. L., 1948, On the nature of the phosphorus-containing lipids of cabbage leaves and their relation to a phospholipid-splitting enzyme contained in these leaves, *J. Biol. Chem.* **172**:191.

Hanahan, D. J., Joseph, M., and Morales, R., 1980, The isolation and characterization of a third or neutral phospholipase A_2 in the venom of *Agkistrodon halys blomhofii, Biochim. Biophys. Acta* **619**:640.

Hansen, C. A., Mah, S., and Williamson, J. R., 1986, Formation and metabolism of inositol 1,3,4,5-tetrakisphosphate in liver, *J. Biol. Chem.* **261**:8100.

Hardt, S. L., 1979, Rates of diffusion controlled reactions in one, two and three dimensions, *Biophys. Chem.* **10**:239.

Harrison, J. S., and Trevelyan, W. E., 1963, Phospholipid breakdown in baker's yeast during drying, *Nature* **200**:1189.

Hattori, H., and Kanfer, J. N., 1984, Synaptosomal phospholipase D: Potential role in providing choline for acetylcholine synthesis, *Biochem. Biophys. Res. Commun.* **124**:945.

Hauser, H., Pascher, I., Pearson, R. H., and Sundell, S., 1981, Preferred conformation and molecular packing of phosphatidylethanolamine and phosphatidylcholine, *Biochim. Biophys. Acta* **650**:21.

Haverkate, F., and van Deenen, L. L. M., 1965, Isolation and chemical characterization of phosphatidylglycerol from spinach leaves, *Biochim. Biophys. Acta* **106**:78.

Hawthorne, J. N., 1982, Inositol phospholipids, in *Phospholipids* (J. N. Hawthorne and G. B. Ansell, eds.), vol. 4, pp. 263–278, Elsevier Medical, New York.

Hazlett, T. L., and Dennis, E. A., 1985, Affinity chromatography of phospholipase A_2 from *Naja naja naja* (Indian cobra) venom, *Toxicon* **23**:457.

Heath, M. F., and Jacobson, W., 1976, Phospholipases A_1 and A_2 in lamellar inclusion bodies of the alveolar epithelium of rabbit lung, *Biochim. Biophys. Acta* **441**:443.

Heinrikson, R. L., Krueger, E. T., and Keim, P. S., 1977, Amino acid sequence of phospholipase A_2-α from the venom of *Crotalus adamanteus:* A new classification of phospholipase A_2 based upon structural determinants, *J. Biol. Chem.* **252**:4913.

Heller, M., 1978, Phospholipase D, *Adv. Lipid Res.* **16**:267.

Heller, M., and Arad, R., 1970, Properties of the phospholipase D from peanut seeds, *Biochim. Biophys. Acta* **210**:276.

Heller, M., and Shapiro, B., 1966, Enzymatic hydrolysis of sphingomyelin by rat liver, *Biochem. J.* **98**:763.

Heller, M., Mozes, N., Peri, I., and Maes, E., 1974, Phospholipase D from peanut seeds: Final purification and some properties of the enzyme, *Biochim. Biophys. Acta* **369**:397.

Heller, M., Mozes, N., and Maes, E., 1975, Phospholipase D from peanut seeds. EC 3.1.4.4. Phosphatidylcholine phosphatidohydrolase, *Methods Enzymol.* **35B**:229.

Heller, M., Mozes, N., and Peri (Abramovitz), I., 1976, Interactions of phospholipase D with 1,2-diacyl-*sn*-glycerol-3-phosphorylcholine dodecyl sulfate and Ca^{2+}, *Lipids* **11**:604.

Heller, M., Greenzaid, P., and Lichtenberg, D., 1978, The activity of phospholipase D on aggregates of phosphatidylcholine, dodecyl sulfate and Ca^{2+}, in *Enzymes of Lipid Metabolism* (S. Gatt and L. Freysz, eds.), p. 213, Plenum Press, New York.

Hendrickson, H. S., and Dennis, E. A., 1984, Kinetic analysis of the dual phospholipid model for phospholipase A_2 action, *J. Biol. Chem.* **259**:5734.

Hendrickson, H. S., and Rauk, P. N., 1981, Continuous fluorometric assay of phospholipase A_2 with pyrene-labeled lecithin as a substrate, *Anal. Biochem.* **116**:553.

Hendrickson, H. S., Trygstad, W. M., Loftness, T. L., and Sailer, S. L., 1981, Phospholipase A_2 activity on mixed lipid monolayers: Inhibition and activation of phospholipases A_2 from porcine pancreas and *Crotalus adamanteus* by anionic and neutral amphiphiles, *Arch. Biochem. Biophys.* **212:**508.

Herman, E. M., and Chrispeels, M. J., 1980, Characteristics and subcellular localization of phospholipase D and phosphatidic acid phosphatase in mung bean cotyledons, *Plant Physiol.* **66:**1001.

Hershberg, R. D., Reed, G. H., Slotboom, A. J., and deHaas, G. H., 1976, Nuclear magnetic resonance studies of the aggregation of dihexanoyllecithin and of dipeptanoyllecithin in aqueous solutions, *Biochim. Biophys. Acta* **424:**73.

Hetland, O., and Prydz, H., 1982, Phospholipase C from *Bacillus cereus* has sphingomyelinase activity, *Scand. J. Clin. Lab. Invest.* **42:**57.

Hille, J. D. R., Donne-Op den Kelder, G. M., Sauve, P., deHaas, G. H., and Egmond, M. R., 1981, Physicochemical studies on the interaction of pancreatic phospholipase A_2 with a micellar substrate analogue, *Biochemistry* **20:**4068.

Hille, J. D. R., Egmond, M. R., Dijkman, R., van Oort, M. G., Jirgensons, B., and deHaas, G. H., 1983a, Aggregation of porcine pancreatic phospholipase A_2 and its zymogen induced by submicellar concentrations of negatively charged detergents, *Biochemistry* **22:**5397.

Hille, J. D. R., Egmond, M. R., Dijkman, R., van Oort, M. G., Sauve, P., and deHaas, G. H., 1983b, Unusual kinetic behavior of porcine pancreatic (pro)phospholipase A_2 on negatively charged substrates at submicellar concentrations, *Biochemistry* **22:**5343.

Hirasawa, K., Irvine, R. F., and Dawson, R. M. C., 1981a, The catabolism of phosphatidylinositol by an EDTA-insensitive phospholipase A_1 and calcium-dependent phosphatidylinositol phosphodiesterase in rat brain, *Eur. J. Biochem.* **120:**53.

Hirasawa, K., Irvine, R. F., and Dawson, R. M. C., 1981b, The hydrolysis of phosphatidylinositol monolayers at an air/water interface by the calcium-ion-dependent phosphatidylinositol phosphodiesterase of pig brain, *Biochem. J.* **193:**607.

Hirasawa, K., Irvine, R. F., and Dawson, R. M. C., 1982a, Proteolytic activation can produce a phosphatidylinositol phosphodiesterase highly sensitive to Ca^{2+}, *Biochem. J.* **206:**675.

Hirasawa, K., Irvine, R. F., and Dawson, R. M. C., 1982b, Heterogeneity of the calcium-dependent phosphatidylinositol phosphodiesterase in rat brain, *Biochem. J.* **205:**437.

Hirata, F., 1984, Roles of lipomodulin: A phospholipase inhibitory protein in immunoregulation, in *Advances in Inflammation Research* (I. Otterness, R. Capetola, and S. Wong, eds.), vol. 7, Raven Press, New York.

Hirata, F., and Iwata, M., 1983, Role of lipomodulin, a phospholipase inhibitory protein, in immunoregulation by thymocytes, *J. Immunol.* **130:**1930.

Hirata, F., Schiffmann, E., Venkatasubramanian, K., Salomon, D., and Axelrod, J., 1980, A phospholipase A_2 inhibitory protein in rabbit neutrophils induced by glucocorticoids, *Proc. Natl. Acad. Sci. USA* **77:**2533.

Hirata, F., Matsuda, K., Notsu, Y., Hattori, T., and Del Carmine, R., 1984, Phosphorylation at a tyrosine residue of lipomodulin in mitogen-stimulated murine thymocytes, *Proc. Natl. Acad. Sci. USA* **81:**4717.

Ho, R. J., 1970, Radiochemical assay of long chain fatty acids using ^{63}Ni as tracer, *Anal. Chem.* **36:**105.

Ho, R. J., and Meng, H. C., 1969, A simple and ultrasensitive method for determination of free fatty acid by radiochemical assay, *Anal. Biochem.* **31:**426.

Hofmann, S. L., and Majerus, P. W., 1982a, Identification and properties of two distinct phosphatidylinositol-specific phospholipse C enzymes from sheep seminal vesicular glands, *J. Biol. Chem.* **257:**6461.

Hofmann, S., and Majerus, P., 1982b, Modulation of phosphatidylinositol-specific phospholipase C activity by phospholipid interactions, diglycerides, and calcium ions, *J. Biol. Chem.* **257:**14359.

Hokin, L. E., 1985, Receptors and phosphoinositide-generated second messengers, in *Annual Review of Biochemistry* (C. C. Richardson, P. D. Boyer, I. B. Dawid, and A. Meister, eds.), vol. 54, p. 205, Annual Reviews Inc., Palo Alto, CA.

Hokin-Neaverson, M., 1977, Metabolism and role of phosphatidylinositol in acetylcholine-stimulated membrane function, *Adv. Exp. Biol. Med.* **83:**429.

Hokin, M. R., and Hokin, L. E., 1953, Enzyme secretion and the incorporation of ^{32}P into phospholipides of pancreas slices, *J. Biol. Chem.* **203**:967.

Hokin, M. R., and Hokin, L. E., 1960, The role of phosphatidic acid and phosphoinositide in transmembrane transport elicited by acetylcholine and other humoral agents, *Int. Rev. Neurobiol.* **2**:99.

Holmsen, H., Dangelmaier, C. A., and Holmsen, H.-K., 1981, Thrombin-induced platelet responses differ in requirement for receptor occupancy, *J. Biol. Chem.* **256**:9393.

Homma, H., Kobayashi, T., Ito, Y., Kudo, I., Inoue, K., Ikeda, H., Sekiguchi, M., and Nojima, S., 1983, Identification and cloning of the gene coding for lysophospholipase L$_2$ of *E. coli* K-12, *J. Biochem.* **94**:2079.

Homma, H., Kobayashi, T., Chiba, N., Karasawa, K., Mizushima, H., Kudo, I., Inoue, K., Ikeda, H., Sekiguchi, M., and Nojima, S., 1984*a*, The DNA sequence encoding pldA gene, the structural gene for detergent-resistant phospholipase A of *E. coli*, *J. Biochem.* **96**:1655.

Homma, H., Chiba, N., Kobayashi, T., Kudo, I., Inoue, K., Ikeda, H., Sekiguchi, M., and Nojima, S., 1984*b*, Characteristics of detergent-resistant phospholipase A overproduced in *E. coli* cells bearing its cloned structural gene, *J. Biochem.* **96**:1645.

van't Hooft, F. M., van Gent, T., and van Tol, A., 1981, Turnover and uptake by organs of radioactive serum high-density lipoprotein cholesteryl esters and phospholipids in the rat *in vivo, Biochem. J.* **196**:877.

Hortnagl, Heide, Winkler, H., and Hortnagl, H., 1969, The subcellular distribution of lysophospholipase in bovine adrenal medulla, *Eur. J. Biochem.* **10**:243.

Hostetler, K. Y., and Hall, L. B., 1980, Phospholipase C activity of rat tissues, *Biochem. Biophys. Res. Commun.* **96**:388.

Hostetler, K. Y., Yazaki, P. J., and van den Bosch, H., 1982, Purification of lysosomal phospholipase A, *J. Biol. Chem.* **257**:13367.

Houtsmuller, U. M. T., and van Deenen, L. L. M., 1963, Studies on the phospholipids and phospholipase from *Bacillus cereus, Proc. Koninkl. Ned. Akad. Wetensch. B* **64**:528.

Hsueh, W., Kuhn, C., III, and Needleman, P., 1979, Relationship of prostaglandin secretion by rabbit alveolar macrophages to phagocytosis and lysosomal enzyme release, *Biochem. J.* **184**:345.

Hsueh, W., Desai, U., Gonzalez-Crussi, F., Lamb, R., and Chu, A., 1981, Two phospholipase pools for prostaglandin synthesis in macrophages, *Nature* **290**:710.

Hsueh, W., Lamb, R., and Gonzalez-Crussi, F., 1982, Decreased phospholipase A$_2$ activity and prostaglandin biosynthesis in *Bacillus Calmette-Guerin*-activated alveolar macrophages, *Biochim. Biophys. Acta* **710**:406.

Huang, C., and Lee, L. P., 1973, Diffusion studies on phosphatidylcholine vesicles, *J. Am. Chem. Soc.* **95**:234.

Hubscher, G., 1962, Metabolism of phospholipids. VI. The effect of metal ions on the incorporation of L-serine into phosphatidylserine, *Biochim. Biophys. Acta* **57**:555.

Hughes, A., 1935, The action of snake venoms on surface films, *Biochem. J.* **29**:437.

Hui, D. Y., and Harmony, J. A. K., 1980, Phosphatidylinositol turnover in mitogen-activated lymphocytes: Suppression by low-density lipoproteins, *Biochem. J.* **192**:91.

Hulsmann, W. C., Oerlemans, M. C., and Jansen, H., 1980, Activity of heparin-releasable liver lipase dependency on the degree of saturation of the fatty acids in the acyl glycerol substrates, *Biochim. Biophys. Acta* **618**:364.

Humes, J. L., Burger, S., Galvage, M., Kuehl, F. A., Wightman, P. D., Dahlgren, M. E., Davies, P., and Bonney, R. J., 1980, The diminished production of arachidonic acid oxygenation products by elicited mouse peritoneal macrophages: Possible mechanisms, *J. Immunol.* **124**:2110.

Hunt, G. R. A., and Jones, I. C., 1984, Application of ^{1}H-NMR to the design of liposomes for oral use: Synergistic activity of bile salts and pancreatic phospholipase A$_2$ in the induced permeability of small unilamellar phospholipid vesicles, *J. Microencapsulation* **1**:113.

Hysmith, R. M., and Franson, F. C., 1982*a*, Elevated levels of cellular and extracellular phospholipases from pathogenic *Naegleria fowleri, Biochim. Biophys. Acta* **711**:26.

Hysmith, R. M., and Franson, R. C., 1982*b*, Degradation of human myelin phospholipids by phospholipase-enriched culture media of pathogenic *Naegleria fowleri, Biochim. Biophys. Acta* **712**:698.

Ichimasa, M., Morooka, T., and Niimura, T., 1984, Purification and properties of a membrane-bound phospholipase B from baker's yeast (*Saccharomyces cerevisiae*), *J. Biochem.* **95:**137.

Ikezawa, H., Yamanegi, M., Taguchi, R., Miyashita, T., and Ohyabu, T., 1976, Studies on phosphatidylinositol phosphodiesterase (phospholipase C type) of *Bacillus cereus*. I. Purification, properties and phosphatase-releasing activity, *Biochim. Biophys. Acta* **450:**154.

Ikezawa, H., Mori, M., Ohyabu, T., and Taguchi, R., 1978, Studies on sphingomyelinase of *Bacillus cereus*. I. Purification and properties, *Biochim. Biophys. Acta* **528:**247.

Ikezawa, H., Nakabayashi, T., Suzuki, K., Nakajima, M., Taguchi, T., and Taguchi, R., 1983, Complete purification of phosphatidylinositol-specific phospholipase C from a strain of *Bacillus thuringiensis*, *J. Biochem.* **93:**1717.

Imamura, S., and Horiuti, Y., 1979, Purification of *Streptomyces chromofuscus* phospholipase D by hydrophobic affinity chromatography on palmitoylated cellulose, *J. Biochem.* **85:**79.

Imamura, S., and Horiuti, Y., 1980, Purification of phospholipase B from *Penicillium notatum* by hydrophobic chromatography on palmitoyl cellulose, *J. Lipid Res.* **21:**180.

Irvine, R. F., and Dawson, R. M. C., 1979, Transfer of arachidonic acid between phospholipids in rat liver microsomes, *Biochem. Biophys. Res. Commun.* **91:**1399.

Irvine, R. F., and Dawson, R. M. C., 1983, Phosphatidylinositol phosphodiesterase of rat brain: Ca^{2+}-dependency, pH optima and heterogeneity, *Biochem. J. Lett.* **215:**431.

Irvine, R. F., Hemington, N., and Dawson, R. M. C., 1978, The hydrolysis of phosphatidylinositol by lysosomal enzymes of rat liver and brain, *Biochem. J.* **176:**475.

Irvine, R. F., Hemington, N., and Dawson, R. M. C., 1979, The calcium-dependent phosphatidylinositol-phosphodiesterase of rat brain: Mechanisms of suppression and stimulation, *Eur. J. Biochem.* **99:**525.

Ishii, H., Connolly, T. M., Bross, T. E., and Majerus, P. W., 1986, Inositol cyclic triphosphate [inositol 1,2(cyclic)-4,5-triphosphate] is formed upon thrombostimulation of human platelets, *Proc. Natl. Acad. Sci. USA* **83:**6397.

Ispolatovskaya, M. V., 1971, Type A *Clostridium perfringens* toxin, in *Microbial Toxins* (S. Kadis, T. C. Montie, and S. J. Ajl, eds.), vol. IIA: *Bacterial Protein Toxins*, pp. 109–158, Academic Press, New York.

Itaya, K., 1977, A more sensitive and stable colorimetric determination of free fatty acids in blood, *J. Lipid Res.* **18:**663.

Itonaga, M., Aida, Y., and Onoue, K., 1984, Structural studies of Fc receptors. IV. Structure required for phospholipids for reconstitution of the delipidated Fc receptor of macrophages, *J. Biochem.* **95:**1145.

Jackowski, S., Rettenmier, C. W., Sherr, C. J., and Rock, C. O., 1986, A guanine nucleotide-dependent phosphatidylinositol 4,5-diphosphate phospholipase C in cells transformed by the v-*fms* and v-*fes* oncogenes, *J. Biol. Chem.* **261:**4978.

Jahn, C. E., Osborne, J. C., Jr., Schaefer, E. J., and Brewer, H. B., Jr., 1983, Activation of the enzymic activity of hepatic lipase by apolipoprotein A-II: Characterization of a major component of high density lipoprotein as the activating plasma component *in vitro*, *Eur. J. Biochem.* **131:**25.

Jain, M. K., and Apitz-Castro, R. C., 1978, Lag phase during the action of phospholipase A$_2$ on phosphatidylcholine bilayers containing alkanols, *J. Biol. Chem.* **253:**7005.

Jain, M. K., and Cordes, E. H., 1973a, Phospholipases. I. Effect of *n*-alkanols on the rate of hydrolysis of egg phosphatidylcholine, *J. Membr. Biol.* **14:**101.

Jain, M. K., and Cordes, E. H., 1973b, Phospholipases. II. Effect of sonication and addition of cholesterol to the rate of hydrolysis of various lecithins, *J. Membr. Biol.* **14:**119.

Jain, M. K., and deHaas, G. H., 1983, Activation of phospholipase A$_2$ by freshly added lysophospholipids, *Biochim. Biophys. Acta* **736:**157.

Jain, M. K., van Echteld, C. J. A., Ramirez, F., deGier, J., deHaas, G. H., and van Deenen, L. L. M., 1980, Association of lysophosphatidylcholine with fatty acids in aqueous phase to form bilayers, *Nature* **284:**486.

Jansen, H., and Hulsmann, W. C., 1980, Heparin-releasable (liver) lipase(s) may play a role in the uptake of cholesterol by steroid-secreting tissues, *Trends Biochem. Sci.* **5:**265.

Jansen, H., Meyer, H., deHaas, G. H., and Kaptein, R., 1978a, A photochemically induced dynamic nuclear polarization study of pancreatic phospholipase A$_2$, *J. Biol. Chem.* **253:**6346.

Jansen, H., van Berkel, T. J. C., and Hulsmann, W. C., 1978*b*, Binding of liver lipase to parenchymal and non-parenchymal rat liver cells, *Biochem. Biophys. Res. Commun.* **85**:148.

Jansen, H., Kalkman, C., Zonneveld, A. J., and Hulsmann, W. D., 1979, Secretion of triacylglycerol hydrolase activity by isolated parenchymal rat liver cells, *FEBS Lett.* **98**:299.

Jansen, H., van Tol, A., and Hulsmann, W. C., 1980, On the metabolic function of heparin-releasable liver lipase, *Biochem. Biophys. Res. Commun.* **92**:53.

Janssen, L. H. M., deBruin, S. H., and deHaas, G. H., 1972, Hydrogen-ion-titration studies of pancreatic phospholipase A and its zymogen, *Eur. J. Biochem.* **28**:156.

Jarvis, A. A., Cain, C., and Dennis, E. A., 1984, Purification and characterization of a lysophospholipase from human amnionic membranes, *J. Biol. Chem.* **259**:15188.

Jeng, T. W., and Fraenkel-Conrat, H., 1978, Chemical modification of histidine and lysine residues of crotoxin, *FEBS Lett.* **87**:291.

Jensen, G. L., and Bensadoun, A., 1981, Purification, stabilization, and characterization of rat hepatic triglyceride lipase, *Anal. Biochem.* **113**:246.

Jensen, G. L., Daggy, B., and Bensadoun, A., 1982, Triacylglycerol lipase, monoacylglycerol lipase and phospholipase activities of highly purified rat hepatic lipase, *Biochim. Biophys. Acta* **710**:464.

Jesse, R. L., and Franson, R. C., 1979, Modulation of purified phospholipase A_2 activity from human platelets by calcium and indomethacin, *Biochim. Biophys. Acta* **575**:467.

Jiang, K.-T., Shyy, Y.-J., and Tsai, M.-D., 1984, Phospholipids chiral at phosphorus: Absolute configuration of chiral thiophospholipids and stereospecificity of phospholipase D, *Biochemistry* **23**:1661.

Jimeno-Abendano, J., and Zahler, P., 1979, Purified phospholipase A_2 from sheep erythrocyte membrane: Preferential hydrolysis according to polar groups and 2-acyl chains, *Biochim. Biophys. Acta* **573**:266.

Johnson, C. E., and Bonventre, P. F., 1967, Lethal toxin of *Bacillus cereus*. I. Relationships and nature of toxin, hemolysin, and phospholipase, *J. Bacteriol.* **94**:306.

Johnston, J. M., 1977, Gastrointestinal tissue, in *Lipid Metabolism in Mammals* (F. Snyder, ed.), pp. 151–187, Plenum Press, New York.

Jones, L. M., and Michell, R. H., 1975, The relationship of calcium to receptor-controlled stimulation of phosphatidylinositol turnover: Effects of acetylcholine, adrenaline, calcium ions, cinchocaine and a bivalent cation ionophore on rat parotid-gland fragments, *Biochem. J.* **148**:479.

Jones, L. M., and Michell, R. H., 1978, Enhanced phosphatidylinositol breakdown as a calcium-independent response of rat parotid fragments to substance P, *Biochem. Soc. Trans.* **6**:1035.

Jones, L. M., Cockcroft, S., and Michell, R. H., 1979, Stimulation of phosphatidylinositol turnover in various tissues by cholinergic and adrenergic agonists, by histamine and by caerulein, *Biochem. J.* **182**:669.

deJong, J. G. N., van den Bosch, H., Rijken, D., and van Deenen, L. L. M., 1974, Studies on lysophospholipases. III. The complete purification of the two proteins with lysophospholipase activity from beef liver, *Biochim. Biophys. Acta* **369**:50.

deJong, J. G. N., van den Besselaar, A. M. H. P., and van den Bosch, H., 1976, Studies on lysophospholipases. VIII. Immunochemical differences between two phospholipases from beef liver, *Biochim. Biophys. Acta* **441**:221.

Joubert, F. J., and Haylett, T., 1981, Snake venoms—purification, some properties and amino acid sequence of a phospholipase A_2 (DE-1) from *Trimeresurus okinavensis* (Hime-habu) venom, *Z. Physiol. Chem.* **362**:997.

op den Kamp, J. A. F., 1981, The asymmetric architecture of membranes, in *Membrane Structure* (Finean/Michell, eds.), ch. 3, p. 83, Elsevier/North-Holland Biomedical Press, Amsterdam.

op den Kamp, J. A. F., deGier, J., and van Deenen, L. L. M., 1974, Hydrolysis of phosphatidylcholine liposomes by pancreatic phospholipase A_2 at the transition temperature, *Biochim. Biophys. Acta* **345**:253.

Kanfer, J. N., 1980, The base exchange enzymes and phospholipase D of mammalian tissue, *Can. J. Biochem.* **58**:1370.

Kanfer, J. N., and Kennedy, E. P., 1964, Metabolism and function of bacterial lipids. II. Biosynthesis of phospholipids in *Escherichia coli*, *J. Biol. Chem.* **239**:1720.

Kannagi, R., and Koizumi, K., 1979, Effect of different physical states of phospholipid substrates on partially purified platelet phospholipase A_2 activity, *Biochim. Biophys. Acta* **556**:423.

Kannan, K. K., Lovgren, S., Cid-Dresdner, H., Petef, M., and Eaker, D., 1977, Crystallization and crystallographic data of notexin: A neurotoxic basic phospholipase A from the venom of Australian tiger snake *Notechis scutatus scutatus, Toxicon* **15**:435.

Kaplan-Harris, L., and Elsbach, P., 1980, The antiinflammatory activity of analogs of indomethacin correlates with their inhibitory effects on phospholipase A_2 of rabbit polymorphonuclear leukocytes, *Biochim. Biophys. Acta* **618**:318.

Kaplan, L., Weiss, J., and Elsbach, P., 1978, Low concentrations of indomethacin inhibit phospholipase A_2 of rabbit polymorphonuclear leukocytes, *Proc. Natl. Acad. Sci. USA* **75**:2955.

Karasawa, K., Kudo, I., Kobayashi, T., Sa-Eki, T., Inoue, K., and Nojima, S., 1985, Purification and characterization of lysophospholipase L_2 of *Escherichia coli* K-12, *J. Biochem.* **98**:1117.

Kater, L. A., Goetzl, E. J., and Austen, K. F., 1976, Isolation of human eosinophil phospholipase D, *J. Clin. Invest.* **57**:1173.

Kates, M., 1953, Lecithinase activity of chloroplasts, *Nature* **172**:814.

Kates, M., 1954a, Lecithinase systems in sugar beet, spinach, cabbage, and carrot, *Can. J. Biochem. Physiol.* **32**:571.

Kates, M., 1954b, Hydrolysis of lecithin by plant plastid enzymes, *Can. J. Biochem. Physiol.* **33**:575.

Kates, M., 1956, Hydrolysis of glycerophosphatides by plastid phospholipase C, *Can. J. Biochem. Biophys.* **34**:967.

Kates, M., 1957, Effects of solvents and surface-active agents on plastid phospholipase C activity, *Can. J. Biochem. Physiol.* **35**:127.

Kates, M., 1970, Plant phospholipids and glycolipids, *Adv. Lipid Res.* **8**:225.

Kates, M., 1972, Methods in lipidology, in *Laboratory Techniques in Biochemistry and Molecular Biology* (T. S. Worh and E. Worh, eds.), vol. 3, pp. 269–579, North Holland/American Elsevier, New York.

Kates, M., and Gorham, P. R., 1957, Coalescence as a factor in solvent stimulation of plastid phospholipase C activity, *Can. J. Biochem. Physiol.* **35**:119.

Kawahara, Y., Takai, Y., Minakuchi, R., Sano, K., and Nishizuka, Y., 1980, Possible involvement of Ca^{2+}-activated, phospholipid-dependent protein kinase in platelet activation, *J. Biochem.* **88**:913.

Kawasaki, N., and Saito, K., 1973, Purification and some properties of lysophospholipase from *Pencillium notatum, Biochim. Biophys. Acta* **296**:426.

Kawasaki, N., Sugatani, J., and Saito, K., 1975, Studies on a phospholipase B from *Penicillium notatum, J. Biochem.* **77**:1233.

Kawauchi, S., Iwanaga, S., Samejima, N., and Suzuki, T., 1971, Isolation and characterization of two phospholipase A's from the venom of *Agkistrodon halys blomhoffii, Biochim. Biophys. Acta* **236**:142.

Keith, C., Feldman, D. S., Deganello, S., Glick, J., Ward, K. B., Jones, E. O., and Sigler, P. B., 1981, The 2.5 Å crystal structure of a dimeric phospholipase A_2 from the venom of *Crotalus atrox, J. Biol. Chem.* **256**:8602.

Kemp, C. L., and Howatson, A. F., 1966, Action of phospholipase C on erythrocyte membranes and Rauscher virus, *Virology* **30**:147.

Kensil, C. R., and Dennis, E. A., 1979, Action of cobra venom phospholipase A_2 on the gel and liquid crystalline states of dimyristoyl and dipalmitoyl phosphatidylcholine vesicles, *J. Biol. Chem.* **254**:5843.

Kensil, C. R., and Dennis, E. A., 1985, Action of cobra venom phospholipase A_2 on large unilamellar vesicles: Comparison with small unilamellar vesicles and multibilayers, *Lipids* **20**:80.

Kihara, H., Ishimaru, K., and Ohno, M., 1981, Cleavage of *Trimeresurus flavoviridis* phospholipase A_2 with cyanogen bromide: Sequence of the short peptide fragment and formation of a noncovalently bonded complex from the fragments, *J. Biochem.* **90**:363.

Kim, S., and Martin, G. M., 1981, Preparation of cell-size unilamellar liposomes with high capture volume and defined size distribution, *Biochim. Biophys. Acta* **646**:1.

King, E. J., and Nolan, M., 1933, The enzymic hydrolysis of phosphatides. II. Lysolecithin, *Biochem. J.* **27**:403.

King, T. P., Sobotka, A. K., Alagon, A., Kochoumian, L., and Lichtenstein, L. M., 1978, Protein allergens of white-faced hornet, yellow hornet, and yellow jacket venoms, *Biochemistry* **17:**5165.

Kinnunen, P. K. J., and Ehnholm, C., 1976, Effect of serum and C-apoproteins from very low density lipoproteins on human postheparin plasma hepatic lipase, *FEBS Lett.* **65:**354.

Kirk, C. J., Michell, R. H., and Hems, D. A., 1981, Phosphatidylinositol metabolism in rat hepatocytes stimulated by vasopressin, *Biochem. J.* **194:**155.

Kito, M., Narita, H., Takamura, H., Park, H. J., Matsuura, T., and Tanaka, K., 1986, Dissociation of Ca^{2+} mobilization from breakdown of phosphatidylinositol 4,5-bisphosphate in activated human platelets, *J. Biochem.* **99:**1277.

Kobayashi, T., Homma, H., Natori, Y., Kudo, I., Inoue, K., and Nojima, S., 1984, Isolation of two kinds of *E. coli* K-12 mutants for lysophospholipase L_2: One with an elevated level of the enzyme and the other defective in it, *J. Biochem.* **96:**137.

Kobayashi, T., Kudo, I., Karasawa, K., Mizushima, H., Inoue, K., and Nojima, S., 1985a, Nucleotide sequence of the pldB gene and characteristics of deduced amino acid sequence of lysophospholipase L_2 in *Escherichia coli*, *J. Biochem.* **98:**1017.

Kobayashi, T., Kudo, I., Homma, H., Karasawa, K., Inoue, K., Ikeda, H., and Nojima, S., 1985b, Gene organization of pldA and pldB, the structural genes for detergent-resistant phospholipase A and lysophospholipase L_2 of *Escherichia coli*, *J. Biochem.* **98:**1007.

Kokke, R., Hooghwinkel, J. M., Booij, H. L., van den Bosch, H., Zelles, L., Mulder, E., and van Deenen, L. L. M., 1963, Metabolism of lysolecithin and lecithin in a yeast supernatant, *Biochim. Biophys. Acta* **70:**351.

Kondo, K., Narita, K., and Lee, C.-Y., 1978, Chemical properties and amino acid composition of β_1-bungarotoxin from the venom of *Bungarus multicinctus* (formosan banded krait), *J. Biochem.* **83:**91.

Kondo, K., Toda, H., Narita, K., and Lee, C.-Y., 1982a, Amino acid sequence of β_2-bungarotoxin from *Bungarus multicinctus* venom: The amino acid substitutions in the B chains, *J. Biochem.* **91:**1519.

Kondo, K., Toda, H., Narita, K., and Lee, C.-Y., 1982b, Amino acid sequences of three β-bungarotoxins (β_3-, β_4-, and β_5-bungarotoxins) from *Bungarus multicinctus* venom: Amino acid substitutions in the A chains, *J. Biochem.* **91:**1531.

Kornberg, A., and McConnell, H. M., 1971, Lateral diffusion of phospholipid in a vesicle membrane, *Proc. Natl. Acad. Sci. USA* **68:**2564.

Kornberg, A., Spudick, J. A., Nelson, D. L., and Deutscher, M. P., 1968, Origin of proteins in sporulation, *Annu. Rev. Biochem.* **37:**51.

Kothary, M. H., and Kreger, A. S., 1985, Purification and characterization of an extracellular cytolysin produced by *Vibrio damsela, Infect. Immun.* **49:**25.

Kovatchev, S., and Eibl, H., 1978, The preparation of phospholipids by phospholipase D, *Adv. Exp. Med. Biol.* **101:**221.

Kramer, R. M., Wuthrich, C., Bollier, C., Allegrini, P. R., and Zahler, P., 1978, Isolation of a phospholipase A_2 from sheep erythrocyte membranes in the presence of detergents, *Biochim. Biophys. Acta* **507:**381.

Kramer, R. M., Pritzker, C. R., and Deykin, D., 1984, Coenzyme A-mediated arachidonic acid transacylation in human platelets, *J. Biol. Chem.* **259:**2403.

Krug, E. L., and Kent, C., 1984, Phospholipase C from *Clostridium perfringens:* Preparation and characterization of homogeneous enzyme, *Arch. Biochem. Biophys.* **231:**400.

Kubo, M., Matsuzawa, Y., Yokoyama, S., Tajima, S., Ishikawa, K., Yamamoto, A., and Tarui, S., 1982, Mechanism of inhibition of hepatic triglyceride lipase from human postheparin plasma by apolipoproteins A-I and A-II, *J. Biochem.* **92:**865.

Kurioka, S., and Liu, P., 1967, Effect of the hemolysin of *Pseudomonas aeruginosa* on phosphatides and on phospholipase C activity, *J. Bacteriol.* **93:**670.

Kurpiewski, G., Forrester, L. J., Barrett, J. T., and Campbell, B. J., 1981, Platelet aggregation and sphingomyelinase D activity of a purified toxin from the venom of *Loxosceles reclusa*, *Biochim. Biophys. Acta* **678:**467.

Kurup, P. A., 1966, Phospholipase A of the venom of South Indian scorpion, *Heterometrus scaber*, *Indian J. Biochem.* **3:**164.

Kushner, D. J., 1962, Formation and release of lecithinase activity by growing cultures of *Bacillus cereus, Can. J. Microbiol.* **8:**673.

Kuusi, T., Kinnunen, P. K. J., Ehnholm, C., and Nikkila, E. A., 1979*a*, A simple purification procedure for rat hepatic lipase, *FEBS Lett.* **98:**314.

Kuusi, T., Bry, K., Nikkila, E. A., and Kinnunen, P. K. J., 1979*b*, Modification of the substrate specificity of rat liver hepatic lipase by collagenase treatment, *Med. Biol.* **57:**192.

Kuusi, T., Kinnunen, P. K. J., and Nikkila, E. A., 1979*c*, Hepatic endothelial lipase antiserum influences rat liver low and high density lipoproteins *in vivo, FEBS Lett.* **104:**384.

Kuusi, T., Nikkila, E. A., Virtanen, I., and Kinnunen, P. K. J., 1979*d*, Localization of the heparin-releasable lipase *in situ* in the rat liver, *Biochem. J.* **181:**245.

Kuusi, T., Saarinen, P., and Nikkila, E. A., 1980, Evidence for the role of hepatic endothelial lipase in the metabolism of plasma high density lipoprotein$_2$ in man, *Atherosclerosis* **36:**589.

Kyes, P., 1902, Ueber die wirkungsweise des cobragiftes, I,II, *Berlin Klin. Wochenschr.* **39:**886,918.

Kyes, P., 1907, Ueber die lecithide des schlangengiftes, *Biochem. Z.* **IV:**99.

Kyte, J., and Doolittle, R. F., 1982, A simple method for displaying the hydropathic character of a protein, *J. Mol. Biol.* **157:**105.

Lal, A. A., and Garg, N. K., 1979, *Hartmanella culbertsoni:* Biochemical changes in the brain of the meningoencephalitic mouse, *Exp. Parasitol.* **48:**331.

Lands, W. E. M., and Crawford, C. G., 1976, Enzymes of membrane phospholipid metabolism, in *The Enzymes of Biological Membranes* (A. Martonosi, ed.), pp. 3–85, Plenum Press, New York.

Lands, W. E. M., and Hart, P., 1965, Metabolism of plasmalogen, *Biochim. Biophys. Acta* **98:**532.

Lang, V., Pryhitka, G., and Buckley, J. T., 1977, Effect of neomycin and ionophore A23187 on ATP levels and turnover of polyphosphoinositides in human erythrocytes, *Can. J. Biochem.* **55:**1007.

Lanni, C., and Franson, R. C., 1981, Localization and partial purification of a neutral-active phospholipase A$_2$ from BCG-induced rabbit alveolar macrophages, *Biochim. Biophys. Acta* **658:**54.

Lapetina, E. G., Schmitges, C. J., Chandrabose, K., and Cuatrecasas, P., 1977, Cyclic adenosine 3′,5′-monophosphate and prostacyclin inhibit membrane phospholipase activity in platelets, *Biochem. Biophys. Res. Commun.* **76:**828.

Lapetina, E. G., Chandrabose, K. A., and Cuatrecasas, P., 1978, Ionophore A23187- and thrombin-induced platelet aggregation: Independence from cyclooxygenase products, *Proc. Natl. Acad. Sci. USA* **75:**818.

Lapetina, E. G., Billah, M. M., and Cuatrecasas, P., 1981*a*, The initial action of thrombin on platelets: Conversion of phosphatidylinositol to phosphatidic acid preceding the production of arachidonic acid, *J. Biol. Chem.* **256:**5037.

Lapetina, E. G., Billah, M. M., and Cuatrecasas, P., 1981*b*, The phosphatidylinositol cycle and the regulation of arachidonic acid production, *Nature* **292:**367.

Laychock, S. G., Franson, R. C., Weglicki, W. B., and Rubin, R. P., 1977, Identification and partial characterization of phospholipases in isolated adrenocortical cells, *Biochem. J.* **164:**753.

Lehmann, V., 1972, Properties of phospholipase C from *Acinetobacter calcoaceticus, Acta Pathol. Microbiol. Scand.* **B80:**827.

Lehmann, V., 1973*a*, Haemolytic activity of various strains of *Acinetobacter, Acta Pathol. Microbiol. Scand.* **B81:**427.

Lehmann, V., 1973*b*,, The nature of the phospholipase C from *Acinetobacter calcoaceticus:* Effects on whole cells and red cell membranes, *Acta Pathol. Microbiol. Scand.* **B81:**419.

Lehmann, V., and Whg, J. N., 1972, Leukolytic activity of *Acinetobacter calcoaceticus, Acta Pathol. Microbiol. Scand.* **B80:**823.

Leibovitz-Ben Gershon, Z., and Gatt, S., 1976, Lysolecithinase activity in subcellular fractions of rat organs, *Biochem. Biophys. Res. Commun.* **69:**592.

Leitersdorf, E., Stein, O., and Stein, Y., 1984, Synthesis and secretion of triacyglycerol lipase by cultured rat hepatocytes, *Biochim. Biophys. Acta* **794:**261.

Lenard, J., and Singer, S. J., 1968, Structure of membranes: Reaction of red blood cell membranes with phospholipase C, *Science* **159:**738.

Levene, P. A., and Rolf, I. P., 1923, Lysolecithins and lysocephalins, *J. Biol. Chem.* **55**:743.

Levene, P. A., Rolf, I. P., and Simms, H. S., 1923–1924, Lysolecithins and lysocephalins. II. Isolation and properties of lysolecithins and lysocephalins, *J. Biol. Chem.* **58**:859.

Lewis, J. C., and de la Lande, I. S., 1967, Pharmacological and enzymic constituents of the venom of an Australian "bulldog" ant *Myrmecia pyriformis, Toxicon* **4**:225.

Lichtenberg, D., Romero, G., Menashe, M., and Biltonen, R. L., 1986, Hydrolysis of dipalmi-toylphosphatidylcholine large unilamellar vesicles by porcine pancreatic phospholipase A_2, *J. Biol. Chem.* **261**:5334.

Linder, R., and Bernheimer, A. W., 1978, Effect of sphingomyelin-containing liposomes of phospholipase D from *Corynebacterium ovis* and the cytolysin from *Stoichactis helianthus, Biochim. Biophys. Acta* **530**:236.

Lindgren, J. A., Claesson, H.-E., and Hammarstrom, S., 1978, Stimulation of arachidonic acid release and prostaglandin production in 3T3 fibroblasts by adenosine 3′:5′-monophosphate, in *Advances in Prostaglandin and Thromboxane Research* (C. Galli *et al.*, eds.), vol. 3, Raven Press, New York.

Lippiello, P. M., Dijkstra, J., van Galen, M., Scherphof, G., and Waite, M., 1981, Uptake and metabolism of chylomicron-remnant lipids by non-parenchymal cells in perfused liver and by Kupffer cells in culture, *J. Biol. Chem.* **256**:7454.

Lippiello, P. M., Sisson, P., and Waite, M., 1985, The uptake and metabolism of chylomicron-remnant lipids by rat liver parenchymal and non-parenchymal cells *in vitro, Biochem. J.* **232**:395.

Little, C., 1977a, The histidine residues of phospholipase C from *Bacillus cereus, Biochem. J.* **167**:399.

Little, C., 1977b, Phospholipase C from *Bacillus cereus:* Action on some artificial lecithins, *Acta Chem. Scand.* **B31**:267.

Little, C., 1978, Conformational studies on phospholipase C from *Bacillus cereus:* The effect of urea on the enzyme, *Biochem. J.* **175**:977.

Little, C., 1981a, Phospholipase C from *Bacillus cereus.* EC 3.1.4.3. Phosphatidylcholine choline-phosphohydrolase, *Methods Enzymol.* **71**:725.

Little, C., 1981b, The effect of metal-ion substitution on the conformational stability of *Bacillus cereus* phospholipase C, *Biochem. Soc. Trans.* **9**:448.

Little, C., and Aurebekk, B., 1977, Inactivation of phospholipase C from *Bacillus cereus* by a carboxyl group modifying reagent, *Acta Chem. Scand.* **B31**:273.

Little, C., and Johansen, S., 1979, Unfolding and refolding of phospholipase C from *Bacillus cereus* in solutions of guanidinium chloride, *Biochem. J.* **179**:509.

Little, C., Aurebekk, B., and Otnaess, A.-B., 1975, Purification by affinity chromatography of phospholipase C from *Bacillus cereus, FEBS Lett.* **52**:175.

Little, C., Aakre, S.-E., Rumsby, M. G., and Gwarsha, K., 1982, Effect of Co^{2+}-substitution on the substrate specificity of phospholipase C from *Bacillus cereus* during attack on two membrane systems, *Biochem. J.* **207**:117.

Lloveras, J., and Douste-Blazy, L., 1973, Hydrolysis of [^{3}H,^{14}C]phosphatidylethanolamines by acid phospholipases A of subcellular fractions of rat spleen, *Eur. J. Biochem.* **33**:567.

Lombardo, D., and Dennis, E. A., 1985a, Cobra venom phospholipase A_2 inhibition by manoalide: A novel type of phospholipase inhibitor, *J. Biol. Chem.* **260**:7234.

Lombardo, D., and Dennis, E. A., 1985b, Immobilized phospholipase A_2 from cobra venom: Prevention of substrate interfacial and activator effects, *J. Biol. Chem.* **260**:16114.

Long, C., Odavic, R., and Sargent, E. J., 1967, The chemical nature of the products obtained by the action of cabbage-leaf phospholipase D on lysolecithin: The structure of lysolecithin, *Biochem. J.* **102**:221.

Longmore, W. J., Oldenberg, V., and van Golde, L. M. G., 1979, Phospholipases A_2 in rat lung microsomes; substrate specificity towards endogenous phosphatidylcholines, *Biochim. Biophys. Acta* **572**:452.

Lory, S., and Tai, P. C., 1983, Characterization of the phospholipase C gene of *Pseudomonas aeruginosa* clones in *Escherichia coli, Gene* **22**:95.

Low, M., 1981, Phosphatidylinositol-specific phospholipase C from *Staphylococcus aureus, Methods Enzymol.* **71**:741.

Low, M. G., and Finean, J. B., 1977, Modification of erythrocyte membranes by a purified phosphatidylinositol-specific phospholipase C (*Staphylococcus aureus*), *Biochem. J.* **162**:235.

Low, M. G., and Finean, J. B., 1978, Specific release of plasma membrane enzymes by a phosphatidylinositol-specific phospholipase C, *Biochim. Biophys. Acta* **508**:565.

Low, M. G., and Weglicki, W., 1983, Resolution of myocardial phospholipase C into several forms with distinct properties, *Biochem. J.* **215**:325.

Low, M. G., Carroll, R. C., and Weglicki, W. B., 1984, Multiple forms of phosphoinositide-specific phospholipase C of different relative molecular masses in animal tissues: Evidence for modification of the platelet enzyme by Ca^{2+}-dependent proteinase, *Biochem. J.* **221**:813.

Lumb, R. H., and Allen, K. F., 1976, Properties of microsomal phospholipases in rat liver and hepatoma, *Biochim. Biophys. Acta* **450**:175.

Lysenko, O., 1973, Phospholipase C of *Pseudomonas chlororaphis* and its toxicity for insects, *Folia Microbiol.* **18**:125.

Macfarlane, M. G., and Knight, B. C. J. G., 1941, The biochemistry of bacterial toxins. I. The lecithinase activity of *Cl. welchii* toxins, *Biochem. J.* **35**:884.

Magee, W. L., Gallai-Hatchard, J., Sanders, H., and Thompson, R. H. S., 1962, The purification and properties of phospholipase A_2 from human pancreas, *Biochem. J.* **83**:17.

Magnusson, B. J., Doery, H. M., and Gulasekharam, J., 1962, Phospholipase activity of *Staphylococcus* toxin, *Nature* **196**:270.

Mahadevappa, V. G., and Holub, B. J., 1983, Degradation of different molecular species of phosphatidylinositol in thrombin-stimulated human platelets, *J. Biol. Chem.* **258**:5337.

Maheswaran, S. K., and Lindorfer, R. K., 1967, Staphylococcal β-hemolysin. II. Phospholipase C activity of purified β-hemolysin, *J. Bacteriol.* **94**:1313.

Majerus, P. W., Wilson, D. B., Connolly, T. M., Bross, T. E., and Neufeld, E. J., 1985, Phosphoinositide turnover provides a link in stimulus-response coupling, *Trends Biochem. Sci.* **10**:168.

Malmqvist, M., Malmqvist, T., and Mollby, R., 1978, Phospholipids immobilized on beaded agarose by hydrophobic interaction as hydrophilic substrates for phospholipase C, *FEBS Lett.* **90**:243.

Malmqvist, T., 1981, A screening method for bacterial production of sphingomyelinase, *FEMS Microbiol. Lett.* **10**:91.

Malmqvist, T., and Mollby, R., 1981, Effects of staphylococcal β-haemolysin on immobilized sphingomyelin and on the sheep erythrocyte membrane, in *Staphylococci and Staphylococcal Infections* (J. Jeljaszewicz, ed.), pp. 253–259, Gustav Fischer Verlag, Stuttgart.

Mansback, C. M., 1984, Ontogeny of intestinal phospholipase A_2 in the rat, *Biochim. Biophys. Acta* **794**:484.

Mansback, C. M., Pieroni, G., and Verger, R., 1982, Intestinal phospholipase, a novel enzyme, *J. Clin. Invest.* **69**:368.

Maraganore, J. M., and Heinrikson, R. L., 1985, The role of lysyl residues of phospholipases A_2 in the formation of the catalytic complex, *Biochem. Biophys. Res. Commun.* **131**:129.

Maraganore, J. M., and Heinrikson, R. L., 1986, The lysine-49 phospholipase A_2 from the venom of *Agkistrodon p. piscivorus*: Relation of structure and function to other phospholipases A_2, *J. Biol. Chem.* **261**:4797.

Maraganore, J. M., Merutka, G., Cho, W., Welches, W., Kezdy, F. J., and Heinrikson, R. L., 1984, A new class of phospholipases A_2 with lysine in place of aspartate 49: Functional consequences for calcium and substrate binding, *J. Biol. Chem.* **259**:13839.

Maraganore, J. M., Poorman, R. A., and Heinrikson, R. L., 1986a, Phospholipases A_2: Variations on a structural theme and their implications as to mechanism, *J. Protein Chem.*, in press.

Maraganore, J. M., Poorman, R. A., and Heinrikson, R. L., 1986b, Identification of functionally-relevant sequence homology between the phospholipases A from *E. coli* and those from venom and pancreatic sources, *Biochem. J.*, in press.

Marinetti, G., 1964, Hydrolysis of cardiolipin by snake venom phospholipase A, *Biochim. Biophys. Acta* **84**:55.

Marples, E. A., and Thompson, R. H. S., 1960, The distribution of phospholipase B in mammalian tissue, *Biochem. J.* **74**:123.

Martin, J. K., Luthra, M. G., Wells, M. A., Watts, R. P., and Hanahan, D. J., 1975, Phospholipase

A_2 as a probe of phospholipid distribution in erythrocyte membranes: Factors influencing the apparent specificity of the reaction, *Biochemistry* **14**:5400.

Mason, R. J., Stossel, T. P., and Vaughan, M., 1972, Lipids of alveolar macrophages, polymorphonuclear leukocytes, and their phagocytic vesicles, *J. Clin. Invest.* **51**:2399.

Matsuzawa, Y., and Hostetler, K. Y., 1980, Properties of phospholipase C isolated from rat liver lysosomes, *J. Biol. Chem.* **255**:646.

Matsuzawa, Y., Poorthuis, B. J. H. M., and Hostetler, K. Y., 1978, Mechanism of phosphatidylinositol stimulation of lysosomal bis(monoacylglyceryl)phosphate synthesis, *J. Biol. Chem.* **253**:6650.

Mauco, G., Chap, H., and Douste-Blazy, L., 1979, Characterization and properties of a phosphatidylinositol phosphodiesterase (phospholipase C) from platelet cytosol, *FEBS Lett.* **100**:367.

Mauco, G., Dangelmaier, C. A., and Smith, J. B., 1984, Inositol lipids, phosphatidate and diacylglycerol share stearoylarachidonoylglycerol as a common backbone in thrombin-stimulated human platelets, *Biochem. J.* **224**:933.

Mavis, R. D., Bell, R. M., and Vagelos, P. R., 1972, Effect of phospholipase C hydrolysis of membrane phospholipids on membranous enzymes, *J. Biol. Chem.* **247**:2835.

Menashe, M., Lichtenberg, D., Gutierrez-Merino, C., and Biltonen, R. L., 1981, Relationship between the activity of pancreatic phospholipase A_2 and the physical state of the phospholipid substrate, *J. Biol. Chem.* **256**:4541.

Menashe, M., Romero, G., Biltonen, R. L., and Lichtenberg, D., 1986, Hydrolysis of dipalmitoylphosphatidylcholine small unilamellar vesicles by porcine pancreatic phospholipase A_2, *J. Biol. Chem.* **261**:5328.

Meyer, H., Verhoef, H., Hendriks, F. F. A., Slotboom, A. J., and deHaas, G. H., 1979a, Comparative studies of tyrosine modification in pancreatic phospholipases. 1. Reaction of tetranitromethane with pig, horse, and ox phospholipases A_2 and their zymogens, *Biochemistry* **16**:3582.

Meyer, H., Puijk, W. C., Dijkman, R., Foda-van der Hoorn, M. M. E. L., Pattus, F., Slotboom, A. J., and deHaas, G. H., 1979b, Comparative studies of tyrosine modification in pancreatic phospholipases. 2. Properties of the nitrotyrosyl, aminotyrosyl, and dansylaminotyrosyl derivatives of pig, horse, and ox phospholipases A_2 and their zymogens, *Biochemistry* **18**:3589.

Michel, G. P. F., and Starka, J., 1979, Phospholipase A activity with integrated phospholipid vesicles in intact cells of an envelope mutant of *Escherichia coli*, *FEBS Lett.* **108**:261.

Michell, R. H., 1979, Mechanisms of cell surface receptors for hormone and neurotransmitters, in *Companion to Biochemistry* (A. T. Bull, J. R. Lagnudo, J. O. Thomas, and K. F. Tipton, eds.), vol. 2, pp. 205–228, Longmans, London.

Michell, R. H., Kirk, C. J., Jones, L. M., Downes, C. P., and Creba, J. A., 1981, The stimulation of inositol lipid metabolism that accompanies calcium mobilization in stimulated cells, defined characteristics and unanswered questions, *Philos. Trans. R. Soc. Lond. (Biol.)* **296**:123.

Miller, C. H., Parce, J. W., Sisson, P., and Waite, M., 1981, Specificity of lipoprotein lipase and hepatic lipase toward monoacylglycerols varying in the acyl composition, *Biochim. Biophys. Acta* **665**:385.

Misaki, H., and Matsumoto, M., 1978, Purification of lysophospholipase of *Vibrio parahaemolyticus* and its properties, *J. Biochem.* **83**:1395.

Misiorowski, D. L., and Wells, M. A., 1974, The activity of phospholipase A_2 in reversed micelles of phosphatidylcholine in diethyl ether: Effect of water and cations, *Biochemistry* **13**:4921.

Mollay, C., and Kreil, G., 1974, Enhancement of bee venom phospholipase A_2 activity by melittin, direct lytic factor from cobra venom and polymixin B, *FEBS Lett.* **46**:141.

Mollby, R., 1978, Bacterial phospholipases, in *Bacterial Toxins and Cell Membranes* (J. Jeljaszewicz and T. Wadstrom, eds.), pp. 367–424, Academic Press, New York.

Mollby, R., and Wadstrom, T., 1973, Purification of phospholipase C (alpha-toxin) from *Clostridium perfringens*, *Biochim. Biophys. Acta* **321**:569.

Molnar, D. M., 1962, Separation of the toxin of *Bacillus cereus* into two components and nonidentity of the toxin with phospholipase, *J. Bacteriol.* **84**:147.

Moore, R. B., and Appel, S. H., 1984, Calcium-dependent hydrolyses of polyphosphoinositides in human erythrocyte membranes, *Can. J. Biochem.* **62**:363.

Moskowitz, N., Schook, W., and Puszkin, S., 1982, Interaction of brain synaptic vesicles induced by endogenous Ca^{2+}-dependent phospholipase A_2, *Science* **216**:305.

Mueller, H. W., Purdon, A. D., Smith, J. B., and Wykle, R. L., 1985, 1-0-Alkyl-linked phospho-glycerides of human platelets: Distribution of arachidonate and other acyl residues in the ether-linked and diacyl species, *Lipids* **18**:814.

Mukerjee, P., and Mysels, K. J., 1971, Critical micelle concentrations of aqueous surfactant systems, National Bureau of Standards, NSRDS-NBS, 36.

Murofushi, M., Sato, T., and Fujii, T., 1973, Protective effect of calcium ions on hypotonic hemolysis of human erythrocytes, *Chem. Pharm. Bull.* (Tokyo) **21**:1364.

Myrnes, B. J., and Little, C., 1980, A simple purification scheme yielding crystalline phospholipase C from *Bacillus cereus*, *Acta Chem. Scand.* **B34**:375.

Myrnes, B. J., and Little, C., 1981, Identification of the apparently essential lysine residues in phospholipase C (*Bacillus cereus*), *Biochem. J.* **193**:805.

Nachbaur, J., Colbeau, A., and Vignais, P. M., 1972, Distribution of membrane-confined phos-pholipases A in the rat hepatocyte, *Biochim. Biophys. Acta* **274**:426.

Nagano, K., 1977, Triplet information in helix prediction applied to the analysis of super-secondary structures, *J. Mol. Biol.* **109**:251.

Nalbone, G., and Hostetler, K. Y., 1985, Subcellular localization of the phospholipases A of rat heart: Evidence for a cytosolic phospholipase A_1, *J. Lipid Res.* **26**:104.

Nalbone, G., Lairon, D., Charbonnier-Augeire, M., Vigne, J.-L., Leonardi, J., Chabert, C., Hau-ton, J. C., and Verger, R., 1980, Pancreatic phospholipase A_2 hydrolysis of phosphatidyl-cholines in various physicochemical states, *Biochim. Biophys. Acta* **620**:612.

Nantel, G., Baraff, G., and Proulx, P., 1978, The lipase activity of a partially-purified lipolytic enzyme of *Escherichia coli*, *Can. J. Biochem.* **56**:324.

Natori, Y., Nishijima, M., Nojima, S., and Satoh, H., 1980, Purification and properties of a membrane-bound phospholipase A_2 from rat ascites hepatoma 108A cells, *J. Biochem.* **87**:959.

Natori, Y., Karasawa, K., Arai, H., Tamori-Natori, Y., and Nojima, S., 1983, Partial purification and properties of phospholipase A_2 from rat liver mitochondria, *J. Biochem.* **93**:631.

Newkirk, J. D., and Waite, M., 1971, Identification of phospholipase A_1 in plasma membranes, *Biochim. Biophys. Acta* **225**:224.

Newkirk, J. D., and Waite, M., 1973, Phospholipid hydrolysis by phospholipases A_1 and A_2 in plasma membranes of rat liver, *Biochim. Biophys. Acta* **298**:562.

Nieuwenhuizen, W., Steenburg, P., and deHaas, G. H., 1973, The isolation and properties of two prephospholipases A_2 from porcine pancreas, *Eur. J. Biochem.* **40**:1.

Nilson-Ehle, P., Garfinkel, A. S., and Schotz, M. C., 1976, Intra- and extracellular forms of lipoprotein lipase in adipose tissue, *Biochim. Biophys. Acta* **431**:147.

Nishijima, M., Akamatsu, Y., and Nojima, S., 1974, Purification and properties of a membrane-bound phospholipase A_1 from *Mycobacteria phlei*, *J. Biol. Chem.* **249**:5658.

Nishijima, M., Nakaike, S., Tamori, Y., and Nojima, S., 1977, Detergent-resistant phospholipase A of *Escherichia coli* K-12, *Eur. J. Biochem.* **73**:115.

North, E. A., and Doery, H. M., 1958a, Antagonism between the actions of staphylococcal toxin and tiger snake venom, *Nature* **181**:1542.

North, E. A., and Doery, H. M., 1958b, Protective action of venoms containing phospholipase-A against certain bacterial exotoxins, *Nature* **182**:1374.

Nukuni, Z., 1932, *Bull Agr. Chem. Soc. Japan* **8**:104.

O'Flaherty, J. T., 1985, Neutrophil degranulation: Evidence pertaining to its mediation by the combined effects of leukotriene B_4, platelet-activating factor, and 5-HETE, *J. Cell. Physiol.* **122**:229.

O'Flaherty, J. T., Thomas, M. J., Hammett, M. H., Carroll, C., McCall, C. E., and Wykle, R. L., 1983, 5-L-Hydroxy-6,8,11,14-eicosatetraenoate potentiates the human neutrophil degran-ulating action of platelet-activating factor, *Biochem. Biophys. Res. Commun.* **111**:1.

Ohara, O., Tamaki, M., Nakamura, E., Tsuruta, Y., Fujii, Y., Shin, M., Teraoka, H., and Okamoto, M., 1986, Dog and rat pancreatic phospholipases A_2: Complete amino acid sequences de-duced from complementary DNAs, *J. Biochem.* **99**:733.

Ohno, M., Honda, A., Tanaka, S., Mohri, N., Shieh, T. C., and Kihara, H., 1984, Interaction of

Trimeresurus flavoviridis phospholipase A_2 and its fragment with calcium ion, *J. Biochem.* **96**:1183.

Ohyabu, T., Taguchi, R., and Ikezawa, H., 1978, Studies on phosphatidylinositol phosphodiesterase (phospholipase C type) of *Bacillus cereus*. II. *In vivo* and immunochemical studies of phosphatase-releasing activity, *Arch. Biochem. Biophys.* **190**:1.

Okajima, F., and Ui, M., 1984, ADP-ribosylation of the specific membrane protein by islet-activating protein, pertussis toxin, associated with inhibition of a chemotactic peptide-induced arachidonate release in neutrophils: A possible role of the toxin substrate Ca^{2+} mobilizing biosignaling, *J. Biol. Chem.* **259**:13863.

Okawa, Y., and Yamaguchi, T., 1975a, Studies of phospholipases from *Streptomyces*. III. Purification and properties of *Streptomyces hachijoensis* phospholipase C, *J. Biochem.* **78**:537.

Okawa, Y., and Yamaguchi, T., 1975b, Studies of phospholipases from *Streptomyces*. II. Purification and properties of *Streptomyces hachijoensis* phospholipase D, *J. Biochem.* **78**:363.

Okazaki, T., Strauss III, J. F., and Fleckinger, G. L., 1977, Lysosomal phospholipase A activities of rat ovarian tissue, *Biochim. Biophys. Acta* **487**:343.

Okumura, T., Kimura, S., and Saito, K., 1980, A novel purification procedure for *Penicillium notatum* phospholipase B and evidence for a modification of phospholipase B activity by the action of an endogenous protease, *Biochim. Biophys. Acta* **617**:264.

Okumura, T., Sugatani, J., and Saito, K., 1981, Role of the carbohydrate moiety of phospholipase B from *Penicillium notatum* in enzyme activity, *Arch. Biochem. Biophys.* **211**:419.

Okuyama, H., and Nojima, S., 1969, The presence of phospholipase A in *Escherichia coli, Biochim. Biophys. Acta* **176**:120.

Ono, Y., and Nojima, S., 1969, Phospholipases of the membrane fraction of *Mycobacterium phlei*, *Biochim. Biophys. Acta* **176**:111.

Ono, Y., and White, D. C., 1970a, Cardiolipin-specific phospholipase D activity in *Haemophilus parainfluenzae, J. Bacteriol.* **103**:111.

Ono, Y., and White, D. C., 1970b, Cardiolipin-specific phospholipase D activity in *Haemophilus parainfluenzae*: Characteristics and possible significance, *J. Bacteriol.* **104**:712.

van Oort, M. G., Dijkman, R., Hille, J. D. R., and deHaas, G. H., 1985a, Kinetic behavior of porcine pancreatic phospholipase A_2 on zwitterionic and negatively charged single-chain substrates, *Biochemistry* **24**:7987.

van Oort, M. G., Dijkman, R., Hille, J. D. R., and deHaas, G. H., 1985b, Kinetic behavior of porcine pancreatic phospholipase A_2 on zwitterionic and negatively charged double-chain substrates, *Biochemistry* **24**:7993.

Otnaess, A.-B., 1980, The hydrolysis of sphingomyelin by phospholipase C from *Bacillus cereus*, *FEBS Lett.* **114**:202.

Otnaess, A.-B., Little, C., and Prydz, H., 1974, The synthesis of phospholipase C by *Bacillus cereus*. II. A screening method for mutants, *Acta Pathol. Microbiol. Scand.* **B82**:354.

Otnaess, A.-B., Little, C., Sletten, K., Wallin, R., Johnsen, S., Flengsrud, R., and Prydz, H., 1977, Some characteristics of phospholipase C from *Bacillus cereus*, *Eur. J. Biochem.* **79**:459.

Ottolenghi, A., Gollub, S., and Ulin, A., 1961, Studies on phospholipase from *Bacillus cereus*. I. Separation of phospholipolytic and hemolysin activities, *Bacteriol. Proc.*, p. 171.

Owen, M. D., 1979, Chemical components in the venoms of *Ropalidia revolutionalis* and *Polistes humilis* (hymenoptera, vespidae), *Toxicon* **17**:519.

Pace-Asciak, C., and Granstrom, E., 1983, Prostaglandins and related substances, in *New Comprehensive Biochemistry* (A. Neuberger and L. L. M. van Deenen, eds.), vol. 5, Elsevier, Toronto.

Palmer, J. W., Schmid, P. C., Pfeiffer, D. R., and Schmid, H. H., 1981, Lipids and lipolytic enzyme activities of rat heart mitochondria, *Arch. Biochem. Biophys.* **211**:674.

Parker, J., Daniel, L. W., and Waite, M., 1987, Evidence of protein kinase C involvement in phorbol diester stimulated arachidonic acid release and prostaglandin synthesis, *J. Biol. Chem.*, **262**:(in press).

Parthasarathy, S., Steinbrecher, U. P., Barnett, J., Witztum, J. L., and Steinberg, D., 1985, Essential role of phospholipase A_2 activity in endothelial cell-induced modification of low density lipoprotein, *Proc. Natl. Acad. Sci. USA* **82**:3000.

Pasek, M., Keith, C., Feldman, D., and Sigler, P. B., 1975, Characterization of crystals of two venom phospholipases A_2, *J. Mol. Biol.* **97**:395.

Patriarca, P., Beckerdite, S., and Elsbach, P., 1972, Phospholipases and phospholipid turnover in *Escherichia coli* spheroplasts, *Biochim. Biophys. Acta* **260**:593.

Patten, G. M., Cann, S., Brunengraber, H., and Lowenstein, J. M., 1981, Separation of methyl esters of fatty acids by gas chromatography on capillary columns, including the separation of deuterated from nondeuterated fatty acids, *Methods Enzymol.* **72**:8.

Pattus, F., Slotboom, A. J., and deHaas, G. H., 1979a, Regulation of phospholipase A_2 activity by the lipid–water interface: A monolayer approach, *Biochemistry* **18**:2691.

Pattus, F., Slotboom, A. J., and deHaas, G. H., 1979b, Regulation of the interaction of pancreatic phospholipase A_2 with lipid–water interfaces by Ca^{2+} ions: A monolayer study, *Biochemistry* **18**:2698.

Pattus, F., Slotboom, A. J., and deHaas, G. H., 1979c, Amino acid substitutions of the NH_2 terminal Ala^1 of porcine pancreatic phospholipase A_2: A monolayer study, *Biochemistry* **18**:2703.

Paysant, M., Bitran, M., Wald, R., and Polonovski, J., 1970, Phospholipase A des globules rouges chez l'homme. Action sur les phospholipides endogenes et exogenes, *Bull. Soc. Chim. Biol.* **52**:1257.

Pentchev, P. G., Brady, R. O., Gal, A. E., and Hibbert, S. R., 1977, The isolation and characterization of sphingomyelinase from human tissue, *Biochim. Biophys. Acta* **488**:312.

Pepinsky, R. B., Sinclair, L. K., Browning, J. L., Mattaliano, R. J., Smart, J. E., Chow, E. P., Falbel, T. Ribolini, A., Garwin, J. L., and Wallner, B. P., 1986, Purification and partial sequence analysis of a 37-kDa protein that inhibits phospholipase A_2 activity from rat peritoneal exudates, *J. Biol. Chem.* **261**:4239.

Perret, B., Chap, H. J., and Douste-Blazy, L., 1979, Asymmetric distribution of arachidonic acid in the plasma membrane of human platelets: A determination using purified phospholipases and a rapid method for membrane isolation, *Biochim. Biophys. Acta* **556**:434.

dePierre, J. W., 1977, Extraction and quantitation of free fatty acids produced by hydrolysis of microsomal membranes, *Anal. Biochem.* **83**:82.

Pieterson, W. A., Vidal, J. C., Volwerk, J. J., and deHaas, G. H., 1974, Zymogen-catalyzed hydrolysis of monomeric substrates and the presence of a recognition site for lipid–water interfaces in phospholipase A_2, *Biochemistry* **13**:1455.

Pluckthun, A., and Dennis, E. A., 1982a, Acyl and phosphoryl migration in lysophospholipids: Importance in phospholipid synthesis and phospholipase specificity, *Biochemistry* **21**:1743.

Pluckthun, A., and Dennis, E. A., 1982b, Role of monomeric activators in cobra venom phospholipase A_2 action, *Biochemistry* **21**:1750.

Pluckthun, A., and Dennis, E. A., 1985, Activation, aggregation, and product inhibition of cobra venom phospholipase A_2 and comparison with other phospholipases, *J. Biol. Chem.* **260**:11099.

Pluckthun, A., Rohlfs, R., Davidson, F. F., and Dennis, E. A., 1985, Short-chain phosphatidylethanolamines: Physical properties and suceptibility of the monomers to phospholipase A_2 action, *Biochemistry* **24**:4201.

Pluckthun, A., DeBony, J., Fanni, T., and Dennis, E. A., 1986, Conformation of fatty acyl chains in α- and β-phosphatidylcholine and phosphatidylethanolamine derivatives in sonicated vesicles, *Biochim. Biophys. Acta* **856**:144.

dePont, J. J. H. H. M., van Prooijen-van Eeden, A., and Bonting, S. L., 1978, Role of negatively charged phospholipids in highly purified (Na^+ + K^+)-ATPase from rabbit kidney outer medulla: Studies on (Na^+ + K^+)-activated ATPase, XXXIX, *Biochim. Biophys. Acta* **408**:464.

Poon, P. H., and Wells, M. A., 1974, Physical studies of egg phosphatidylcholine in diethyl ether–water solutions, *Biochemistry* **13**:4928.

Porter, N. A., and Weenen, H., 1981, High performance ligand chromatographic separations of phospholipids and phospholipid oxidation products, *Methods Enzymol.* **72**:34.

Proulx, P. R., and van Deenen, L. L. M., 1966, Acylation of lysophosphoglycerides by *Escherichia coli*, *Biochim. Biophys. Acta* **125**:591.

Proulx, P. R., and Fung, C. K., 1969, Metabolism of phosphoglycerides in *E. coli*. IV. The positional specificity and properties of phospholipase A, *Can. J. Biochem.* **47**:1125.

Prpic, V., Blackmore, P. F., and Exton, J. H., 1982, Phosphatidylinositol breakdown induced by vasopressin and epinephrine in hepatocytes is calcium-dependent, *J. Biol. Chem.* **257**:11323.

Puijk, W. C., Verheij, H. M., and deHaas, G. H., 1977, The primary structure of phospholipase A_2 from porcine pancreas: A reinvestigation, *Biochim. Biophys. Acta* **492**:254.

Purdon, A. D., Tinker, D. O., and Spero, L., 1976, The interaction of *Crotalus atrox* phospholipase A₂ with calcium ion and 1-anilinonaphthalene-8-sulfonate, *Can. J. Biochem.* **55**:205.

Quarles, R. H., and Dawson, R. M. C., 1969*a*, The hydrolysis of monolayers of phosphatidyl-[Me-^{14}C]choline by phospholipase D, *Biochem. J.* **113**:697.

Quarles, R. H., and Dawson, R. M. C., 1969*b*, Shift in the optimal pH of phospholipase D produced by activating long-chain anions, *Biochem. J.* **112**:795.

Quarles, R. H., and Dawson, R. M. C., 1969*c*, The distribution of phospholipase D in developing and mature plants, *Biochem. J.* **112**:787.

Raetz, C. R. H., 1978, Enzymology, genetics, and regulation of membrane phospholipid synthesis in *Escherichia coli*, *Microbiol. Rev.* **42**:614.

Rahman, Y. E., Cerny, E. A., and Peraino, C., 1973, Purification and some properties of rat spleen phospholipase A, *Biochim. Biophys. Acta* **321**:526.

Ramesha, C. S., and Thompson, G. A., Jr., 1982, Changes in the lipid composition and physical properties of *Tetrahymena* ciliary membranes following low-temperature acclimation, *Biochemistry* **21**:3612.

Ramesha, C. S., Dickens, B. F., and Thompson, G. A., Jr., 1982, Phospholipid molecular species alterations in *Tetrahymena* ciliary membranes following low-temperature acclimation, *Biochemistry* **21**:3618.

Rao, B. G., and Spence, M. W., 1976, Sphingomyelinase activity at pH 7.4 in human brain and a comparison to activity at pH 5.0, *J. Lipid Res.* **17**:506.

Rao, R. H., Waite, M., and Myrvik, Q. N., 1981, Deacylation of dipalmitoyllecithin by phospholipase A in alveolar macrophages, *Exp. Lung Res.* **2**:9.

Raybin, D. M., Bertsch, L. L., and Kornberg, A., 1972, A phospholipase in *Bacillus megaterium* unique to spores and sporangia, *Biochemistry* **11**:1754.

Record, M., Lloveras, J., Ribes, G., and Douste-Blazy, L., 1977, Phospholipases A₁ and A₂ in subcellular fractions and plasma membranes of Krebs II ascites cells, *Cancer Res.* **37**:4372.

Record, M., Loyter, A., and Gatt, S., 1980, Utilization of membranous lipid substrates by membranous enzymes: Hydrolysis of sphingomyelin in erythrocyte "ghosts" and liposomes by the membranous sphingomyelinase of chicken erythrocyte "ghosts," *Biochem. J.* **187**:115.

Reisman, R. E., Littler, S. J., and Wypych, J. I., 1984, Comparison of the biochemical, immunologic and allergenic properties of vespid venoms collected in early and late summer, *Toxicon* **22**:148.

Reman, F. C., Nieuwenhuizen, W., and Boonders, T. M., 1978, Studies on the surface structure of very low density lipoproteins, *Chem. Phys. Lipids* **21**:223.

Richards, D. E., Irvine, R. F., and Dawson, R. M. C., 1979, Hydrolysis of membrane phospholipids by phospholipases of rat liver lysosomes, *Biochem. J.* **182**:599.

Rietsch, J., Pattus, F. Desnuelle, P., and Verger, R., 1977, Further studies of mode of action of lipolytic enzymes, *J. Biol. Chem.* **252**:4313.

Rillema, J. A., Osmialowski, E. C., and Linebaugh, B. E., 1980, Phospholipase A₂ activity in DMBA-induced mammary tumors of rats, *Biochim. Biophys. Acta* **617**:150.

Rimon, A., and Shapiro, B., 1959, Properties and specificity of pancreatic phospholipase A, *Biochem. J.* **71**:620.

Rittenhouse-Simmons, S., 1979, Production of diglyceride from phosphatidylinositol in activated human platelets, *J. Clin. Invest.* **63**:580.

Rittenhouse-Simmons, S., 1981, Differential activation of platelet phospholipases by thrombin and ionophore A23187, *J. Biol. Chem.* **256**:4153.

Rittenhouse, S. F., 1984, Activation of human platelet phospholipase C by ionophore A23187 is totally dependent upon cyclo-oxygenase products and ADP, *Biochem. J.* **222**:103.

Rittenhouse-Simmons, S., and Deykin, D., 1981, Release and metabolism of arachidonate in human platelets, in *Platelets in Biology and Pathology* (J. L. Gordon, ed.), vol. 2, pp. 349–372, Elsevier/North Holland Biomedical Press, Amsterdam.

Rittenhouse, S. E., and Sasson, J. P., 1985, Mass changes in myoinositol trisphosphate in human platelets stimulated by thrombin: Inhibitory effects of phorbol ester, *J. Biol. Chem.* **260**:8657.

Rnata, G. Q., and Caneva, G., 1901, Della scompoizione delle lecitine, *Ann. Hyg.* **5**:79.

Roberts, M. F., and Dennis, E. A., 1977, Proton nuclear magnetic resonance demonstration of conformationality nonequivalent phospholipid fatty acid chains in mixed micelles, *J. Am. Chem. Soc.* **99**(18):6142.

Roberts, M. F., Deems, R. A., Mincey, T. C., and Dennis, E. A., 1977*a*, Chemical modification of the histidine residue in phospholipase A₂ (*Naja naja naja*): A case of half site reactivity, *J. Biol. Chem.* **252**:2405.

Roberts, M. F., Deems, R. A., and Dennis, E. A., 1977*b*, Spectral perturbations of the histidine and tryptophan in cobra venom phospholipase A₂ upon metal ion and mixed micelle binding, *J. Biol. Chem.* **252**:6011.

Roberts, M. F., Deems, R. A., and Dennis, E. A., 1977*c*, Dual role of interfacial phospholipid in phospholipase A₂ catalysis, *Proc. Natl. Acad. Sci. USA* **74**:1950.

Roberts, M. F., Bothner-By, A. A., and Dennis, E. A., 1978, Magnetic nonequivalence within the fatty acyl chains of phospholipids in membrane models: ^{1}H nuclear magnetic resonance studies of the α-methylene groups, *Biochemistry* **17**:935.

Roberts, M. F., Adamich, M., Robson, R. T., and Dennis, E. A., 1979, Phospholipid activation of cobra venom phospholipase A₂. I. Lipid–lipid or lipid–enzyme interaction, *Biochemistry* **18**:3301.

Robinson, D. R., Bastian, D., Hamer, P. J., Pichiarallo, D. M., and Stephenson, M. L., 1981, Mechanism of stimulation of prostaglandin synthesis by a factor from rheumatoid synovial tissue, *Proc. Natl. Acad. Sci. USA* **78**:5160.

Robinson, M., and Waite, M., 1983, Physical–chemical requirements for the catalysis of substrates by lysosomal phospholipase A₁, *J. Biol. Chem.* **258**:14371.

Robson, R. J., and Dennis, E. A., 1978, Characterization of mixed micelles of phospholipids of various classes and a synthetic, homogeneous analogue of the nonionic detergent Triton X-100 containing nine oxyethylene groups, *Biochim. Biophys. Acta* **508**:513.

Rock, C. O., and Snyder, F., 1975, Rapid purification of phospholipase A₂ from *Crotalus adamanteus* venom by affinity chromatography, *J. Biol. Chem.* **250**:6564.

Rodnight, R., 1956, Cerebral diphosphoinositide breakdown: Activation, complexity, and distribution in animal (mainly nervous) tissues, *Biochem. J.* **63**:223.

Roelofsen, B., and Schatzmann, H. J., 1977, The lipid requirement of the $(Ca^{2+} + Mg^{2+})$-ATPase in the human erythrocyte membrane, as studied by various highly purified phospholipases, *Biochim. Biophys. Acta* **464**:17.

Roholt, O. A., and Schlamowtitz, M., 1961, Studies of the use of dihexanoyllecithin and other lecithins as substrates for phospholipase A, *Arch. Biochem. Biophys.* **94**:364.

Rosenberg, P., Ishay, J., and Gitter, S., 1977, Phospholipases A and B activities of the oriental hornet (*Vespa orientalis*) venom and venom apparatus, *Toxicon* **15**:141.

Rosenthal, A. F., 1975, Chemical synthesis of phospholipids and analogues of phospholipids containing carbon–phosphorus bonds, *Methods Enzymol.* **35**:429.

Ross, T. S., and Majerus, P. W., 1986, Isolation of D-myo-inositol 1 : 2-cyclic phosphate 2-inositolphosphohydrolase from human placenta, *J. Biol. Chem.* **261**:11119.

Ruggiero, M., and Lapetina, E. G., 1985, Leupeptin selectively inhibits human platelet responses induced by thrombin and trypsin: A role for proteolytic activation of phospholipase C, *Biochem. Biophys. Res. Commun.* **131**:1198.

Sahu, S., and Lynn, W. S., 1977, Characterization of phospholipase A from pulmonary secretions of patients with alveolar proteinosis, *Biochim. Biophys. Acta* **489**:307.

Saito, K., and Hanahan, D. J., 1962, A study of the purification and properties of the phospholipase A of *Crotalus adamanteus* venom, *Biochemistry* **1**:521.

Saito, K., and Kates, M., 1974, Substrate specificity of a highly-purified phospholipase B from *Penicillium notatum*, *Biochim. Biophys. Acta* **369**:245.

Saito, M., and Kanfer, J., 1973, Solubilization and properties of a membrane-bound enzyme from rat brain catalyzing a base exchange reaction, *Biochem. Biophys. Res. Commun.* **53**:391.

Saito, M., and Kanfer, J., 1975, Phosphatidohydrolase activity in a solubilized preparation from rat brain particulate fraction, *Arch. Biochem. Biophys.* **169**:318.

Saito, M., Bourque, E., and Kanfer, J., 1974, Phosphatidohydrolase and base-exchange activity of commercial phospholipase D, *Arch. Biochem. Biophys.* **164**:420.

Sato, H., and Murata, R., 1973, Role of zinc in the production of *Clostridium perfringens* alpha toxin, *Infect. Immun.* **8**:360.

Scandella, C. J., and Kornberg, A., 1971, A membrane-bound phospholipase A₁ from *Escherichia coli*, *Biochemistry* **10**:4447.

van Scharrenburg, G. J. M., deHaas, G. H., and Slotboom, A. J., 1980, Regeneration of full enzymatic activity by reoxidation of reduced pancreatic phospholipase A_2, *Hoppe Seylers Z. Physiol. Chem.* **361**:571.

van Scharrenburg, G. J. M., Puijk, W. C., Egmond, M. R., deHaas, G. H., and Slotboom, A. J., 1981, Semisynthesis of phospholipase A_2: Preparation and properties of arginine-6 bovine pancreatic phospholipase A_2, *Biochemistry* **20**:158.

van Scharrenburg, G. J. M., Puijk, W. C., Egmond, M. R., van der Schaft, P. H., deHaas, G. H., and Slotboom, A. J., 1982, Effects of substitution of the absolutely invariant glutamine-4 and phenylalanine-5 in bovine pancreatic phospholipase A_2 on enzymatic activity and substrate binding properties, *Biochemistry* **21**:1345.

van Scharrenburg, G. J. M., Puijk, W. C., deHaas, G. H., and Slotboom, A. J., 1983, Semisynthesis of phospholipase A_2: The effect of substitution of amino-acid residues at positions 6 and 7 in bovine and porcine pancreatic phospholipases A_2 on catalytic and substrate-binding properties, *Eur. J. Biochem.* **133**:83.

van Scharrenburg, G. J. M., Jansen, E. H. J. M., Egmond, M. R., deHaas, G. H., and Slotboom, A. J., 1984, Structural importance of the amino-terminal residue of pancreatic phospholipase A_2, *Biochemistry* **23**:6285.

Schneider, P. B., and Kennedy, E. P., 1967, Sphingomyelinase in normal human spleens and in spleens from subjects with Neimann–Pick disease, *J. Lipid Res.* **8**:202.

Schoonderwoerd, K., Hulsmann, W. C., and Jansen, H., 1983, Regulation of liver lipase. I. Evidence for several regulatory sites, studied in corticotrophin-treated rats, *Biochim. Biophys. Acta* **754**:279.

Schranner, R., and Richter, P. H., 1978, Rate enhancement by guided diffusion chain length dependence of repressor operator association rates, *Biophys. Chem.* **8**:135.

Scow, R. O., and Egelrud, T., 1976, Hydrolysis of chylomicron phosphatidylcholine *in vitro* by lipoprotein lipase, phospholipase A_2 and phospholipase C, *Biochim. Biophys. Acta* **431**:538.

Scow, R. O., Stein, Y., and Stein, O., 1967, Incorporation of dietary lecithin and lysolecithin into lymph chylomicrons in the rat, *J. Biol. Chem.* **242**:4919.

Shakir, K. M. M., 1981, Phospholipase A_2 activity of post-heparin plasma; a rapid and sensitive assay and partial characterization, *Anal. Biochem.* **111**:64.

Shapiro, B., 1953, Purification and properties of a lysolecithinase from pancreas, *Biochem. J.* **53**:663.

Shen, B. W., Tsao, F. H. C., Law, J. H., and Kezdy, F. J., 1975, Kinetic study of the hydrolysis of lecithin monolayers by *Crotalus adamanteus* α-phospholipase A_2: Monomer–dimer equilibrium, *J. Am. Chem. Soc.* **97**:1205.

Shipolini, R. A., Callewaert, G. L., Cottrell, R. S., Doonan, S., Vernon, C. A., and Banks, B. E. C., 1971, Phospholipase A_2 from bee venom, *Eur. J. Biochem.* **20**:459.

Shipolini, R. A., Callewaert, G. L., Cottrell, R. S., and Vernon, C. A., 1974a, The amino-acid sequence and carbohydrate content of phospholipase A_2 from bee venom, *Eur. J. Biochem.* **48**:465.

Shipolini, R. A., Doonan, S., and Vernon, C. A., 1974b, The disulfide bridges of phospholipase A_2 from bee venom, *Eur. J. Biochem.* **48**:477.

Shukla, S. D., and Hanahan, D. J., 1982a, Identification of domains of phosphatidylcholine in human erythrocyte plasma membranes: Differential action of acidic and basic phospholipases A_2 from *Agkistrodon halys blomhoffii*, *J. Biol. Chem.* **257**:2908.

Shukla, S. D., and Hanahan, D. J., 1982b, Membrane alterations in cellular aging: Susceptibility of phospholipids in density (age)-separated human erythrocytes to phospholipase A_2, *Arch. Biochem. Biophys.* **214**:335.

Siess, W., and Binder, H., 1985, Thrombin induces the rapid formation of inositol bisphosphate and inositol trisphosphate in human platelets, *FEBS Lett.* **180**:107.

Siess, W., Siegel, F. L., and Lapetina, E. G., 1983, Arachidonic acid stimulates the formation of 1,2-diacyglycerol and phosphatidic acid in human platelets: Degree of phospholipase C activation correlates with protein phosphorylation, platelet shape change, serotonin release, and aggregation, *J. Biol. Chem.* **258**:11236.

Siess, W., Weber, P. C., and Lapetina, E. G., 1984, Activation of phospholipase C is dissociated from arachidonate metabolism during platelet shape change induced by thrombin or platelet-activating factor: Epinephrine does not induce phospholipase C activation or platelet shape change, *J. Biol. Chem.* **259**:8286.

Slein, M. W., and Logan, G. F., 1963, Partial purification and properties of two phospholipases of *Bacillus cereus*, *J. Bacteriol.* **85**:370.

Slein, M. W., and Logan, G. F., 1965, Characterization of the phospholipases of *Bacillus cereus* and their effects on erythrocytes, bone, and kidney cells, *J. Bacteriol.* **90**:69.

Sloane-Stanley, G. H., 1953, Anaerobic reactions of phospholipins in brain suspensions, *Biochem. J.* **53**:613.

Slotboom, A. J., and deHaas, G. H., 1975, Specific transformations at the N-terminal region of phospholipase A₂, *Biochemistry* **14**:5394.

Slotboom, A. J., Verger, R., Verheij, H. M., Baartmans, P. H. M., van Deenen, L. L. M., and deHaas, G. H., 1976, Application of enantiomeric 2-sn-phosphatidylcholines in interfacial enzyme kinetics of lipolysis, *Chem. Phys. Lipids* **17**:128.

Slotboom, A. J., Jansen, E. H. J. M., Pattus, F., and deHaas, G. H., 1978, Semisynthetic studies on phospholipase A₂, in *Semisynthetic Peptides and Proteins* (R. E. Offord and C. DiBello, eds.), pp. 315–349, Academic Press, London.

Slotboom, A. J., Verheij, H. M., and deHaas, G. H., 1982, On the mechanism of phospholipase A₂, in *Phospholipids* (J. N. Hawthorne and G. B. Ansell, eds.), vol. 4, ch. 10, pp. 359–434, Elsevier Biomedical, Amsterdam.

Slotta, K. H., and Fraenkel-Conrat, H. L., 1938, Schlangengifte III. Mitteil: reinigung und krystallisation des klapperschlangengiftes, *Chem. Ber.* **71**:1076.

Small, D. M., 1986, The physical chemistry of lipids: From alkanes to phospholipids, in *Handbook of Lipid Research*, vol. 4, Plenum Press, New York.

Smith, A. D., and Winkler, H., 1968, Lysosomal phospholipase A₁ and A₂ of bovine adrenal medulla, *Biochem. J.* **108**:867.

Smith, A. D., Gul, S., and Thompson, R. H. S., 1972, The effect of fatty acids and of albumin on the action of a purified phospholipase A₂ from cobra venom on synthetic lecithins, *Biochim. Biophys. Acta* **289**:147.

Smith, C. D., Lane, B. C., Kusaka, I., Verghese, M. W., and Snyderman, R., 1985, Chemoattractant receptor-induced hydrolysis of phosphatidylinositol 4,5-bisphosphate in human polymorphonuclear leukocyte membranes: Requirement for a guanine nucleotide regulatory protein, *J. Biol. Chem.* **260**:5875.

Smith, C. D., Cox, C. C., and Snyderman, R., 1986, Receptor-coupled activation of phosphoinositide-specific phospholipase C by an N protein, *Science* **232**:97.

Smith, D. M., and Waite, M., 1986, Phospholipid metabolism in human neutrophil subfractions, *Arch. Biochem. Biophys.* **246**:263.

Smith, H. R., 1954, The phosphatides of the latex of *Havea brasiliensis*, *Biochem. J.* **56**:240.

Smith, J. B., and Kozlowski, A., 1986, A model for the release of arachidonic acid during platelet activation, *Thrombosis*, submitted.

Smith, J. B., and Silver, M. J., 1973, Phospholipase A₁ of human blood platelets, *Biochem. J.* **131**:615.

Smith, J. B., Dangelmaier, C., and Mauco, G., 1985, Measurement of arachidonic acid liberation in thrombin-stimulated human platelets: Use of agents that inhibit both the cyclooxygenase and lipoxygenase enzymes, *Biochim. Biophys. Acta* **835**:344.

Smith, W. L., and Borgeat, P., 1985, The eicosanoids: Prostaglandins, thromboxanes, leukotrienes, and hydroxy-eicosaenoic acids, in *Biochemistry of Lipids and Membranes* (D. E. Vance and J. E. Vance, eds.), ch. 11, Benjamin-Cummings Publishing, Menlo Park, CA.

Snyderman, R., Smith, C. D., and Verghese, M. W., 1986, A model for leukocyte regulation by chemoattractant receptors: Roles of a guanine nucleotide regulatory protein and polyphosphoinositide metabolism, *J. Leukocyte Biol.*, in press.

Sonoki, S., and Ikezawa, H., 1975, Studies on phospholipase C from *Pseudomonas aureofaciens*. I. Purification and some properties of phospholipase C, *Biochim. Biophys. Acta* **403**:412.

Sonoki, S., and Ikezawa, H., 1976, Studies on phospholipase C from *Pseudomonas aureofaciens*. II. Further studies on the properties of the enzyme. *J. Biochem.* **80**:361.

Soucek, A., and Souckova, A., 1974, Toxicity of bacterial sphingomyelinases D, *J. Hyg. Epidemiol. Microbiol. Immunol.* **18**:327.

Soucek, A., Michalec, C., and Souckova, A., 1971, Identification and characterization of a new enzyme of the group "phospholipase D" isolated from *Corynebacterium ovis, Biochim. Biophys. Acta* **227:**116.

Spence, M. W., and Burgess, J. K., 1978, Acid and neutral sphingomyelinases of rat brain: Activity in developing brain and regional distribution in adult brain, *J. Neurochem.* **30:**917.

Stahl, W. L., 1973, Phospholipase C purification and specificity with respect to individual phospholipids and brain microsomal membrane phospholipids, *Arch. Biochem. Biophys.* **154:**47.

Stanacev, N. Z., Stuhne-Sekalec, L., and Domazet, Z., 1973, The enzymatic formation of cardiolipid from phosphatidylglycerol by the transphosphatidylation mechanism catalyzed by phospholipase D, *Can J. Biochem.* **51:**747.

Steinbrecher, U. P., Parthasarathy, S., Leake, D. S., Witztum, J., and Steinberg, D., 1984, Modification of low density lipoprotein by endothelial cells involves lipid peroxidation and degradation of low density lipoprotein phospholipids, *Proc. Natl. Acad. Sci. USA* **81:**3883.

Stillway, L. W., and Lane, C. E., 1971, Phospholipase in the nematocyst toxin of *Physalia physalis, Toxicon* **9:**193.

Stocks, J., and Dalton, D. J., 1980, Activation of the phospholipase A activity of lipoprotein lipase by apoprotein C11, *Lipids* **15:**186.

Stoffel, W., 1975, Chemical synthesis of choline-labeled lecithins and sphingomyelins, *Methods Enzymol.* **35:**533.

Strauss, H., Leibovitz-Ben Gershon, Z., and Heller, M., 1976, Enzymatic hydrolysis of 1-monoacyl-*sn*-glycero-3-phosphorylcholine (1-lysolecithin) by phospholipases from peanut seed, *Lipids* **11:**442.

Strong, P. N., Wood, J. N., and Ivanyi, J., 1984, Characterization of monoclonal antibodies against β-bungarotoxin and their use as structural probes for related phospholipase A_2 enzymes and presynaptic phospholipase neurotoxins, *Eur. J. Biochem.* **142:**145.

Sugatani, J., Kawasaki, N., and Saito, K., 1978, Studies on a phospholipase B from *Penicillium notatum, Biochim. Biophys. Acta* **529:**29.

Sugatani, J., Okumura, T., and Saito, K., 1980, Studies on a phospholipase B from *Penicillium notatum* substrate specificity and properties of active site, *Biochim. Biophys. Acta* **620:**372.

Sugiura, T., Onuma, Y., Sekiguchi, N., and Waku, K., 1982, Ether phospholipids in guinea pig polymorphonuclear leukocytes and macrophages: Occurrence of high levels of 1-*O*-alkyl-2-acyl-*sn*-glycero-3-phosphocholine, *Biochim. Biophys. Acta* **712:**515.

Sunamoto, J., Kondo, H., Nomura, T., and Okamoto, H., 1980, Liposomal membranes. 2. Synthesis of a novel pyrene labeled lecithin and structural studies on liposomal bilayers, *J. Am. Chem. Soc.* **102:**1146.

Sundaram, G. S., Shakir, K. M. M., Barnes, G., and Margolis, S., 1978, Release of phospholipase A and triglyceride lipase from rat liver, *J. Biol. Chem.* **253:**7703.

Sundler, R., Alberts, A. W., and Vagelos, P. R., 1978, Enzymatic properties of phosphatidylinositol inositolphosphohydrolase from *Bacillus cereus:* Substrate dilution in detergent–phospholipid micelles and bilayer vesicles, *J. Biol. Chem.* **253:**4175.

Suzuki, T., Saito-Taki, T., Sadasivan, R., and Nitta, T., 1982, Biochemical signal transmitted by Fcγ receptors: Phospholipase A_2 activity of Fcγ2b receptor of murine macrophage cell line P388D₁, *Proc. Natl. Acad. Sci. USA* **79:**591.

Suzuki, Y., and Matsumoto, M., 1974, Acid phospholipase A_1 and A_2 in the cells, and subcellular redistribution of their activities in the cells infected with measles virus, *Biochim. Biophys. Res. Commun.* **57:**505.

Suzuki, Y., and Matsumoto, M., 1978, Acid phospholipase A_1 requiring phospholipids or Triton X-100 in the cytosol of cultured cells, *J. Biochem.* **84:**1411.

Taguchi, R., and Ikezawa, H., 1975, Phospholipase C from *Clostridium novyi* Type A.I, *Biochim. Biophys. Acta* **409:**75.

Taguchi, R., and Ikezawa, H., 1977, Phospholipase C from *Clostridium novyi* Type A. II. Factors influencing the enzyme activity, *J. Biochem.* **82:**1217.

Taguchi, R., and Ikezawa, H., 1978, Phosphatidylinositol-specific phospholipase C from *Clostridium novyi* Type A, *Arch. Biochem. Biophys.* **186:**196.

Takahashi, T., Sugahara, T., and Ohsaka, A., 1974, Purification of *Clostridium perfringens* phospholipase C (α toxin) by affinity chromatography on Agarose-linked egg yolk lipoprotein, *Biochim. Biophys. Acta* **351:**155.

Takahashi, T., Sugahara, T., and Ohsaka, A., 1981, Phospholipase C from *Clostridium perfringens*, *Methods Enzymol.* **71**:710.

Takai, Y., Kishimoto, A., Iwasa, Y., Kawahara, Y., Mori, T., and Nishizuka, Y., 1979, Calcium-dependent activation of a multifunctional protein kinase by membrane phospholipids, *J. Biol. Chem.* **254**:3692.

Takenawa, T., and Nagai, Y., 1981, Purification of phosphatidylinositol-specific phospholipase C from rat liver, *J. Biol. Chem.* **256**:6769.

Takenawa, T., and Nagai, Y., 1982, Effect of unsaturated fatty acids and Ca^{2+} on phosphatidylinositol synthesis and breakdown, *J. Biochem.* **91**:793.

Taki, T., and Kanfer, J. N., 1979, Partial purification and properties of a rat brain phospholipase D, *J. Biol. Chem.* **254**:9761.

Tamori, Y., Nishijima, M., and Nojima, S., 1979, Properties of a purified detergent-resistant phospholipase A of *Escherichia coli* K-12, *J. Biochem.* **86**:1129.

Tanford, C., 1976, *The Hydrophobic Effect: Formation of Micelles and Biological Membranes*, John Wiley & Sons, New York.

Tanford, C., 1980, *The Hydrophobic Effect: Formation of Micelles and Biological Membranes*, John Wiley & Sons, New York.

Tausk, R. J. M., van Esch, J., Karmiggelt, J., Voordouw, G., and Overbeek, J. Th. G., 1974, Physical–chemical studies of short-chain lecithin homologues. II. Micellar weights of dihexanoyl- and diheptanoyllecithin, *Biophys. Chem.* **1**:184.

Teramoto, T., Tojo, H., Yamano, T., and Okamoto, M., 1983, Purification and some properties of rat spleen phospholipase A₂, *J. Biochem.* **93**:1353.

Testa, J., Daniel, L. W., and Kreger, A. S., 1984, Extracellular phospholipase A₂ and lysophospholipase produced by *Vibrio vulnificus*, *Infect. Immun.* **45**:458.

Thakkar, J. K., Sperelakis, N., Pang, D., and Franson, R. C., 1983, Characterization of phospholipase A₂ activity in rat aorta smooth muscle cells, *Biochim. Biophys. Acta* **750**:134.

Thannhauser, S. J., and Reichel, M., 1940, Studies on animal lipids. XVI. The occurrence of sphingomyelin as a mixture of sphingomyelin fatty acid ester and free sphingomyelin, demonstrated by enzymatic hydrolysis and mild saponification, *J. Biol. Chem.* **135**:1.

Thomas, A. P., Alexander, J., and Williamson, J. R., 1984, Relationship between inositol polyphosphate production and the increase of cytosolic free Ca^{2+} induced by vasopressin in isolated hepatocytes, *J. Biol. Chem.* **259**:5574.

Thompson, W., and Dawson, R. M. C., 1964, The triphosphoinositide phosphodiesterase of brain tissue, *Biochem. J.* **91**:237.

Thuren, T., Vainio, P., Virtanen, J. A., Somerharju, P., Blomqvist, K., and Kinnunen, P. K. J., 1984, Evidence for the control of the action of phospholipases A by the physical state of the substrate, *Biochemistry* **23**:5129.

Tinker, D. O., Low, R., and Lucassen, M., 1980, Heterogeneous catalysis by phospholipase A₂: Mechanism of hydrolysis of gel phase phosphatidylcholine, *Can. J. Biochem.* **58**:898.

Tojo, H., Teramoto, T., Yamano, T., and Okamoto, M., 1984, Purification of intracellular phospholipase A₂ from rat spleen supernatant by reverse-phase high-performance liquid chromatography, *Anal. Biochem.* **137**:533.

van Tol., A., van Gent, T., and Jansen, H., 1980, Degradation of high density lipoprotein by heparin-releasable liver lipase, *Biochem. Biophys. Res. Commun.* **94**:101.

Tomita, M., Taguchi, R., and Ikezawa, H., 1982, Molecular properties and kinetic studies on sphingomyelinase of *Bacillus cereus*, *Biochim. Biophys. Acta* **704**:90.

Tookey, H. L., and Balls, A. K., 1956, Plant phospholipase D, *J. Biol. Chem.* **218**:213.

Traynor, J. R., and Authi, K. S., 1981, Phospholipase A₂ activity of lysosomal origin secreted by polymorphonuclear leucocytes during phagocytosis or on treatment with calcium, *Biochim. Biophys. Acta* **665**:571.

Trugnan, G., Bereziat, G., Manier, M.-C., and Polonovski, J., 1979, Phospholipase activities in subcellular fractions of human platelets, *Biochim. Biophys. Acta* **573**:61.

Tsai, T.-C., Hart, J., Jiang, R.-T., Bruzik, K., and Tsai, M.-D., 1985, Phospholipids chiral at phosphorus: Use of chiral thiophosphatidylcholine to study the metal-binding properties of bee venom phospholipase A₂, *Biochemistry* **24**:3180.

Tsao, F. H. C., Cohen, H., Snyder, W. R., Kezdy, F. J., and Law, J. H., 1973, Multiple forms of porcine pancreatic phospholipase A₂: Isolation and specificity, *J. Supramol. Struct.* **1**:490.

Tsujita, T., Nakagawa, A., Shirai, K., Saito, Y., and Okuda, H., 1984, Methyl butyrate-hydrolyzing activity of hepatic triglyceride lipase from rat post-heparin plasma, *J. Biol. Chem.* **259:**11215.

Tu, A. T., 1977, in *Venoms: Chemistry and Molecular Biology,* pp. 257–293, John Wiley & Sons.

Tzur, R., and Shapiro, B., 1972, Purification of phospholipase D from plants, *Biochim. Biophys. Acta* **280:**290.

Uhing, R. J., Prpic, V., Jiang, H., and Exton, J. H., 1986, Hormone-stimulated polyphosphoino-sitide breakdown in rat liver plasma membranes: Roles of guanine nucleotides and calcium, *J. Biol. Chem.* **261:**2140.

Ulitzur, S., and Heller, M., 1981, Bioluminescent assay for lipase phospholipase A_2 and phospholipase C, *Methods Enzymol.* **72:**338.

Upreti, G. C., and Jain, M. K., 1978, Effect of the state of phosphatidylcholine on the rate of its hydrolysis by phospholipase A_2, *Arch. Biochem. Biophys.* **188:**364.

Upreti, G. C., and Jain, M. K., 1980, Action of phospholipase A_2 on unmodified phosphatidyl-choline bilayers: Organizational defects are preferred sites of action, *J. Membr. Biol.* **55:**113.

Upreti, G. C., Rainier, S., and Jain, M. K., 1980, Intrinsic differences in the perturbing ability of alkanols in bilayer: Action of phospholipase A_2 on the alkanol-modified phospholipid bilayers, *J. Membr. Biol.* **55:**97.

Vasil, M. L., Berka, R. M., Gray, G. L., and Pavlovskis, O. R., 1985, Biochemical and genetic studies of iron-regulated (exotoxin A) and phosphate-regulated (hemolysin phospholipase C) virulence factors of *Pseudomonas aeruginosa, Antibiot. Chemother.* **36:**23.

Vaskovsky, V. E., and Khotimchenko, S. V., 1983, Micro-chromatographic test of transphos-phatidylic activity of phospholipase D in algae and other plants, *J. Chromatogr.* **261:**324.

Vensel, L. A., and Kantrowitz, E. R., 1980, An essential arginine in porcine phospholipase A_2, *J. Biol. Chem.* **255:**7306.

Verger, R., 1980, Enzyme kinetics of lipolysis, *Methods Enzymol.* **64B:**340.

Verger, R., and deHaas, G. H., 1973, Enzyme reactions in a membrane model. 1. A new technique to study enzyme reactions in monolayers, *Chem. Phys. Lipids* **10:**127.

Verger, R., and deHaas, G. H., 1976, Interfacial enzyme kinetics of lipolysis, *Ann. Rev. Biophys. Bioenerg.* **5:**77.

Verger, R., and Pattus, F., 1982, Lipid–protein interactions in monolayers, *Chem. Phys. Lipids* **30:**189.

Verger, R., and Pieroni, G., 1986, Monomolecular layers: A bio-topology in the past, present and future, in *Lipids and Membranes: Past, Present and Future* (J. A. F. op den Kamp, B. Roelofsen, and K. W. A. Wirtz, eds.), ch. 7, p. 153, Elsevier Science Publishers, Amsterdam.

Verger, R., Mieras, M. C. E., and deHaas, G. H., 1973, Action of phospholipase A at interfaces, *J. Biol. Chem.* **248:**4023.

Verger, R., Ferrato, F., Mansbach, C. M., and Pieroni, G., 1982, A novel intestinal phospholipase A_2: Purification and some molecular characteristics, *Biochemistry* **21:**6883.

Verghese, M. W., Smith, C. D., and Snyderman, R., 1985, Potential role for a guanine nucleotide regulatory protein in chemoattractant receptor mediated polyphosphoinositide metabolism, Ca^{++} mobilization and cellular responses by leukocytes, *Biochem. Biophys. Res. Commun.* **127:**450.

Verheij, H. M., Volwerk, J. J., Jansen, E. H. J. M., Puijk, W. C., Dijkstra, B. W., Drenth, J., and deHaas, G. H., 1980a, Methylation of histidine-48 in pancreatic phospholipase A_2: Role of histidine and calcium ion in the catalytic mechanism, *Biochemistry* **19:**743.

Verheij, H. M., Boffa, M.-C., Rothen, C., Bryckaert, M.-C., Verger, R., and deHaas, G. H., 1980b, Correlation of enzymatic activity and anticoagulant properties of phospholipase A_2, *Eur. J. Biochem.* **112:**25.

Verheij, H. M., Egmond, M. R., and deHaas, G. H., 1981a, Chemical modification of the α-amino group in snake venom phospholipases A_2: A comparison of the interaction of pancreatic and venom phospholipases with lipid–water interfaces, *Biochemistry* **20:**94.

Verheij, H. M., Slotboom, A. J., and deHaas, G. H., 1981b, Structure and function of phospholipase A_2, *Rev. Physiol. Biochem. Pharmacol.* **91:**91.

Verheij, H. M., Westerman, J., Sternby, B., and deHaas, G. H., 1983, The complete primary structure of phospholipase A_2 from human pancreas, *Biochim. Biophys. Acta* **747:**93.

Verkleij, A. J., Zwaal, R. F. A., Roelofsen, B., Comfurius, P., Kastelijn, P., and van Deenen, L. L. M., 1973, The asymmetric distribution of phospholipids in human red cell membrane—

a combined study using phospholipases and freeze-etch electron microscopy, *Biochim. Biophys. Acta* **323**:178.

Victor, M., Weiss, J., Klempner, M. S., and Elsbach, P., 1981, Phospholipase A_2 activity in the plasma membrane of human polymorphonuclear leukocytes, *FEBS Lett.* **136**:298.

Victoria, E. J., and Korn, E. D., 1975a, Enzymes of phospholipid metabolism in the plasma membrane of *Acanthamoeba castellanii, J. Lipid Res.* **16**:54.

Victoria, E. J., and Korn, E. D., 1975b, Plasma membrane and soluble lysophospholipases of *Acanthamoeba castellanii, Arch. Biochem. Biophys.* **171**:255.

Victoria. E. J., van Golde, L. M. G., Hostetler, K. Y., Scherphof, G. L., and van Deenen, L. L. M., 1971, Some studies on the metabolism of phospholipids in plasma membranes from rat liver, *Biochim. Biophys. Acta* **239**:443.

Viljoen, C. C., and Botes, D. P., 1979, Influence of pH on the kinetic and spectral properties of phospholipase A_2 from *Bitis gabonica* (gaboon adder) snake venom, *Toxicon* **17**:77.

Viljoen, C. C., Botes, D. P., and Schabort, J. C., 1975, Spectral properties of *Bitis gabonica* venom phospholipase A_2 in the presence of divalent metal ion, substrate and hydrolysis products, *Toxicon* **13**:343.

Viljoen, C. C., Visser, L., and Botes, D. P., 1976, An essential tryptophan in the active site of phospholipase A_2 from the venom of *Bitis gabonica, Biochim. Biophys. Acta* **438**:424.

Vogel, W. C., Ryan, W. G., Koppel, J. L., and Olwin, J. H., 1965, Post-heparin phospholipase and fatty acid transesterification in human plasma, *J. Lipid Res.* **6**:335.

Vogel, W. C., Pluckthun, A., Miller-Eberhard, H. J., and Dennis, E. A., 1981, Hemolytic assay for venom phospholipase A_2, *Anal. Biochem.* **118**:262.

Volwerk, J. J., Pieterson, W. A., and deHaas, G. H., 1974, Histidine at the active site of phospholipase A_2, *Biochemistry* **13**:1446.

Volwerk, J. J., Dedieu, A. G. R., Verheij, H. M., Dijkman, R., and deHaas, G. H., 1979, Hydrolysis of monomeric substrates by porcine pancreatic (pro)phospholipase A_2; the use of a spectrophotometric assay, *Recl. Trav. Chim. Pays-Bas* **98**:214.

Volwerk, J. J., Jost, P. C., deHaas, G. H., and Griffith, O. H., 1984, Evidence that the zymogen of phospholipase A_2 binds to a negatively charged lipid–water interface, *Chem. Phys. Lipids* **36**:101.

Vos, M. M., op den Kamp, J. A. F., Beckerdite-Quagliata, S., and Elsbach, P., 1978, Acylation of monoacylglycerophosphoethanolamine in the inner and outer membranes of the envelope of an *Escherichia coli* K12 strain and its phospholipase A-deficient mutant, *Biochim. Biophys. Acta* **508**:165.

Wadstrom, T., and Mollby, R., 1971, Studies on extracellular proteins from *Staphylococcus aureus.* VI. Production and purification of β-haemolysin in large scale, *Biochim. Biophys. Acta* **242**:288.

Waite, M., 1969, Isolation of rat liver mitochondrial membrane fractions and localization of the phospholipase A, *Biochemistry* **8**:2536.

Waite, M., and van Deenen, L. L. M., 1967, Hydrolysis of phospholipids and glycerides by rat liver preparations, *Biochim. Biophys. Acta* **137**:498.

Waite, M., and Sisson, P., 1971, Partial purification and characterization of the phospholipase A_2 from rat liver mitochondria, *Biochemistry* **10**:2377.

Waite, M., and Sisson, P., 1972, Effect of local anesthetics on phospholipases from mitochondria and lysosomes: A probe into the role of Ca^{2+} in phospholipid hydrolysis, *Biochemistry* **11**:3098.

Waite, M., and Sisson, P., 1973a, Utilization of neutral glycerides and phosphatidylethanolamine by the phospholipase A_1 of the plasma membrane of rat liver, *J. Biol. Chem.* **248**:7985.

Waite, M., and Sisson, P., 1973b, Solubilization by heparin of the phospholipase A_1 from the plasma membranes of rat liver, *J. Biol. Chem.* **248**:7201.

Waite, M., and Sisson, P., 1974, Studies on the substrate specificity of the phospholipase A_1 of the plasma membrane of rat liver, *J. Biol. Chem.* **249**:6401.

Waite, M., and Sisson, P., 1976a, Utilization of serum lipoprotein lipids by the monoacylglycerol acyltransferase, *Biochim. Biophys. Acta* **450**:301.

Waite, M., and Sisson, P., 1976b, Mode of action of the plasmalemma phospholipase from rat liver, in *Lipids* (R. Paoletti, G. Porcelletti, and G. Jacini, eds.), pp. 127–139, Raven Press, New York.

Waite, M., Scherphof, G. L., Boshouwers, F. M. G., and van Deenen, L. L. M., 1969, Differentiation of phospholipases A in the mitochondria and lysosomes of rat liver, *J. Lipid. Res.* **10**:411.

Waite, M., Sisson, P., and Blackwell, E., 1970, A comparison of mitochondrial with microsomal acylation of monoacylphosphoglycerides, *Biochemistry* **9**:746.

Waite, M., Parce, B., Morton, R., Cunningham, C., and Morris, H. P., 1977, The deacylation and reacylation of phosphoglycerides in microsomes of Morris hepatoma 7777 and host rat liver, *Cancer Res.* **37**:2092.

Waite, M., Sisson, P., and El-Maghrabi, R., 1978, Comparison of the lipolytic activities in liver perfusates and liver plasma membranes from rats, *Biochim. Biophys. Acta* **530**:292.

Waite, M., DeChatelet, L. R., King, V., and Shirley, P. S., 1979, Phagocytosis-induced release of arachidonic acid from human neutrophils, *Biochem. Biophys. Res. Commun.* **90**:984.

Waite, M., Rao, R. H., Griffin, H., Franson, R., Miller, C., Sisson, P., and Frye, J., 1981, Phospholipases A_1 from lysosomes and plasma membranes of rat liver. EC 3.1.1.32 Phosphatidate 1-acylhydrolase, *Methods Enzymol.* **71**:674.

Waku, K., and Nakazawa, Y., 1972, Hydrolyses of 1-*O*-alkyl-, 1-*O*-alkenyl-, and 1-acyl-2-[1-^{14}C]-linoleoyl-glycero-3-phosphorylcholine by various phospholipases, *J. Biochem.* **72**:149.

Walenga, R., Vanderhoek, J. Y., and Feinstein, M. B., 1980, Serine esterase inhibitors block stimulus-induced mobilization of arachidonic acid and phosphatidylinositide-specific phospholipase C activity in platelets, *J. Biol. Chem.* **255**:6024.

Wallace, M. A., and Fain, J. N., 1985, Guanosine 5'-*O*-thiotriphosphate stimulates phospholipase C activity in plasma membranes of rat hepatocytes, *J. Biol. Chem.* **260**:9527.

Wallach, F. H., 1969, *Membrane Proteins*, Proceedings of the Symposium of the New York Heart Association, Little, Brown Publishers, Boston, MA.

Wallner, B. P., Mattaliano, R. J., Hession, C., Cate, R. L., Tizard, R., Sinclair, L. K., Foeller, C., Chow, E. P., Browning, J. L., Ramachandran, K. L., and Pepinsky, R. B., 1986, Cloning and expression of human lipocortin, a phospholipase A_2 inhibitor with potential anti-inflammatory activity, *Nature* **320**:77.

Walsh, C. E., Waite, M., Thomas, M. J., and DeChatelet, L. R., 1981, Release and metabolism of arachidonic acid in human neutrophils, *J. Biol. Chem.* **256**:7228.

Weglicki, W., Waite, M., Sisson, P., and Shohet, S., 1971, Myocardial phospholipase A of microsomal and mitochondrial fractions, *Biochim. Biophys. Acta* **231**:512.

Weglicki, W., Waite, M., and Stam, A., 1972, Association of phospholipase A with a myocardial membrane preparation containing the $(Na^+ + K^+)$-Mg^{2+}-ATPase, *J. Mol. Cell. Cardiol.* **4**:195.

Weglicki, W. B., Ruth, R. C., Owens, K., Griffin, H. D., and Waite, M., 1974, Changes in lipid composition of Triton-filled lysosomes during lysis, *Biochim. Biophys. Acta* **337**:145.

Weiss, J., Franson, R. C., Beckerdite, S., Schmeidler, K., and Elsbach, P., 1975, Partial characterization and purification of a rabbit granulocyte factor that increases permeability of *Escherichia coli*, *J. Clin. Invest.* **55**:33.

Weiss, J., Beckerdite-Quagliata, S., and Elsbach, P., 1979, Determinants of the action of the phospholipases A on the envelope phospholipids of *Escherichia coli*, *J. Biol. Chem.* **254**:11010.

Wells, M. A., 1971, Evidence for *O*-acyl cleavage during hydrolysis of 1,2-diacyl-*sn*-glycero-3-phosphorylcholine by the phospholipases A_2 of *Crotalus adamanteus* venom, *Biochim. Biophys. Acta* **248**:80.

Wells, M. A., 1972, A kinetic study of the phospholipase A_2 (*Crotalus adamanteus*) catalyzed hydrolysis of 1,2-dibutyryl-*sn*-glycero-3-phosphorylcholine, *Biochemistry* **11**:1030.

Wells, M. A., 1973*a*, Spectral perturbations of *Crotalus adamanteus* phospholipase A_2 induced by divalent cation binding, *Biochemistry* **12**:1080.

Wells, M. A., 1973*b*, Effects of chemical modification on the activity of *Crotalus adamanteus* phospholipase A_2: Evidence for an essential amino group, *Biochemistry* **12**:1086.

Wells, M. A., 1974, The mechanism of interfacial activation of phospholipase A_2, *Biochemistry* **13**:2248.

Wells, M. A., 1975, A simple and high yield purification of *Crotalus adamanteus* phospholipase A_2, *Biochim. Biophys. Acta* **380**:357.

Wells, M. A., and Hanahan, D. J., 1969, Studies on phospholipase A. I. Isolation and characterization of two enzymes from *Crotalus adamanteus* venom, *Biochemistry* **8**:414.

van Wezel, F. M., and deHaas, G. H., 1975, Phospholipase A₂ isoenzyme from porcine pancreas: Purification and some properties, *Biochim. Biophys. Acta* **410:**299.

van Wezel, F. M., Slotboom, A. J., and deHaas, G. H., 1976, Studies on the role of methionine in porcine pancreatic phospholipase A₂, *Biochim. Biophys. Acta* **452:**101.

Wightman, P. D., Humes, J. L., Davies, P., and Bonney, R. J., 1981a, Identification and characterization of two phospholipases A₂ activities in resident mouse peritoneal macrophages, *Biochem. J.* **195:**427.

Wightman, P., Dahlgren, M. E., Davies, P., and Bonney, R. J., 1981b, The selective release of phospholipase A₂ by resident mouse peritoneal macrophages, *Biochem. J.* **200:**441.

Wightman, P. D., Dahlgren, M. E., and Bonney, R. J., 1982, Protein kinase activation of phospholipase A₂ in sonicates of mouse peritoneal macrophages, *J. Biol. Chem.* **257:**6650.

Williams, R. J. P., 1981, Studies of larger motions within proteins by nuclear magnetic resonance spectroscopy, *Biochem. Soc. Symp.* **46:**57.

Willman, C., and Hendrickson, H. S., 1978, Positive surface charge inhibition of phospholipase A₂ in mixed monolayer systems, *Arch. Biochem. Biophys.* **191:**298.

Willstatter, R., and Ludecke, K., 1904, Zur kenntniss des lecithins, *Ber. Chem. Ges.* **XXXVII:**3753.

Wilschut, J. C., Regts, J., Westenberg, H., and Scherphof, G., 1978, Action of phospholipases A₂ on phosphatidylcholine bilayers; effects of the phase transition, bilayer curvature and structural defects, *Biochim. Biophys. Acta* **508:**185.

Wilson, D. B., Prescott, S. M., and Majerus, P. W., 1982, Discovery of an arachidonoyl coenzyme A synthetase in human platelets, *J. Biol. Chem.* **257:**3510.

Windler, E., Chao, Y.-s., and Havel, R. J., 1980, Determinants of hepatic uptake of triglyceride-rich lipoproteins and their remnants in the rat, *J. Biol. Chem.* **255:**5475.

deWinter, J. M., Vianen, G. M., and van den Bosch, H., 1982, Purification of rat liver mitochondrial phospholipase A₂, *Biochim. Biophys. Acta* **712:**332.

deWinter, J. M., Korpancova, J., and van den Bosch, H., 1984, Regulatory aspects of rat liver mitochondrial phospholipase A₂: Effects of calcium ions and calmodulin, *Arch. Biochem. Biophys.* **234:**243.

Wiseman, G. M., 1975, The hemolysins of *Staphylococcus aureus, Bacteriol. Rev.* **39:**317.

Wiseman, G. M., and Caird, J. D., 1967, The nature of staphylococcal beta hemolysin. I. Mode of action, *Can. J. Microbiol.* **13:**369.

Witt, W., Bruller, H.-J., Falker, G., and Fuhrmann, G. F., 1982, Purification and properties of a phospholipid acyl hydrolase from plasma membranes of *Saccharomyces cerevisiae, Biochim. Biophys. Acta* **711:**403.

Witt, W., Schweingruber, M. E., and Mertsching, A., 1984a, Phospholipase B from the plasma membrane of *Saccharomyces cerevisiae:* Separation of two forms with different carbohydrate content, *Biochim. Biophys. Acta* **795:**108.

Witt, W., Mertsching, A., and Konig, E., 1984b, Secretion of phospholipase B from *Saccharomyces cerevisiae, Biochim. Biophys. Acta* **795:**117.

Woelk, H., Rubley, N., Arienti, G., Gaiti, A., and Porcellati, G., 1982, Occurrence and properties of phospholipases A₁ of plasma membranes prepared from neuronal- and glial-enriched fractions of the rabbit cerebral cortex, *J. Neurochem.* **36:**875.

Wolf, C., Sagaert, L., Bereziat, G., and Polonovski, J., 1981, The deacylation of rat platelet phospholipids during thrombin-induced aggregation studied by a fluorescence method, *FEBS Lett.* **135:**285.

Wolf, R. A., and Gross, R. W., 1985, Identification of neutral active phospholipase C which hydrolyzes choline glycerophospholipids and plasmalogen selective phospholipase A₂ in canine myocardium, *J. Biol. Chem.* **260:**7296.

Wunderlich, F., Ronai, A., Speth, V., Seelig, J., and Blume, A., 1975, Thermotropic lipid clustering in tetrahymena membranes, *Biochemistry* **14:**3730.

Wurster, N., Elsbach, P., Rand, J., and Simon, E. J., 1971, Effects of levorphanol on phospholipid metabolism and composition in *Escherichia coli, Biochim. Biophys. Acta* **248:**282.

Wykle, R. L., and Schremmer, J. M., 1974, A lysophospholipase D pathway in the metabolism of ether-linked lipids in brain microsomes, *J. Biol. Chem.* **249:**1742.

Wykle, R. L., Kraemer, W. F., and Schremmer, J. M., 1977, Studies on the lysophospholipase D of rat liver and other tissues, *Arch. Biochem. Biophys.* **184:**149.

Wykle, R. L., Kraemer, W. F., and Schremmer, J. M., 1980, Specificity of lysophospholipase D, *Biochim. Biophys. Acta* **619:**58.

Wykle, R. L., Olson, S. C., and O'Flaherty, J. T., 1986, Biochemical pathways of platelet activating factor synthesis and breakdown, *Adv. Inflammation Res.* **11:**71.

Yadgar, S., and Gatt, S., 1980, Enzymic hydrolysis of sphingomyelin in the presence of bile salts, *Biochem. J.* **185:**749.

Yamakawa, Y., and Ohsaka, A., 1977, Purification and some properties of phospholipase C (alpha-toxin) of *Clostridium perfringens, J. Biochem.* **81:**115.

Yamanaka, T., Hanada, E., and Suzuki, K., 1981, Acid sphingomyelinase of human brain: Improved purification procedures and characterization, *J. Biol. Chem.* **256:**3884.

Yang, C. C., 1974, Chemistry and evolution of toxins in snake venoms, *Toxicon* **12:**1.

Yang, C. C., and King, K., 1980, Chemical modification of the histidine residue in phospholipase A_2 from the *Hemachatus haemachatus* snake venom, *Toxicon* **18:**529.

Yang, S. F., Freer, S., and Benson, A. A., 1967, Transphosphatidylation by phospholipase D, *J. Biol. Chem.* **242:**477.

Yoshida, H., Kudo, T., Shinkai, W., and Tamiya, N., 1979, Phospholipase A of sea snake *Laticauda semifasciata* venom: Isolation and properties of novel forms lacking tryptophan, *J. Biochem.* **85:**379.

Zeller, E. A., 1951, Enzymes as essential components of bacterial and animal toxins, *The Enzymes* **1:**986.

Zhelkovskii, A. M., D'yakov, V. D., Ginodman, L. M., and Antonov, V. K., 1978, Active center of phospholipase A_2 from the venom of the Central Asian cobra: A catalytically active carboxy group, *Bioorgh. Khim.* **4:**1665.

Zurini, M., Hugentobler, G., and Gazzotti, P., 1981, Activity of phospholipase A_2 in the inner membrane of rat-liver mitochondria, *Eur. J. Biochem.* **119:**517.

Zwaal, R. F. A., 1978, Membrane and lipid involvement in blood coagulation, *Biochim. Biophys. Acta* **515:**163.

Zwaal, R. F. A., Roelofsen, B., Comfurius, P., and van Deenen, L. L. M., 1971, Complete purification and some properties of phospholipase C from *Bacillus cereus, Biochim. Biophys. Acta* **233:**474.

Zwaal, R. F. A., Roelofsen, B., and Colley, C. M., 1973, Localization of red cell membrane constituents, *Biochim. Biophys. Acta* **300:**159.

Zwaal, R. F. A., Fluckiger, R., Moser, S., and Zahler, P., 1974, Lecithinase activities at the external surface of ruminant erythrocyte membranes, *Biochim. Biophys. Acta* **373:**416.

Zwaal, R. F. A., Roelofsen, B., Comfurius, P., and van Deenen, L. L. M., 1975, Organization of phospholipids in human red cell membranes as detected by the action of various purified phospholipases, *Biochim. Biophys. Acta* **406:**83.

Index